European Communities Oil and Gas Technological Development Projects

Third Status Report

Compiled by
R. DE BAUW
E. MILLICH
J.P. JOULIA
D. VAN ASSELT
J.W. BRONKHORST

Commission of the European Communities
Directorate-General for Energy,
Brussels

Published by
Graham & Trotman
for the Commission of the European Communities

Published in 1987 by

Graham & Trotman Limited
Sterling House
66 Wilton Road
London SW1V 1DE
United Kingdom

Graham & Trotman Inc.
Kluwer Academic Publishers Group
101 Philip Drive
Assinippi Park
Norwell, MA 02061 USA

for the Commission of the European Communities,
Directorate-General Telecommunications, Information Industries and
Innovation

EUR 11 163

© ECSC, EEC, EAEC, Brussels and Luxembourg, 1987

ISBN 0 86010 976 3

Printed in Great Britain at the Alden Press, Oxford

C O N T E N T S

PREFACE

The 1973 oil crisis highlighted the dependency of the Community on imported hydrocarbons to satisfy its energy demand.

Therefore, in order to improve security of supply the Community has developed since 1973 a programme assisting the oil industry to develop new technologies required for exploiting oil and gas resources outside and inside the Community territories.

This programme (Regulations 3056/73 and 3639/85) has allowed remarkable achievements in a sector where innovation is needed to take up the challenge of producing oil and gas in difficult environments.

This report shows the achievements of the Community programme. It gives evidence of the high technical level which has already been attained by the companies in the oil and gas sector with the support of the Community.

Continued promotion of innovative energy technologies through research development and demonstration has been agreed upon the Community as an objective for 1995. The programme to support technological development in the hydrocarbon sector, which aims to improve research action towards reduction of costs, greater efficiency and improvement of safety forms an important part in the realisation of this objective.

Nic MOSAR

INTRODUCTION

The implementation of a Community energy strategy is one of the objectives which the Community has set itself. Due to the importance of hydrocarbons in the Community's energy supplies and the Community's dependence on imports, the fundamental objective of this policy is to ensure long-term security of supply. With this aim in view the Commission of the European Communities, through to Directorate General for Energy, is involved in a programme to support the development of new technologies in the hydrocarbon sector.

The development of appropriate technologies for oil and gas exploration, production, storage and transportation contributes to reduce the Community's dependency in imported energy sources. Therefore, the Council of Ministers introduced in 1973 a first programme to provide financial assistance to the development of new processes, techniques and tools in the hydrocarbons sector. This programme, based on Regulation 3056/73, operated for 12 years until 1985. After a thorough evaluation of its results the Council adopted a Commission proposal to introduce a second programme governed by a new Regulation - 3639/85 - valid for a limited period of 4 years.

The purpose of this report is to present the technological development carried out under contract in the framework of the 2 Regulations governing the programme. The introduction begins by examining briefly the programme in view of the 1995 energy objectives adopted by the Community. It then presents a summary of the programme's content, its implementation and supervision structure and some of the main results achieved till now. Finally, a detailed summary of each of the projects from the seventh (1981) up to the twelfth (1986) round is presented. This summary is presented by subject area covered by the programme. Projects completed before 1984 which were included in the previous status report* have not been considered as well as 1986 projects for which contracts have not already been signed.

COMMUNITY ENERGY RESEARCH AND DEVELOPMENT STRATEGY

The 1973 oil crisis highlighted the dependency of the Community on imported energy sources to satisfy its energy demand. This situation, although improved by Community action, still exists. In 1986, total primary energy consumption in the EEC reached 1033.6 m toe. Oil and gas, which are mainly imported, amounted to 465.9 m toe and 187 m toe respectively.

Early in the 1970's this dependency on oil and gas led the Commmunity to set up instruments to enable its energy demand to be met with the best efficiency. The Community energy research and development strategy was an important component towards the achievement of this objective.

---

* European Communities,Oil and Gas Technological Development Projects - Second Status Report - 1984

The Community has adopted for 1995 new energy objectives, among them continued promotion of innovative energy technologies through research, development and demonstration. The programme for technological development in oil and gas sector has shown important contribution towards the satisfaction of this objective.

The recent fall in the price of crude oil may have created the impression that the need to pursue this effort is less justified than in the past. Nevertheless, even if the short term effect on European offshore production may appear limited, the development of oil and gas fields in the North Sea and in other difficult areas inside and outside the Community, as well as the expertise level of the oil-related industry may suffer greatly as a result of this situation, endangering long-term security of supply.

Obviously, reduction of costs, increase of efficiency and improvement of safety, priorities which were established in the framework of the Regulation 3639/85, will remain the major technical trends. Therefore continuity in the technological development effort is a necessity.

PROGRAMME CHARACTERISTICS

The financial support provided with Regulations 3056/73 and 3639/86 is intended to promote technological development directly related to those activities in exploration, production, transportation and storage of hydrocarbons which are likely to improve the security of Community oil and gas supplies. The support granted is in the form of a subsidy which is repayable by the beneficiary in case of commercial exploitation of the project. This support does not exceed 40% of the eligible costs.

The projects in the hydrocarbon sector which can be considered within the programme, are those which satisfy the following conditions:

- develop innovatory techniques, processes or products or exploit a new application of techniques, processes or products for which the research stage is completed;

- offer prospects of industrial, economic and commercial viability;

- present difficulties concerning financing because of considerable technical and economic risks involved, so that most probably they would not be carried out without Community financial support.

In order to improve the efficiency of the programme, it is foreseen that the Commission, after consulting the Advisory Committee, will draw up priorities for the selection of the project. These priorities are referred to in the invitations to submit projects.

In the selection, a preference is given to projets involving associations of at least two independent companies established in different member states, provided the contribution of these undertakings is effective and significant, and to projects submitted by small and medium-sized firms.

IMPLEMENTATION AND SUPERVISION STRUCTURE

The Commission has several responsibilities as far as the implementation of the Regulations governing the programme is concerned. Each year, the Commission issues invitations to submit projects which lay down priorities for the selection of projects drafted with the assistance of an Advisory Committee.

The Commission evaluates the proposals and after consulting an Advisory Committee, decides whether or not to grant support to projects. When a support has been granted, the services of the Commission are then responsible for the negotiation and conclusion of the contracts.

During the project execution, the Commission scrutinises progress of the work both technically and financially. The technical coordination of the programme is the responsibility of Directorate C, "Hydrocarbons", of the Directorate General for Energy, while the contract division, Directorate A, "Energy policy, analyses, forecasts and contracts" is responsible for all administrative matters. Reports concerning the state of advancement of the programme are made every 2 years to the European Parliament and the Council.

STATUS OF IMPLEMENTATION

From 1974 to 1986, 185 companies submitted 805 projects in response to the Commission's annual invitations under Regulations 3056/73 and 3639/85. The Council and the Commission granted financial support – totalling 422 millions of units of account – to 509 projects which results in 460 contracts. The support attributed to individual projects reached an average of 35% of the eligible costs.

The following major technical achievements, among others, have been registered :

- the development of new systems for collecting, processing and interpretating geophysical data;

- the development of drilling installations on board dynamically positoned vessels, capable of drilling for oil and gas at depths of as much as 1,880 meters;

- the completion in the Adriatic Sea of the first horizontal drilling operation;

- the development of new platform concepts including tension leg platforms (TLP);

- the study of various floating production systems and among them the SWOPS production system for exploiting small accumulations and early production operations;

- the implementation of pilot projects relating to the methods used for enhanced recovery of oil and gas;

- the development of submarine vehicles for underwater works;

- the pipelaying trials in Sicilian waters leading to the construction of the first subsea gas-line linking Africa to Europe, which is now conveying 12 billion of cubic meters per year of Algerian natural gas to Italy.

Looking at the commercialisation of the developed techniques, it appears that of the 144 projects supported during the first 5 rounds of the programme, from 1975 to 1979, 47 projects reached a commercialisation stage. The average repayment rate for these five rounds is 31.7% with a high value for the first. During the following 5 rounds 21 projects reached commercialisation. Considering that a long period is required to commercialise a development on the market, there is no doubt that this last figure will increase in the future.

DIFFUSION OF KNOWLEDGE AND RESULTS

The diffusion of knowledge and results of projects in the Community hydrocarbon project scheme is mainly the prerogative of the contractor. Where results have been positive they have been presented as papers in technical conferences and articles in specialised newspapers, and companies have promoted them in the main exhibitions of the oil industry in order to achieve commercialisation of the development.

Nevertheless, the Commission has an important role to play in the diffusion of knowledge:

a) In April 1979 and December 1984, Symposia were held in Luxembourg where the results were presented and discussed by Community contractors and other interested oil and gas industry representatives. Attended by several hundred participants, they were the occasion to review programme achievement. In March 1988, a third Symposium will be held in Luxembourg again.

b) Regularly, technical conferences are organized. These meetings encourage the cooperation between Community contractors working on similar problems and act as forum for the presentation of the results and their assessment.

c) A data base, called "SESAME", has been set up which – when open to the public in the near future – will provide anyone interested with details of the projects. It will be a permanent means to make contact with specialists and companies in order to achieve a better spreading of the technology developed with the support of the Community.

d) Reports on the status of the programme similar to this one are published.

INFORMATION FOR FUTURE PROPONENTS

The Commission normally seeks to attract research and development proposals for subvention by the Community by issuing an annual call for tender. Since the provision of subvention may interest a great many proponents, the call for tenders is published in the Official Journal of the Communities.

It has been the practice to publish the call for tenders in July with a closing date for proposals end of December. Research proposals should be submitted in the form required by the call for tender.

Proposals should be sent to:

    Directorate for Hydrocarbons
    Directorate General for Energy
    Commission of the European Communities
    rue de la Loi, 200
    B - 1049 Brussels

Further information on the programme may also be obtained from the above address.

BREAKDOWN OF SUPPORT BY SECTOR OF ACTIVITY FROM 1974 TO 1986

TOTAL: 422.6 MILLIONS OF ECUS

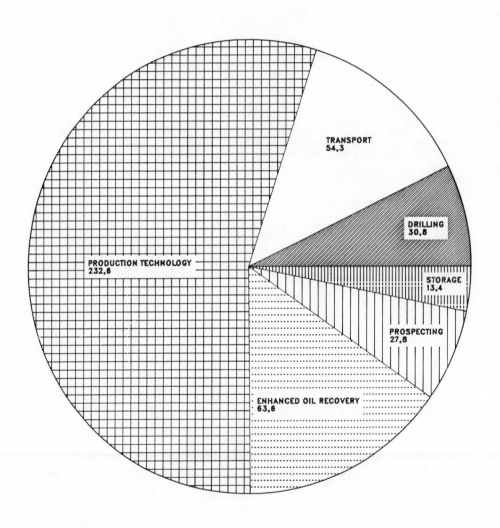

BREAKDOWN OF PROJECTS BY SECTOR OF ACTIVITY FROM 1974 TO 1986

TOTAL: 509 PROJECTS.

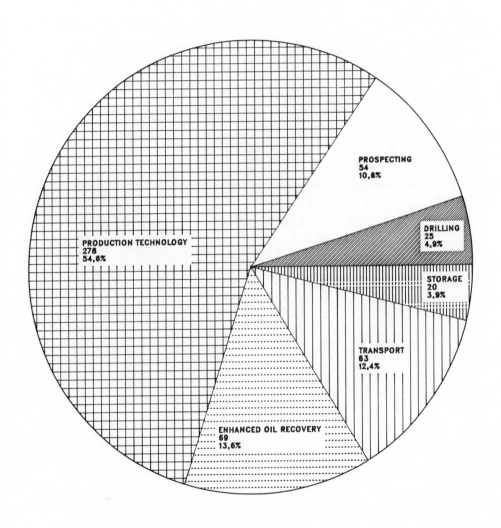

PROSPECTING
54
10,6%

DRILLING
25
4,9%

PRODUCTION TECHNOLOGY
278
54,6%

STORAGE
20
3,9%

TRANSPORT
63
12,4%

ENHANCED OIL RECOVERY
69
13,6%

GEOPHYSICS AND PROSPECTING

```
***********************************************************************************
*  TITLE : FURTHER DEVELOPMENT AND TESTING OF THE    *      PROJECT NO          *
*          MAGNETOTELLURIES REMOTE REFERENCE         *                          *
*          TECHNIQUE.                                *   TH./01027/81/DE/..      *
*                                                    *                          *
***********************************************************************************
*  CONTRACTOR :                                      *  PROGRAM :               *
*    B.G.R. HANNOVER                                 *    HYDROCARBONS          *
*    ALFRED-BENTZ-HAUS         TEL 0511/643-0        *                          *
*    POSTFACH 510153           TLX 923730 BGR HAD    *                          *
*    D - 3000 HANNOVER 51                            *  SECTOR :                *
*                                                    *    GEOPHYSICS            *
*  PERSON TO CONTACT FOR FURTHER INFORMATION :       *                          *
*    DR. W. LOSECKE                                  *                          *
*                                                    *                          *
*                                                    *                          *
***********************************************************************************
```

VERSION : 12/02/87

AIM OF THE PROJECT :
****
The Magnetotelluric Remote Reference Technique (RRMT) is to be developed
further for routine application to hydrocarbon exploration. An increase in the
reliability of the data is obtained with RRMT by reduction of the influence of
noise level.
The work is concentrated on two areas :
(a) further development of the measuring equipment, the data processing
programs, and the interpretation techniques;
(b) development and testing of an exploration strategy for specific types of
geological structures.

PROJECT DESCRIPTION :
****
The RRMT method uses two precisely synchronised MT equipments placed at two
sites A and B: the data of site A are processed together with those of site B
for improving the results for site A and vice-versa.
This method is routinely applied in hydrocarbon exploration. The improvements
granted by the RRMT method are tested in such areas as volcanic or alpine
layers or thick evaporite beds. These activities are accompanied by the
development of the corresponding hard and software.
RRMT measurements are carried out along
- 1 profile in the study area of East Holstein (W-GERMANY)
- 2 profiles in the study area of Nienburg-Versmold-Lippstadt (W-GERMANY)
- 2 profiles in the study area of the Alps (W-GERMANY/AUSTRIA)

STATE OF ADVANCEMENT :
****
Completed

RESULTS :
****
Values for parameters have been established on the basis of measurements and processing results obtained so far. By standardisation of the processing methods, the program processes the data four times faster than the earlier versions. Software to calculate confidence limits was developed for the single and Remote Reference processing program package.

Extensive measurements in three highly industrialised test areas of FRG showed that the quality of the sounding curves is significantly improved by the RRMT technique relative to the single-site magnetotelluric technique.

RRMT makes it possible to obtain usable results in areas where the sounding curves obtained with "classical" MT methods cannot be interpreted or only with a great expenditure of time and effort.

The testing of the RRMT technique yielded the following conclusions:
- Electromagnetic noise that correlates over large distances must be
    expected in the period range of T < 10 s. Therefore, the distance
    between the base and reference stations should be at least 20 km.
- The two stations should be sited consecutively along the profile or along
    neighbouring profiles and survey time must be as short as possible.
- In some parts of the North German basin, good-conducting sediments were
    found but thickness resulted in periods as high as 10000 s., in order
    to determine basement depths.
- "H" solution is preferable for the processing of the data.
- Period range T < 10 s must be given careful attention during measuring and
processing. Sorting out poor data requires great number of measurements.
Furthermore exploration, geophysical and geological data were obtained for the study areas.

In the Eastern Holstein test area, a very good-conducting layer was found in the Mesozoic. In the pre-Permian no good-conducting layers of significant thickness were found, but structures with relatively high resistivities are prominent.

In contrast to the East Holstein area, the RRMT results in large parts of the Nienburg-Versmold-Lippstadt study area show a thick, good-conducting layer in the pre-Permian. The depth to the basement decreases from 15 km at Hodenhagen to about 10 km in the Wiehen mountains to the south-west. Large lateral changes in conductivity were observed along a profile between Duemmer and Lippstandt.

The RRMT data for the Alpine study area shows low resitivities for the Molasse basin and high resistivities for the Calcereous Alps. The good-conducting Molasse layers can be traced below the Alpine nappe.

REFERENCES :
****
RODEMANN, H., LOSECKE, W., KNOEDEL, K.* THE MT-EQUIPMENT OF THE FEDERAL INSTITUTE FOR GEOSCIENCES AND NATURAL RESOURCES. -
J. GEOMAG. GEOELECTR., 35, 399-405, 1983.
LOSECKE, W., KNOEDEL, K., MUELLER, W., RODEMANN, H.: TECHNOLOGICAL DEVELOPMENT AND TESTING OF REMOTE REFERENCE MAGNETOTELLURICS. -
IN: NEW TECHNOLOGIES FOR THE EXPLORATION AND EXPLOITATION OF OIL AND GAS RESOURCES, EDITED BY R. DE BAUW, E. MILLICH,
J.P. JOULIA AND D. VAN ASELT, VOL. 1, 70-76, GRAHAM & TROTMAN, LONDON 1985.
KNOEDEL, K., LOSECKE, W., GRINAT, M., MUELLER, W.: FURTHER DEVELOPMENT AND TESTING OF THE MAGNETOTELLURIC REMOTE REFERENCE TECHNIQUE. - FINAL REPORT ON RESEARCH PROJECT EC/BGR CONTRACT NO. THO21.27/81, HANNOVE

```
********************************************************************************
* TITLE : NEW POSITION FIXING SYSTEM FOR  REMOTE  *       PROJECT NO          *
*         SENSING OF HYDROCARBONS                 *                           *
*                                                 *                           *
*                                                 *    TH./01030/82/UK/..      *
*                                                 *                           *
********************************************************************************
* CONTRACTOR :                                    * PROGRAM :                 *
*   S.S.L.                                        *   HYDROCARBONS            *
*   PO BOX 36                    TEL 029573746    *                           *
*   UK - BANBURRY OXON OX15JB  TLX 24224 REF 1885 *                           *
*                                                 * SECTOR :                  *
*                                                 *   GEOPHYSICS              *
* PERSON TO CONTACT FOR FURTHER INFORMATION :     *                           *
*   MR. B. LONDSDALE                              *                           *
*                                                 *                           *
*                                                 *                           *
********************************************************************************
```

VERSION : 12/03/87

AIM OF THE PROJECT :
****

The aim of the project is to evaluate the new ratio position fixing system
"navstar" for its suitability as an aid to offshore hydrocarbon exploration.
Navstar is a satellite based system funded by the U.S. Dept. of Defence. It
provides a 24 hour global position to an accuracy of approximately 100 metres
together with velocity and atomic time output.

PROJECT DESCRIPTION :
****

There are three major phases of the project.
A) selection of receivers and integration into test platforms (aircraft, boat).
   Construction of special purpose data recording systems.
B) operational trials.
   Establishment of a test range overland using microwave transponders.
   Conduct flight trials over test range over water.
   Conduct marine test.
C) analysis of data.
   Computer printout of tracks and comparisons with other inputs such as mini-
ranger loran c.

STATE OF ADVANCEMENT :
****

Integration test with precision navigation comparator have been concluded and
the project is now at a stage where flight/marine tests could commence.
Nevertheless deployment of navstar satellites has not proceeded according to
plan due to the Challenger space shuttle loss which implies a delay in the
project completion.

RESULTS :
****

An instrumented bell jetranger helicopter has been flown with an onboard data
aquisition system in conjunction with a microwave trisponder precision
navigation aid. Loran c was also tested for it suitability as a comparator
system.
The results from the trisponder system indicated an overall accuracy of plus or
minus two metres and the loran c indicated plus or minus one hundred metres.
The loran c data will be used as an aid when post processing data as the output
is in direct latitude/longitude rather than range/range as in the case of the
trisponder. The tests completed so far confirm that the comparison system has
sufficient resolution and accuracy to make possible accurate tests on the gps
receiver when circumstances permit.

```
*********************************************************************************
* TITLE : HIGH PERFORMANCE SEISMIC STREAMER.       *      PROJECT NO          *
*                                                  *                          *
*                                                  *   TH./01031/83/UK/..     *
*                                                  *                          *
*********************************************************************************
* CONTRACTOR :                                     * PROGRAM :                *
*   BRITOIL PLC                                    *   HYDROCARBONS           *
*   ST. VINCENT STREET 150     TEL 041 204 25 66   *                          *
*   UK - GLASGOW G2 5LJ        TLX 77 62 68        *                          *
*                                                  * SECTOR :                 *
*                                                  *   GEOPHYSICS             *
* PERSON TO CONTACT FOR FURTHER INFORMATION :      *                          *
*   MR. J. ANDERSON                                *                          *
*                                                  *                          *
*                                                  *                          *
*********************************************************************************
```

VERSION : 01/03/86

AIM OF THE PROJECT :
****
To develop a 1024 channel seismic streamer and computer controlled seismic data
acquisition system. data acquisition system is located within the streamer and
communication of the data with the seismic vessel is obtained through a high-
speed telemetry system.

PROJECT DESCRIPTION :
****
The phases of the development program are the following :
1. Design and feasibility study
2. Electronic development and acoustic trials
3. Development production and testing of half-length 512 channel
   streamer and computer system
4. Development, production and testing of full 1024 channel
   streamer and computer system
5. Production evaluation and use of full system in seismic surveys.
In the frame of the project TH 01031/83 only phases 1,2 and partially phase 3
were supported. The remaining part of the phase 3 is supported under the
project TH 01041/84.

STATE OF ADVANCEMENT :
****
The project was completed in June 1985.

RESULTS :
****
The full definition of all the packages in the nodes has been completed. This
means data acquisition unit, telemetry control unit and power supply unit. The
telemetry system has been designed. All physical units are of a standard
construction. The acoustic sections, isolators, tail unit and tow cable have
been studied and specified.

```
******************************************************************************
* TITLE : INTEGRATED GEOPHYSICAL PROSPECTING FOR    *      PROJECT NO         *
*         HYDROCARBONS.EXPLORATION OF A DEEP         *                         *
*         EVAPORIC BASIN IN AITOLOAKARMANIA;         *    TH./01032/83/HE/..   *
*         WESTERN GREECE.                            *                         *
******************************************************************************
* CONTRACTOR :                                       * PROGRAM :               *
*   PUBLIC PETROLEUM CORP. OF GREECE                 *   HYDROCARBONS          *
*   19, KIFISSIAS AVE, MAROUSS TEL 8069301-9         *                         *
*   HE - 151-24 ATHENS          TLX 221583 DEP GR    *                         *
*                                                    * SECTOR :                *
*                                                    *   GEOPHYSICS            *
* PERSON TO CONTACT FOR FURTHER INFORMATION :        *                         *
*   DR. M. L. MYRIANTHIS                             *                         *
*                                                    *                         *
*                                                    *                         *
******************************************************************************
```

VERSION : 15/04/87

AIM OF THE PROJECT :
****
The delineation of the deep evaporatic basin of Aitoloakarnania which is part
of the Western Greece extended geosyncline. The final modelling derived through
interractive interpretation of seismic data and computer-aided integration of
all other data.
The objectives of this study have centered around the depth identification of
the preevaporitic basal unit, especially in those areas where it lays within
economic drilling depth. The interpretation approach is quantitive and
synergetic (integrated).

PROJECT DESCRIPTION :
****
1. PHASE 1
A. Acquisition of 329 km seismic profiles using Vibroseis source and 165 km of
conventional dynamite seismic data. Additionally, 567 km of marine seismic
profiles were also recorded using a powerful air gun (all records down to 10
seconds two way travelitime).
B. Acquisition of 1523 land gravity and magnetic points together with 156
magnetotellouric soundings.
C. Acquisition of 22680 marine gravity and magnetic points.
D. Acquisition of 2214 km of aeromagnetic profiles.
PHASE 2
Acquisition of said data.
PHASE 3
Interpretation of all geophysical data together with computer aided integrated
studies.
PHASE 4
Additional work.
The additional work that have helped to better understand the geology of the
area can be divided into:

1. GEOLOGICAL WORK:
a) geological mapping (1:50000) and stratigraphic study of the flysch clastic
sediments over a large area in the Eastern part of Aitoloakarnania.
b) geological mapping (1:50000) in selected areas which exhibit complex
tectonic structure and thrusting fronts, for a better definition of the
tectonic model.
c) detailed geological mapping (1:50000) of smaller areas within the above
complex geological settings.
2. GEOCHEMICAL ANALYSES:
a) analysis of samples from surface mesozoic carbonates, which will help to
understand the source of the hydrocarbon shows.
b) analysis of surface samples from other sequences in the area was also
carried out to ensure a better determination of the geochemical model, since
the palaeotectonic activity in that region has complicated the hydrocarbon
migration path.
3. INTEGRATED STUDIES:
a) integrated study of the geological evolution model of the Alpine chain
between Eurasia, Arabia-Africa, India during the mesozoic and cenozoic periods.
b) integration of the new surface geological data acquired from regional
studies of Greece (data from volcanics and sediment formations with the data
from the Perimediterranean regions).
In addition of all these works a considerable number of micropaleontological
analyses have been performed for the completion of the corresponding
chronostratigraphic model.

STATE OF ADVANCEMENT :
****
Completed

RESULTS :
****
1. The integrated geophysical and geological studies provide the means for an
optimal structural evaluation of tectonically complex areas like Western Greece.
 The proper exploration strategy can now be finilized, based on the experience
from this study, which could serve as a well documented case-study for similar
thrusted environments encountered in Europe or elsewhere.
2. The integrated interpretation of all data made possible the identification
of the Preevaporatic formation and resulted in the recognition of three areas
offering oil exploration potential.
Evaluation of the above areas by further detailed study is recommended,
including extension of studies to 3D solution eventual consideration of a
stratigraphic well, following complementary acquisition of seismic data
assisted by wide angle profiling in land and offshore (OBS).

7

```
****************************************************************************
* TITLE : OBLIQUE SEISMIC PROFILE                 *       PROJECT NO      *
*                                                 *                       *
*                                                 *   TH./01033/83/FR/..  *
*                                                 *                       *
****************************************************************************
* CONTRACTOR :                                    * PROGRAM :             *
*   GERTH                                         *   HYDROCARBONS        *
*   AVENUE DE BOIS PREAU 4       TEL 1 47.52.61.39 *                      *
*   FR - 92502 RUEIL-MALMAISON TLX 203050         *                       *
*                                                 * SECTOR :              *
*                                                 *   GEOPHYSICS          *
* PERSON TO CONTACT FOR FURTHER INFORMATION :     *                       *
*   MR. NICOLETIS                TEL 1 42.91.40.00 *                      *
*                                TLX 615 700      *                       *
*                                                 *                       *
****************************************************************************
                                                      VERSION : 01/01/87
```

AIM OF THE PROJECT :
****
Development of a new improved and less costly method for the study of
reservoirs.

PROJECT DESCRIPTION :
****
The method which is studied makes use of: drilled wells in which seismographs
are installed, and a variable surface source. The project was carried out in 3
phases :
1. Data acquisition
2. Processing
3. Interpretation

STATE OF ADVANCEMENT :
****
Completed.

RESULTS :
****
Two field campaigns have been carried out. Data obtained on the Faulquemont and
Saint Just sites have been processed with the traditional surface seismic
methods.
Acquisition, conventional processing and interpretation of OSPs recorded on
different mineworking and oil sites. Acquisition provided the data base
indispensable for continuation of the project and have improved definition of
the limits and optimum conditions of application of this method. Conventional
batch processing proved to be poorly adapted to this new configuration. leading
to the second phase of the project.
Study and design of a new software using interactive graphics and set-up on a
minicomputer. This software, known as GIPSO (graphical interactive processing
software for offset VSP) which will in the end replace conventional processing
is now being tested. Its modularity and numerous graphic outputs on screen
provide time savings whilst ensuring better quality than anything currently
available.

```
****************************************************************************
* TITLE : GEOLOGICAL MODEL OF RIFTS              *      PROJECT NO        *
*                                                *                        *
*                                                *   TH./01034/83/FR/..   *
*                                                *                        *
****************************************************************************
* CONTRACTOR :                                   * PROGRAM :              *
*   GERTH                                        *   HYDROCARBONS         *
*   AVENUE DE BOIS PREAU 4      TEL 1 47.52.61.39 *                       *
*   FR - 92502 RUEIL-MALMAISON TLX 203050        *                        *
*                                                * SECTOR :               *
*                                                *   GEOPHYSICS           *
* PERSON TO CONTACT FOR FURTHER INFORMATION :    *                        *
*   MR. MASSE                   TEL 59.83.40.00  *                        *
*                               TLX 560 804      *                        *
*                                                *                        *
****************************************************************************
```

VERSION : 01/01/87

AIM OF THE PROJECT :
****
The aim of this project is to build a geological model of the continental rift
to guide the exploration of the deepest strata of the sedimentary basins, based
on the observation of the East African rifts system. This series was chosen
because the East African rifts system with is today perhaps the only one in the
world enabling sedimentation in a rift to be observed at the different
successive stages of its evolution and for a variety of conditions of climate.
Petroleum potential and economic interest related to these deepest lying strata
of the sedimentary basins are considerable. Furthermore, their geological
exploration is generally very difficult owing to their great depths and the
frequent presence of thick covering layers of salt, forming an obstacle to
seismic prospection.

PROJECT DESCRIPTION :
****
The work programme that is being carried out comprises two parts:
- a structural part, covering operations the main purpose of which is addressed
at improving understanding of the dynamics controlling the genesis and
evolution with time of the various basins and chasms along the rift,
- a sedimentological part, covering operations the main purpose of which is to
provide details as to the nature and conditions of sedimentation in these
basins and chasms, at the successive stages of their evolution and under
various conditions of climate.
The execution of this programme is divided into three phases:
PHASE 1: Acquisition of basic data on the geodynamics of the rift and
architecture of its sedimentary deposits
PHASE 2: Study of several sedimentary mechanisms representative of the
successive stages of evolution
PHASE 3: Building an exploration model.

STATE OF ADVANCEMENT :
****
The project was completed mid 1986. Neverthless, work is still going on.

RESULTS :
****
Studies on this East African rifts system covered a wide area across the lakes
Tanganyika, Rukwa and Malawi.
This project has provided the means for a detailed analysis of the origin and
the evolution of this type of basin. This analysis proved notably that the
evolution could be divided into three stages within global drawing tectonics:
- an initial drawing stage
- a drawing stage with tilting of blocks
- a final landslip stage
Different field and laboratory analysis methods were therefore developed : high
resolution seismics, coring heat flow measurement, field structural analysis,
remote detection, geochemical analyses of organic compounds, etc...
On the structural level, we were able to specify the tectonic evolution and
show the importance of these tectonic phenomena on the layout of sediments,
host rocks, reservoirs and roofs.
Hence, we were able to define the exploration subjects to be treated within
this type of sedimentary basins. Nonetheless, additional studies remain
necessary to specify some of the phenomena that were seen.

REFERENCES :
****
"REMOTE DETECTION SUPPORTED BY OBSERVATIONS ON THE GROUND LEVEL OF ACCIDENTS NW-
SE IN THE TANGANYIKA, RUKWA AND MALAWI AREA OF THE EAST AFRICA RIFT.
J. CHROROWICZ, J. LE FOURNIER, C. LE MUT, JP RICHERT, FL SPY ANDERSON AND JJ
TIERCELIN 1983
"GEOLOGICAL CARTOGRAPHY FROM SPACE PHOTOGRAPHIES OF THE TANGANYIKA, RUKWA AND
MALAWI AREA OF THE EAST AFRICAN RIFT.
J. CHROROWICZ 1984
"THE TANGANYIKA LAKE BASIN: TECTONIC AND SEDIMENTARY EVOLUTION"
J. LE FOURNIER, J. CHRROROWICZ, C. THOUIN, F. BALZER, PY CHENET, JP HENRIET, D.
MASSON, A. MONDEGUER, B. ROSENDAHL, FL SPY ANDERSON, JJ TIERCELIN 1985

```
*****************************************************************************
* TITLE : ACOUSTIC MEASUREMENTS ON RESERVOIR ROCK.   *      PROJECT NO     *
*                                                    *                     *
*                                                    *    TH./01035/83/NL/..*
*                                                    *                     *
*****************************************************************************
* CONTRACTOR :                                       * PROGRAM :           *
*    TECHNISCHE HOGESCHOOL DELFT, AFD MIJNBOUWKUNDE   *    HYDROCARBONS      *
*    POSTBUS 5028                TEL 015 78 13 28     *                     *
*    NL - 2600 GA DELFT          TLX 38151           *                     *
*                                                    * SECTOR :            *
*                                                    *    GEOPHYSICS        *
* PERSON TO CONTACT FOR FURTHER INFORMATION :        *                     *
*    IR. J.P. VAN BAAREN                              *                     *
*                                                    *                     *
*                                                    *                     *
*****************************************************************************
```

VERSION : 01/01/87

AIM OF THE PROJECT :
****
The objective of the research is to determine more rock properties from the
microseismogram obtained via borehole measurements with the Sonic tool.

PROJECT DESCRIPTION :
****
Microseismograms are measured under atmospheric pressure and room temperature
on artificially created rock. The synthetic rock closely resembles "natural"
sandstone but it has the advantage that all rock parameters are known exactly
and can be changed independently. Thus by changing one rock parameter it is
attempted to establish a correlation between that particular parameter and the
received microseismogram. To investigate the influence of pressure and
temperature on the recorded wave train the acoustic measurements will also be
performed in a triaxial cell under conditions of up to 500 bar pressure and 300
deg. C.

STATE OF ADVANCEMENT :
****
Ongoing.
A thorough literature survey has been made to be up to date with the state of
the art of acoustic measurements on rock and the wave equation theory.
The triaxial cell is under construction.

RESULTS :
****
Two modelling programs are available simulating three dimensionally the pulsed
acoustic wave motion in a two media configuration with a plane interface using
Biot's theory. One model eliminates the influence of attenuation and dispersion
resulting in a very considerable reduction in required computer-time compared
to the other model. Both models accurately predict the wave velocity
encountered in the measurements.
Transmission measurements through thin slices of synthetic rock indicated the
presence of a slow compressional wave. This strongly points to the fact that
the assumptions made in the models are at least qualitatively applicable to the
artificial rock samples.
A laboratory routine has been established to construct the artificial sandstone
samples.
- Some studies on actual borehole data have been made.
- The data-acquisition system has been completed including a x-, y-, z-
positioning system wich allows directional scanning of a rock sample.
- Reflection measurements are performed under atmospheric pressure and room
temperature.
- Transmission measurements a la Plona are performed to obtain a good acoustic
characterisation of the rock samples.
- The influence of the incorporation of attenuation and dispersion as given by
Biot's theory in acoustic modelling is investigated.

REFERENCES :
****
VAN BAAREN, J.P., VISSER, R. AND HELLER, H.K.J., 1986, CONSTRUCTION OF
RESERVOIR ROCK SAMPLES FOR ACOUSTIC RESEARCH, TRANSACTIONS PAPER FF, SPWLA 10TH
EUROP. EVAL. SYMP., ABERDEEN 22-25 APRIL.

```
******************************************************************************
* TITLE : NEW METHODOLOGY AIMED TO ENHANCE DEEP    *     PROJECT NO         *
*         SEIMIC REFLECTIONS DEFINING THE STATIC   *                        *
*         CORRECTIONS DUE TO A COMPLEX SHALLOW      *     TH./01037/84/IT/.. *
*         GEOLOGY                                   *                        *
******************************************************************************
* CONTRACTOR :                                      * PROGRAM :              *
*   AGIP SPA                                        *   HYDROCARBONS         *
*   ELGE DEPT                 TEL 02 - 52023227     *                        *
*   C.P. 12069                TLX 310246            *                        *
*   IT - 20120 MILANO                               * SECTOR :               *
*                                                   *   GEOPHYSICS           *
* PERSON TO CONTACT FOR FURTHER INFORMATION :       *                        *
*   ING. D. BILGERI                                 *                        *
*                                                   *                        *
*                                                   *                        *
******************************************************************************
                                                    VERSION : 01/01/87
```

AIM OF THE PROJECT :
****
The goal of this research project consists in increasing the resolutive power
of seismic data by searching and developing new methodologies for the
determination of both field and residual static corrections and for the
broadening of the frequency content of the data.

PROJECT DESCRIPTION :
****
The project will develop through different steps of studies and experimental
works in the filed and in the processing :
1. Data acquisition
2. Data processing
3. Special processing (surface static correction)
4. Synthesis of the tests to get an estimate of velocities and thickness and
   to draw conclusions from the seismic results and from the developed
   methodology.
STATE OF ADVANCEMENT :
****
Data acquisition consisting of eight Hydrapulse, four Vibroseis, two Dynamite
seismic lines with the corresponding basic data processing and three geological
surveys have been done. A comparative analysis of field static computation and
special processing of high resolution data are under development.
RESULTS :
****
In field data acquisition new techniques have been experimented with
satisfactory results. Hydrapulse lines seem suitable to define the geophysical
properties of the near surface geological picture.

REFERENCES :
****
- HAGEDOORN, J.G. 1959, THE PLUS-MINUS METHOD...: G.PROSP.,7,158-182.- PALMER,
D., 1980, THE GENERALIZED RECIPROCAL METHOD....:SOC.EXPLOR.GEOPHYS.
-SCHNEIDER, W.A., AND SHIN-YEN KUO, 1985, REFRACTION MODELING FOR STATIC
CORRECTIONS :
  55TH SEG,WASHINGTON,D.C.
- FOSTER, M.R.AND GUINZY,J.,1967,THE COEFFICIENT OF COHERENCE...GEOPHYSICS 32,
602-616.-WALDEN,A.T. AND WHITE, R.E.,1984,ON ERRORS OF FIT AND ACCURACY IN
MATCHING.G.PROSP.32,871-891
-WHITE,R.E.,1980,PARTIAL COHERENCE MATCHING...,G.PROSP.28,333-358.

```
******************************************************************************
* TITLE : 3-D SEISMIC SURVEY IN THE GAS-OIL FIELD    *       PROJECT NO      *
*         OFFSHORE KATAKOLON (WESTERN GREECE)        *                       *
*         DEVELOPING LOW COST TECHNIQUES.            *    TH./01038/84/HE/..  *
*                                                    *                       *
******************************************************************************
* CONTRACTOR :                                       * PROGRAM :             *
*   PUBLIC PETROLEUM CORP. OF GREECE                 *   HYDROCARBONS         *
*   KIFISSIAS AVE 19, MAROUSSI TEL 8069301-9         *                       *
*   HE - 15124 ATHENS          TLX 221583            *                       *
*                                                    * SECTOR :              *
*                                                    *   GEOPHYSICS          *
* PERSON TO CONTACT FOR FURTHER INFORMATION :        *                       *
*   DR. M.L. MYRIANTHIS                              *                       *
*                                                    *                       *
*                                                    *                       *
******************************************************************************
```

VERSION : 21/04/87

AIM OF THE PROJECT :
****
The 3-D seismic survey project covered the area of the oil and gas field
discovered in 1980 by Public Petroleum Corporation. The field name "West
Katakolon" is located 3 miles offshore Cape Katakolon in Western Greece.
The producing horizon is the anticlinal Cretaceous carbonates unconformably
covered by clastic Neogene sediments. A prominent feature SE from the structure
is an evaporitic diaphir which cuts through the carbonates and the younger
Neogene strata.
The results from the tests in the 3 wells that followed the discovery proved
that the field is classified as marginal mainly due to the weater depth in the
area which is about 230 m.
The need for increased confidence in the estimation of reserves led towards the
planning of a "low cost" 3-D seismic survey which is considered as the
appropriate approach for financially marginal oil fields. Other alternative is
the conduct of appraisal wells a far dearer solution than specially dedsigned 3-
D surveys.

PROJECT DESCRIPTION :
****
To meet the requirements of detailed structural delineation high resolution in
the final product was necessary.
Close line spacing in the acquisition stage preserves higher frequencies in
recording the seismic wave field. Generally the higher the frequencies recorded
the higher the resolution achieved.
In the calculation of the desired line spacing enter data such as the desired
frequency the velocity of the reflector and its maximum dips. A line interval
of 37.5m. was considered necessary to assure the desired high frequency to be
recorded. For such a close line spacing requirement, mainly due to steep dips
present in West Katakolon the conventional technique of collecting one line of
sub-surface per vessel pass would make the survey very expensive to be applied
in such marginal field.
To meet the requirements for a low-cost technique the following innovation was
designed and applied. Instead of the conventional one source of seismic waves,
two sources were used separated by a system of paravanes and booms to allow
collection of two seismic lines in one pass of the vessel. This reduced the
cost of data collection by approximately 40%.

15

The center of the two sources were 75 m. apart the number of airguns used per source array was 17 with a volume of 2125 c.in. per arrays. The streamer was positioned in the middle between thw two source arrays. Its length was 2400 m. (96 channels) and the near offset was approx. 200 m. from each array. The variation of this length never exceeded 1-2 m.

Generally the performance of the above innovation system (novel technique) was problem-free and proved to be a real time-money saver.

As a whole the processing package that applied in the processing of the 3-D data is considered to be at the state of the art level. One of the applied techniques though sold be especially emphasised because it greatly reduces the cost of the 3-D survey. This is the Line Interpolation technique.

For the interpolation to be applied, the stacked data were first migrated in-line. Then the data were recorded into cross-lines and interpolated to a line spacing of 18.5 m. using Contractor's Intelligent Interpolation programme. After the interpolation the data were finally migrated in the cross-line direction. Interpolation generates a new trace between every two existing traces.

The interpolation technique was essential to reduce the actual shot line spacing from 37.5 m. to 18.5 m. allowing the high desired resolution to the final product by preserving higher frequencies ranging in this case from 37 to 54 Hz for cross-line dips ranging from 30 to 20 degrees compared to 26-28 Hz max. frequency preserved for non-interpolated data.

This novel technique generates additional lines for input the 3-D migration process that otherwise should be actually shot. The two to one interpolation represents 50% cost saving from lines not required to be actually shot. At the same time the max. frequency preserved in 3-D Migration were increased by 50%.

STATE OF ADVANCEMENT :
****
Completed

RESULTS :
****
The need for increased confidence in the structural interpretation in the case of marginal fields led towards the planning of a 3-D seismic survey which could provide a better basis for reserves estimation.

The area of the 3-D coverage is rectangular in shape with dimensions 13 x 7.5 km. its long axis running NW-SE.

The two novel techniques successfully applied in the Acquisition and processing of the data dramatically reduced the cost and proved the 3-D survey applied in this project to be a cost effective technique for the development of marginal oil fields such as the W. Katakolon.

Following the interactive interpretation it appears that.

1. The evaportic diaphir is now interpreted over a more restricted zone allowing space for mapping the base of Neogene unconformity over a more extended area compared to previous results (2-D seismic).

This unconformity being the top of the producing horizon is a key target for oil exploration in Western Greece.

2. The configuration of the W. Katakolon structure is now different presenting two isolated highs west of the main structure. Also the peak of the main structure appreared at a shallower time contour, i.e. the structure is thicker.

3. An independent high SE of the evaporite is revealed extended over an area comparable in sizesize to the W. Katakolon structure.

4. Conversion of time to depths is considered to be much more accurate than previous attempts. The water bottom and two Neogene horizons A and B were mapped. Velocities from well log data of W. KA-2 were assigned to the layers in between the mapped horizons. The depth maps produced are free of the water bottom effect. This resulted in a slightly different configuration of the structure when compared with the time map. One isolated high instead of two is now observed west of the main structure.

More accurate volumetric calculations are now possible on the basis of the depth maps. The figures of interpretation supplied by G.S.I. show an increase of 15% approximately for the volume of the Reservoir rock. The new volumetric data along with the revealed individual highs are now being studied by our reservoir engineers and a more accurate reserve estimation is expected.

```
*******************************************************************************
* TITLE : HIGHER RESOLUTION MARINE SEISMIC DATA      *       PROJECT NO      *
*         ACQUISITION AND PROCESSING.                *                       *
*                                                    *    TH./01039/84/NL/..  *
*                                                    *                       *
*******************************************************************************
* CONTRACTOR :                                       * PROGRAM :             *
*   TECHNISCHE HOGESCHOOL DELFT, AFD MIJNBOUWKUNDE    *   HYDROCARBONS        *
*   POSTBUS 5028               TEL 015 785190         *                       *
*   NL - 2600 GA DELFT         TLX 38151              *                       *
*                                                    * SECTOR :              *
*                                                    *   GEOPHYSICS          *
* PERSON TO CONTACT FOR FURTHER INFORMATION :         *                       *
*   PROF. A.M. ZIOLKOWSKI                             *                       *
*                                                    *                       *
*                                                    *                       *
*******************************************************************************
                                                    VERSION : 01/01/87
```

AIM OF THE PROJECT :
****
The aim of the project is to improve the resolution of marine seismic data by
improved data acquisition methods and corresponding improvements in the data
processing. In particular, an accurate description of the angular-dependent
source wavefield must be provided and new data processing software developed to
utilize this information.

PROJECT DESCRIPTION :
****
A data set already exists consisting of a seismic line shot over a logged well
in the North Sea. The line has been shot with a number of different source
configurations and in each case the source wavefield can be determined. The
well logs provide a check on the processing of the data in two ways: first,
synthetic seismic data can be calculated from the well logs and compared with
the real data; second, the real data can be deconvolved and inverted to obtain
density and velocity logs that can be compared with the real well logs.

STATE OF ADVANCEMENT :
****
We have almost completed the project. The angular dependent source wavefield
deconvolution and subsequent standard processing of 8 Km of seismic data is now
complete. Reports of this work will be ready in March 1987. The computer coding
of the forward modelling of the offset-dependent synthetic seismograms for both
the acoustic and elastic layered earth models is complete. The reports of the
forward modelling will be ready in June 1987, together with the final report.

RESULTS :
****
Angular-dependent deconvolution of CMP data in the wavenumber-frequency domain
improved the resolution of the data at the target depth of 3000 m. The far
field approximation used in the method reduced the quality of the shallow data
in the first 300 m. A spin-off of this research was the development of an
interactive data-adaptive method for the design of a desired signature for the
deconvolution. The forward-modelling is done by calculating individual plane
wave responses using the reflectivity method and a matrix extension for the
elastic case, superposing the plane waves to create the point source response,
and then superposing the point source responses using the national source
concept. In this scheme the biggest problem has been the superposition of the
plane waves via the fourier-bessel transform. Six different schemes have now
been developed with various trade-offs between accuracy and speed.
Inversion of the deconvolved data for density and velocity is a logical
extension of the original project, but is too big a task to be included. It is
now regarded as the goal of a new research project.

```
****************************************************************************
* TITLE : LONG RANGE HIGH ACCURACY NAVIGATION    *      PROJECT NO        *
* *        SYSTEM.                               *                        *
* *                                              *    TH./01040/84/IR/..   *
* *                                              *                        *
****************************************************************************
* CONTRACTOR :                                   * PROGRAM :              *
*    SEA SURVEY LTD                              *    HYDROCARBONS        *
*    RATHMACULLIG WEST         TEL 021 962600    *                        *
*    BALLYGARVAN               TLX 75850         *                        *
*    IR - CO. CORK                               * SECTOR :               *
* *                                              *    GEOPHYSICS          *
* PERSON TO CONTACT FOR FURTHER INFORMATION :    *                        *
*    MR. T. O'SHEA                               *                        *
* *                                              *                        *
* *                                              *                        *
****************************************************************************
```

VERSION : 20/02/87

AIM OF THE PROJECT :
****
To extend the range of the High Accuracy 2 MHZ Positioning System, know as
"Hyper-Fix" to provide coverage over the frontier areas of the E.E.C.'s
economic zone, i.e. Rockall Trough, Porcupine Bank, West Hebrides, Northern
Norway, Southwest approaches.

PROJECT DESCRIPTION :
****
In order to extend the present range of the Hyperfix system (20 - 250 kms for
24 hour operation) the project envisaged the deployment of a buoy fitted with a
Hyperfix transmitter. The buoy deployed in existing Hyperfix chain coverage
would then retransmit positioning data to the user's vessel. In addition to the
line of position (L.O.P.) generated by the buoy, the user would also receive
from the buoy updated buoy positions as observed by the buoy from the received
L.O.P.'s from the Hyperfix shore stations. To prove that the changing buoy
position was accurately monitored by the Hyperfix, independent checks on the
buoys position would be made using Micro-wave and acoustic positioning systems.

STATE OF ADVANCEMENT :
****
Completed. The project as described above took place during the summer and
autumn months of 1986. The buoy was deployed in the North Sea and within
coverage of the Racal Survey operated the Forth Hyperfix chain.

RESULTS :
****
The evaluation of all the data i.e. monitor records, systems comparisons and
the sea trial yielded the following conclusions:-
The buoy provided a viable L.O.P. to the user vessel.
The sea trial proved that good data at ranges in excess of 185 km was obtained
from the buoy.
The buoy's movement in all sea conditions was accurately reported to the
monitor station and the user vessel.
The system accuracy was within the Industry's acceptable limits i.e. 15-20
meters.
The bouy's power supply system (Gas Generator) initially proved unreliable.
The buoy when deployed was very difficult to service even in the relatively
shallow water close to shore.
Given the current state of the industry with oil companies cutting exploration
budgets, especially in frontier areas where a system like Hybuoy might be used,
it is felt that the expense involved of building and deploying two, possibly
three, such buoy's would prove uneconomical. Therefore further research in this
area has been shelved.

```
*********************************************************************************
* TITLE : HIGH PERFORMANCE SEISMIC STREAMER         *        PROJECT NO        *
*                                                   *                          *
*                                                   *     TH./01041/84/UK/..   *
*                                                   *                          *
*********************************************************************************
* CONTRACTOR :                                      * PROGRAM :                *
*   BRITOIL PLC                                     *   HYDROCARBONS           *
*   BOLTON STREET 29            TEL 01 409 25 25    *                          *
*   UK - LONDON W1Y 8 BN        TLX 8812071         *                          *
*                                                   * SECTOR :                 *
*                                                   *   GEOPHYSICS             *
* PERSON TO CONTACT FOR FURTHER INFORMATION :       *                          *
*   MR. J. ANDERSON                                 *                          *
*                                                   *                          *
*                                                   *                          *
*********************************************************************************
                                                              VERSION : 01/03/86
```

AIM OF THE PROJECT :
****
The project undertakes the work necessary to develop an advanced seismic
streamer and computer system capable of supporting high resolution marine
seismic surveying. This project will improve the capabilities in :
- marginal fields exploration
- identification of subtle trapping mechanisms
- reservoir development.
PROJECT DESCRIPTION :
****
This project is part of a development programme to produce a seismic streamer.
The 2 first phases of this programme have been carried out under the contract
TH 01031/83. The present phase is devoted to the trials of a prototype 512
channel, 1,5 Km streamer, in various North Sea weather conditions. The main
characteristics of the streamer are :
- 100 m acoustically active sections with each section containing
  32 hydrophone groups;
- between any 2 sections, 4 data acquisition modules (DAU) with
  telemetry interface module and power distribution module will form a node.
Each node will :
- absorb data from 32 groups (16 from each adjacent section)
- undertake signal condition (high pass, low pass filtering and calibration)
- perform digitisation (instantaneous floating point at 0.5 msec)
- merge the data on to the telemetry highway.
All auxiliary channels such as compasses, depth sensors, depth controllers etc..
. will be controlled by the telemetry system which itself will interface to a
general purpose computer.
Under software control the computer will undertake :
- data stream organisation   - data validation
- error detection and fault location   - real time processing
- committal to final archival storage.
STATE OF ADVANCEMENT :
****
The project is just completed.
RESULTS :
****
The prototype unit was found to be comparable with a conventional system when
similarly configured but when used at its higher channel capacity, the
prototype gave encouraging results indicating enhanced resolution and lack of
spatial aliasing.

```
*****************************************************************************
* TITLE : INVERSION OF SEISMOGRAMS                 *        PROJECT NO       *
*                                                  *                         *
*                                                  *    TH./01042/84/FR/..   *
*                                                  *                         *
*****************************************************************************
* CONTRACTOR :                                     * PROGRAM :               *
*   GERTH                                          *   HYDROCARBONS          *
*   AVENUE DE BOIS PREAU 4      TEL 1 47 52 61 39  *                         *
*   FR - 92502 RUEIL-MALMAISON TLX 203 050         *                         *
*                                                  * SECTOR :                *
*                                                  *   GEOPHYSICS            *
* PERSON TO CONTACT FOR FURTHER INFORMATION :      *                         *
*   MR. LEGRAND                 TEL 59 83 40 00    *                         *
*                               TLX 560 804        *                         *
*                                                  *                         *
*****************************************************************************
```

VERSION : 01/01/87

AIM OF THE PROJECT :
****
This project is involved with the inversion of reflection seismic data, in
other words attempting to find the subsurface geological parameters that best
explain the surface recorded seismic traces.
The subsurface is approximated as an acoustic medium where wave propagation is
described by only two parameters: velocity of propagation and acoustic
impedance (for 1D-models only the latter is involved).
These two parameters are the unknowns for the inverse problem.
The first objective of this project is to study the feasibility of several
inversion type problems (depending on the geometrical hypotheses : 1D, 2D
horizontally stratified, heterogeneous 2D).
The second goal is to write software for processing large field data volume.

PROJECT DESCRIPTION :
****
The reflection seismic method can be briefly described by a source near the
surface produces a propagation wave in the substratum. The resulting
perturbation which is reflected, refracted or diffracted is measured at the
surface by geophones (Gi) as pressure or displacement Pobs (Gi,t) at a function
of time.
The forward problem involves the computation of synthetic seismograms Pm (Gi,t)
that correspond to a given subsurface model m. Acoustic wave propagation is
assumed.
The inverse problem, briefly explained, will involve the search of an earth
model m that produces seismograms Pm (Gi,t) as close as possible to the
observed seismograms Pobs (Gi,t).
The first step is then construct a tool that efficiency resolve the forward
problem, inversion requiring the computation of J(m) for a large number of
models m.
The next step will require the construction of a second basic tool which can
calculate the gradient of J with respect to the model m.
An optimization strategy must then be defined, entirely automatic or at the
contrary interactive, conviently integrating a priori information of the model.

A sensitivity study has to be done, with two objectives: one is to determine the space of admissible models in which the search of the solution is stable; the second objective is to find the best optimization variables along with a correct preconditioning of the problem to increase stability and the convergence rate.

Many software tests will be required, first with noisy synthetic data, and later with data.

Three inversion problems are to be used:

A 1D inversion : plane waves modelling in an horizontally stratified media. Acoustic impedance as a function of travel-time is sought.

B 2D inversion of horizontally stratified media: waves propagation is 2D, the model to be found is 1D. The redundancy of informations for different offsets allows the search for both velocity and acoustic impedance.

C Heterogeneous 2D inversion : the geometry of the entire problem is 2D, which means that there will be a considerable number of unknowmns, and that a large number of seismic records is required to supply enough information.

This problem is therefore very complex as well as very costly to study.

STATE OF ADVANCEMENT :
****
Ongoing. Problem A is in a monitoring phase, while problems B and C are in a construction and design phase.

RESULTS :
****
Part A of the project, involving 1D inversion, is already well developed. Numerous theoretical as well as pratical results can be found in the literature, describing the possibilities and the difficulties of this problem.

From these results, after the construction of a solid modelling and gradient calculation software package we were able to quickly set a path towards its practical application.

A software package was created for stratigraphic of well data using 1D inversion. A seismic profile of zero offset traces is constructed, after which each trace is inverted by using the low frequency impedance trend of the previous trace. The first inversed trace uses the trend from the well.

This tool appears to be a very useful on for nearly horizontal stratigraphies. Nevertheless, field acquisition problems have been highlighted. These problems have to be solved before the tool can acquire a certain credibility.

Parts B and C of this project related to 2D inversion have needed the invention of new techniques to solve the problems associated with finding a velocity field. These techniques (use of travel time instead of depth, progressive downward continuation in time, progressive increase in frequencies, adapted unknowns) were successfull implemented in the problem B. This has to a satisfying solution for synthetic data of large dimensions.

For heterogeneous 2D media, these new techniques are much too heavy to be implemented at the moment. There is however, one aspect that has been solved in a very successful fashion : that is finding the impedance reflectors when the velocity field is approximately known. The results consist in a refinement of a migration before stack using the wave equation.

```
******************************************************************************
* TITLE : UTILISATION OF DRILLING NOISES IN      *        PROJECT NO        *
*         SEISMICS.                              *                          *
*                                                *     TH./01043/84/FR/..   *
*                                                *                          *
******************************************************************************
* CONTRACTOR :                                   * PROGRAM :                *
*   GERTH                                        *   HYDROCARBONS           *
*   AVENUE DE BOIS PREAU 4      TEL 1 47.52.61.39 *                         *
*   FR - 92502 RUEIL-MALMAISON TLX 203 050       *                          *
*                                                * SECTOR :                 *
*                                                *   GEOPHYSICS             *
* PERSON TO CONTACT FOR FURTHER INFORMATION :    *                          *
*   MR. P. GROS                 TEL 1 47.44.37.32 *                         *
*                               TLX 615 400      *                          *
*                                                *                          *
******************************************************************************
```

VERSION : 01/01/87

AIM OF THE PROJECT :
****

This project is based on the utilisation of the drill bit as an acoustic source.
 Seismic signals emitted during the abrasion of the rock by the drill bit are
recorded via seismic pick-ups placed in a special manner on the surface or in
nearby wells.
After these signals have been processed and interpreted in real time, it is
possible to obtain data comparable to that provided presently off-line by
seismic measurements taken in wells, for example the equivalent of a "speed
log" in relation to the depth, the seismic profile of a well, or a 3D image of
formations crossed in the vicinity of the well. Thus, it is a matter of
studying and finalizing appropriated methods and equipment, thus providing
access to a better instant acknowledgement of the crossed formations, and
allowing to improve drilling procedures.

PROJECT DESCRIPTION :
****

To achieve this goal, it is necessary to use both, the seismic signals received
at the surface on a seismic type apparatus, or on special sensors placed in
nearby wells, and the signals transmitted by the drill string and received at
its head, as the latter signals can be representive of a pseudo-signature of
the seismic signals transmitted by the drill bit.
This project has been divided in five major phases :
Phase 1 - Measures at the drill string
Phase 2 - Measures at the surface geophone
Phase 3 - Combination of both types of measures
Phase 4 - Well seismic source
Phase 5 - Special data acquisition and processing device on the site.
Phase 1, 2 and 3 are closely related, as phase 3 can only be executed once

Phases 1 and 2 have been carried out on the same site and at the same time.
In this regard, measures were made at three different sites in France :
LE MAYET DE MONTAGNE, in July 1984, while drilling in granite for the purpose
of geothermals studies. Measures were taken at a depth between 200 and 800 m,
and recorded with a 47-trace surface device over a length of about 700 m and
placed radially to the well, and on a set of sensors placed in bores near the
well. An accelerometer was fitted at the end of the drill string. Drilling was
executed out in the open with a bottomhole drill hammer.
LACQ SUPERIOR, in May 1985, during a vertical drilling operation, very near a
deviated well. This particularity has allowed us to record with the help of
well geophones, the signals transmitted by the bit drilling at depths of about
400 m, and measured at horizontal distances inferior to 50 m. The purpose was
to compare the signals recorded at the upper end of the drill string with those
recorded at the drill bit.
SOUDRON 116, in October 1985, when drilling a production well on the Soudron
field, south of Chalons sur Marne in the Paris Basin. During this field data
acquisition, measures were taken between 1000 and 2100 m deep (one level of
measures for every string adjunction) on seismic devices comprising sensors
placed, both, at the surface and in bores at a depth of about 40 m.
Phase 3 consisted mainly of the acquisition via numerical processing of seismic
profiles of the transposed and multi-offset wells, or of data on the vertical
seismic speeds by using the records provided by surface devices and drill
strings, and the repeated cross-correlations.
Phase 4 consisted in a feasibility study and in the definition of
specifications in view of designing a prototype seismic source while drilling
in case the feasibility results of phases 1 to 3 proved insufficient at the
level of the signal to noise ratio transmitted only by the drill bit.
Phase 5 of the project depended mainly on the positive results obtained during
the previous phases. The purpose of phase 5 was to study and make an in-situ
acquisition and processing prototype unit, capable of providing in real time
the data required to run the drilling operation and to pursue exploration.

STATE OF ADVANCEMENT :
****
Phases 1, 2 and 3 have been completed while using only the drill bit as seismic
source. For phase 4, a number of manufacturers of drilling equipment, both
French and American, have been investigated. American equipment is not
available and it is not possible, within the scope of this project to study and
develop a seismic source while drilling. Studies interrupted on 31.12.86.

RESULTS :
****
Studied carried out during phase 3, that is the combination of drill string
measurements and surface measurements, have led to the following conclusions :
1. Seismic profiles for transposed wells can be obtained from the sole
recording of drill bit noises, but the quality of the results remain poor, and
depends partly on the hardness of the formation and on the drilling mode
(ordinary rotation or percussion rotation).
2. The signal to noise ratio remaining low, it would be useful to improve it by
using a seismic source on the drill string near the drill bit.
3. The spectrum of seismic signals, emitted at the bit during drilling and
transmitted in the soil is relatively high frequency and may reach 300 to 400
Hz.

4. In most cases, seismic signals recorded at the top of the drill string are not faithfully representative of the signals transmitted at the bit.
For a better correlation with recordings on devices spread at the surface, it is advised, during cross-correlations of traces, to use the "pseudo-signatures" recorded near the bit. However, this raises the problem of their transmission up to the surface, along the drill string.
5. Obtaining an evaluation of the mean speeds of the formation sections between the bit and the surface, and in a correlative manner, of the speeds of seismic waves in vertical sections (section speed) is possible through a statistical study of the curvatures on seismic profiles without the use of plots at the top of the drill string. However, the results obtained like the previous ones are closely linked to the quality of the signal to noise ratio (soft terrains providing hardly any positive result). Their improvement implies the elaboration of a seismic source.
6. Different drilling equipment have been studied both in Europe and in the U.S. A inherent to the study of a seismic source while drilling. There are very few studies on this subject and because of the present context, there is no real incentive for manufacturers to pursue research in this domain. Consequently studies have been interrupted at this stage of the project.

```
******************************************************************************
* TITLE : DEVELOPMENT OF A TECHNIQUE TO EXPLORE      *        PROJECT NO       *
*         SUBSURFACE GAS-WATER CONTACTS.             *                         *
*                                                    *    TH./01046/84/IR/..   *
*                                                    *                         *
******************************************************************************
* CONTRACTOR :                                       * PROGRAM :               *
*   KISH DEVELOPMENT LTD                             *   HYDROCARBONS          *
*   162 CLONTARF ROAD            TEL 01/332211       *                         *
*   CLONTARF                     TLX 33438           *                         *
*   IR - DUBLIN 3                                    * SECTOR :                *
*                                                    *   GEOPHYSICS            *
* PERSON TO CONTACT FOR FURTHER INFORMATION :        *                         *
*   DR. D. NAYLOR                                    *                         *
*                                                    *                         *
*                                                    *                         *
******************************************************************************
                                                             VERSION : 31/12/86
```

AIM OF THE PROJECT :
****
To examine the geological and geophysical problems of detecting and validating
subsurface gas-water interfaces in hydrocarbon reservoirs as an aid to the
direct detection and mapping of gas reserves. A number of known gasfield
examples in Northwest Europe will be examined and compared to other possible,
but unproven, examples. In particular a horizontal subsurface seismic feature
in the Kish Basin, offshore Ireland, will be detailed by a seismic survey, and
later penetrated by an exploration well, at which time geological-geophysical
studies will be made of the critical rock section.

PROJECT DESCRIPTION :
****
Phase 1: carefully monitored seismic acquisition (with large volume water guns)
and processing of reflection seismic data over the Kish Basin example and the
production of time and depth maps of the principal reflecting horizons.
Delineation of the gas water contact. Maturation studies of existing well
samples.
Phase 2: detailed lithological/petrological studies of the two existing Kish
Basin wells and analysis of the possible gas-water contact in the Basin.
Documentation of known gas-water examples from other world-wide. Drilling of a
deep exploration well in the Kish Basin and study of the results.
Phase 3: complete analysis of Kish Basin results.
Studies of three gas-water contacts with reprocessing of seismic data and
detailed analysis of electric log and petrological data.
Phase 4: final report preparation.

STATE OF ADVANCEMENT :
****
Ongoing study of European gas-water contacts. The ability of the seismic method
to detect gas-water contacts has been demonstrated in recent years, although
the precise parameters for optimum recording are not clearly understood. Other
subsurface phenomena may also mimic a gas-water contact on the seismic record.
The project will assess the optimum technical package available to the industry
in gas-prone areas.

RESULTS :
****
Work under Phases 1 and 2 of the project is complete and Phase 3 is now being undertaken.

100km of additinal reflection seismic data were obtained to examine a possible gas-water interface in the Kish Basin. At the same time a maturation study of samples from the existing Amoco 33/22-1 well and related areas was aimed at a study of source potential.

Processing of the Kish Basin lines demonstrated that the reflection was probably not a fluid-contact reflection. Petrographic and electron microscope studies of samples from the nearly Shell 33/21-1 well showed that the potential Triassic reservoir demonstrated high average porosities with only limited secondary mineral development to inhibit permeability..

As part of a wider documentation of gas-water contacts about 20 published and unpublished examples have been collected worldwide. These examples are from a wide variety of geological settings and show that for recognition of a fluid contact reflection it must be a positive reflection, observed in horizontal or near-horizontal attitude, in a trapping position, against a background of moderate dip.

Drilling of the Kish Basin 33/17-1 well took place in May-June 1986.The well was plugged and abandoned as a dry hole at 6,600 feet (2012m).Results from the well substantially confirmed the geological prognosis for the structure and confirmed that the anomalous flat reflector on the prospect was not a fluid-contact reflection.

Permissions have been received from the exploration groups holding licences on the three gas-water contact samples which have been chosen for study. Tapes of the seismic data are being reprocessed in an attempt at fluid-contact optimisation. The first example is from a producing gasfield in the Dutch North Sea, the second a subcommercial gas accumulation from the U.K. Southern North Sea gas province and the third is from the Irish Sea.

```
*******************************************************************************
* TITLE : ACCESS (PHASE 1)                      *      PROJECT NO          *
*                                               *                          *
*                                               *   TH./01047/85/FR/..     *
*                                               *                          *
*******************************************************************************
* CONTRACTOR :                                  * PROGRAM :                *
*   GERTH                                        *   HYDROCARBONS          *
*   AVENUE DE BOIS PREAU 4       TEL 1 47.52.61.39 *                       *
*   FR - 92502 RUEIL-MALMAISON TLX 203 050      *                          *
*                                               * SECTOR :                 *
*                                               *   GEOPHYSICS             *
* PERSON TO CONTACT FOR FURTHER INFORMATION :   *                          *
*   MR. P. LALOUEL               TEL 1 42.91.33.22 *                       *
*                                TLX 615 700     *                          *
*                                               *                          *
*******************************************************************************
                                                        VERSION : 01/01/87
```

AIM OF THE PROJECT :
****
The aim of this project is to develop computer assisted interpretation stations
that are to form the future working tool of explorers. In a word, ACCESS
provides access for all exploration technicians to more comprehensive and
diversified information. It is not only the capability of receiving data, but
also that of communicating analyses being made or completed, either in the head
office or in remote exploration missions.

PROJECT DESCRIPTION :
****
Compared to existing data processing tools, the ACCESS stations will provide
greater flexibility and high capacity, making considerably more efficient
petroleum exploration interpretation, since they will make it possible:
- to manage all the information used or generated at all stage,
- to manipulate this information through their graphical representations,
- to process all this information in integrated and coherent fashion.

STATE OF ADVANCEMENT :
****
Ongoing

RESULTS :
****
Hardware has been selected. Implementation of the core has been achieved and
these tools (man/machine dialog, numerical interpretor, graphical editor...)
can be now used for the development of interpretation applications.
Structure of geological, geophysical and topographic data base has been
designed.
Single well data processing, topographical and geophysical applications are in
"beta test". The cartographical package is being prepared.
All of the first phase of the project has been completed and an independent
structure called "PETROSOFT" will be created in 1987 to commercialise the
product.
The complete package will be available on the market early in 1988.

```
********************************************************************************
* TITLE : SECOND GENERATION MULTIPLEXED STREAMER    *      PROJECT NO        *
*                                                   *                        *
*                                                   *   TH./01049/85/FR/..   *
*                                                   *                        *
********************************************************************************
* CONTRACTOR :                                      * PROGRAM :              *
*   GERTH                                           *   HYDROCARBONS         *
*   4 AVENUE DE BOIS PREAU       TEL 1 47.52.61.39  *                        *
*   FR - 92502 RUEIL-MALMAISON TLX 203 050          *                        *
*                                                   * SECTOR :               *
*                                                   *   GEOPHYSICS           *
* PERSON TO CONTACT FOR FURTHER INFORMATION :       *                        *
*   MR. BEAUDUCEL                TEL 1 47.79.02.14   *                        *
*                                TLX 203 050         *                        *
*                                                   *                        *
********************************************************************************
                                                        VERSION : 01/01/87
```

AIM OF THE PROJECT :
****
Feasibility study of a multiplexed streamer with very high number of traces
(1500), through realisation of subassemblies.

PROJECT DESCRIPTION :
****
Important aspects of this streamer are the following :
- Very low traction noise has been obtained by improving the buoyancy balance
  of the immersed part. This has led to the study of :
  - the weight reduction of boxes containing the electronics and/or the
    distribution of the electronics along the streamer elements
  - the weight reduction and the homogenization of the lineic density of the
    streamer elements.
- An efficient filtering of residual noises :
  - through the utilisation of continuous hydrophones
  - by considering their location at the level of the frequency/number of
    wave (fanshaped filtering).

STATE OF ADVANCEMENT :
****
The following are underway :
- Tests of numerical transmissions on optical fiber
- Measurements on several models of continuous hydrophones
- Study of a complete acquisition subassembly with 8 channels which will be
made in the form of a waterproof module
- Study of a waterproof optical transmission assembly of 30 Mbits/s
subcontracted to SYSOPTIC
- Study of technology to manufacture a continuous flat hydrophone
- Elaboration with potential subcontractors of waterproof module specs.

RESULTS :
****
WATERPROOF MODULES : realisation and measurement of several models of
amplifiers with variable gain (instantaneous Floating Point Amplifier) (the
electronic element contained in the modules)
WEIGHT REDUCED BOXES : drafting specifications for the optical transmission of
30 Mbits/s for which the study is presently being carried out by SYSOPTIC, as
well as specifications for the mixed optical and electrical waterproof
connector.
CONTINOUS HYDROPHONES :
Outlining :
1) the feasibility of such hydrophones
2) the poor rejection versus the acceleration of coaxial hydrophones, thus
implying an evolution towards flat hydrophones
REFERENCES :
****
PATENT INHERENT TO VARIABLE GAIN AMPLIFIERS
3 PATENTS RELATIVE TO CONTINUOUS HYDROPHONES

```
********************************************************************************
* TITLE : MULTIWELL FIELD GEOPHYSICS          *          PROJECT NO          *
*                                             *                              *
*                                             *     TH./01050/85/FR/..       *
*                                             *                              *
********************************************************************************
* CONTRACTOR :                                * PROGRAM :                    *
*   GERTH                                     *   HYDROCARBONS               *
*   AVENUE DE BOIS PREAU 4      TEL 1 47.52.61.39 *                          *
*   FR - 92502 RUEIL-MALMAISON TLX 203 050    *                             *
*                                             * SECTOR :                     *
*                                             *   GEOPHYSICS                 *
* PERSON TO CONTACT FOR FURTHER INFORMATION : *                             *
*   MR. DELVAUX                  TEL 59.83.61.65 *                          *
*                                TLX 560 804   *                             *
*                                             *                              *
********************************************************************************
```

VERSION : 01/01/87

AIM OF THE PROJECT :
****
The object of the project is to develop methods allowing to obtain geophysical
data on the nature of the terrains between different wells within a same field.
The aim is to improve our knowledge on the reservoir layer with the help of
methods complementing those already existing and which are generally
implemented from the surface, such as seismics or electric methods or which are
implemented directly in the wells and presenting a limited investigation range,
such as logs.
The implementation of new methods in wells would provide, at any point of the
reservoir located between these wells, significant and thorough data allowing:
- a better static comprehension of the field :
  new structural definition, location and identification of accidents
  (faults, reefs, etc...), facies variations, etc...
- to obtain data on the dynamic behaviour of the reservoir : preferential
  drains, water inlets, evolution of contacts between fluids.

PROJECT DESCRIPTION :
****
Two major types of methods are being studied :
- methods based on the propagation of acoustic waves between wells
- methods based on the measurement of phenomena provoked like the
electromagnetic measurements between two wells.
One of the particular features of the methodology envisaged within the scope of
this project, resides in the integrated interpretation of results deriving from
the different methods.
Last of all, the latter includes a feasibility phase and a study, execution and
prototype testing phase.
SEISMIC METHOD
Three orientations have been followed, each corresponding to a different
approach of the waves propagation phenomena in the reservoir environment :
- study by transparency :
  while using travel time inversion techniques (first arrivals and selected
reflections). The obtained results are a fine repartition of the geological
speeds between wells. This technique, may, for example, evidence undetectable
speed anomalies thanks to the knowledge of the media crossed at each bore.
- study by reflection :

very high frequency seismics implemented between wells may lead to a very performing visual display of the reservoir. However, due to geometrical particularities at acquisition level, its processing is not yet mastered from the software standpoint. Thus, it is necesary that we prove our ability in emitting and recording seismic signals in a path band according to the objectives of the project, and that we finalize a processing method.
- study on guided propagation :
This phenomenon occurs when a number of conditions concerning speeds, geometry of layers, absence of heterogeneities, are respected. Here we are with an original research orientation which merits exploration (modelling).
ELECTROMAGNETIC METHODS
As the electric properties of a rock depend very closely on its fluids contents, this approach seems, on a conceptual level, paricularly suitable to evidence fluids and contacts between fluids.
However, the transposition of these methods, already used in surface geophysics, remains to be studied more thoroughly within a feasibility phase which is triple :
- study of the reservoir and of the possibilities of these methods (modelling aspect, sensitivity, resolution force)
- drilling environment study
- technological study to draft a prototype pilot chain.
MEANS
All these different approaches mentioned above can only be studied, in a first step, through the modelling tools which will have to be designed or adapted, and can only be achieved once specific equipment will have been implemented in the wells.
- seismic :
three different methods, but a single acquisition chain which should be very wide band (up to 500 to 1000 Hz), of rapid utilisation (important number of traces), non destructive at source level.
- electric :
before setting up the study and execution of a prototype chain, it is necessary to make an overall assessment of technologies used in surface geophysics in order to evaluate the problems.

STATE OF ADVANCEMENT :
****
SEISMIC METHOD : ongoing
Following a study phase on model, prototype tests provided encouraging results.
ELECTRIC METHOD : ongoing
Feasibility is not acquired at all levels.

RESULTS :
****
SEISMIC METHOD
- by transparency : development, adaptations and tests on synthetic data of three 2D or 3D inversion programs (collaboration IFP-SNEA(P)-IRIGM)
- by reflection : familiarize with this type of information thanks to an important volume of synthetic data modelling a complete acquisition between two wells. Utilisation of the latter for special classification, in processing softwares adapted to the seismic profiles of wells with offset.

- guided waves : first modellings which were suspended for a while to improve the computation tools. Resumed early 1987.
- acquisition : an important measurement program with the help of four different well sources, on two different sites. Recording was executed with a conventional 3-component sensor and a wide band 48-channel vertical streamer specially developed for this project. The various sources that have been tested successfully : the shear gun of CPGF, the weight drop on the packer of IFP, the explosive and the sparker of SWRI. Recorded signals have a path band of 500 to 1000 Hz.

The signal to noise ratio is variable and the sensitivity of the streamer needs to be improved ⟨bottom preamplification).
- processing : ongoing.
ELECTROMAGNETIC METHOD
- Studies on models. A number of modelling programs have been written to answer the needs of the field and evaluate the sensitivity of methods versus resistivity variations as well as their spatial resolution. This has allowed us to perform a few studies on synthetic examples. However, these are not sufficient and need to be pursued.
- An environmental study has allowed to confirm that the presence of a casing does not forbid all types of measurements and that a domain can be used between a few hertz (that is 30 - 40 Hz) and a few hundredth of hertz for magnetic measures. It is also possible to use the casing itself as injection electrode.
- A technological survey (investigation, various contacts with different authorities) has allowed to characterize the main pick-ups used in geophysics. Furthermore, very artisanal experimentations have made it possible to orientate the project towards the preparation of a measuring chain.

```
*******************************************************************************
* TITLE : VERTICAL SEISMIC PROFILING AND          *      PROJECT NO         *
*          INTERACTIVE MODELING                    *                         *
*                                                  *   TH./01051/85/DE/..    *
*                                                  *                         *
*******************************************************************************
* CONTRACTOR :                                     * PROGRAM :               *
*   PRAKLA-SEISMOS AG                              *   HYDROCARBONS          *
*   BUCHHOLZER STR. 100        TEL 0511/642 3522   *                         *
*   DE - 3000 HANNOVER 1       TLX 922419 + 922847 PR*                       *
*                                   AKL D          * SECTOR :                *
*                                                  *   GEOPHYSICS            *
* PERSON TO CONTACT FOR FURTHER INFORMATION :      *                         *
*   DR. H.A.K. EDELMANN                            *                         *
*                                                  *                         *
*                                                  *                         *
*******************************************************************************
                                                        VERSION : 25/05/87
```

AIM OF THE PROJECT :
****
The project aims at a better exploitation of VSP data for drillhole planning.
The interactive operation allows to directly include survey data, log data and
VSP data for the decision. The VSP measuring data are directly introduced to
develop a subsurface model.

PROJECT DESCRIPTION :
****
The project consists of two groups of activities. The first group includes the
data acquisition and the onside data processing. The second group includes
sophisticated large computer data processing and interpretation.
Data acquisition consists of two elements, the sonde and the cable which are
lowered into the borehole and the seismic sources, the recording equipment and
onsite data processing computer at the earth surface. The sonde has to be
designed not only to meet the environmental specifications, but also to reduce
the round trip time (lowering the sonde down to the largest depths and
measuring when pulled up back to the surface) to a minimum. For this purpose a
3-component receiver chain, consisting of up to 5 elements with a predetermined
spacing of 25 to 100 m was designed. The number of elements allow to reduce a
round trip time by recording simultaneously at different depths. For deviated
wells with dip angles of more than 30 deg.C against vertical, a gimbled 3-
component geophone receiver was selected for installation in one of the
elements.
The recording time can be appreciably reduced when using multi-source technique.
 This technique can be preferably applied together with the VIBROSEIS method.
Multi-source technique can be applied for multi-offset VSP and for simultaneous
P- and S-wave recording.
The onsite data processing was designed as part of the quality control of the
recording process. The onsite data processing allows to continuously control
the data quality and thus minimize recording time without loss of data quality.
The software to be developed for large computer data processing enables a
careful preparation of the data acquisition by determining the optimum
parameter and by combining all available data and the survey data in a
straightforward manner for later interpretation. For this purpose two ways of
analysis must be used. The first consists of transferring the seismic results
into a 3-dimensional picture of the subsurface in the environment of the well.

This process is called migration. It works sufficiently well in all of those cases in which the layer velocities are known and in wich the layer spacing is not too small compared to the wavelength of the seismic signal. For complicated structures, however, these conditions are normaly not met. In this case an interactive approach must be made starting for the subsurface model constructed from the available data before the VSP measurement has been made. The purpose of the interactive VSP is to refine this model which then becomes a valuable information for further decisions. The two ways of data processing from migration and modelling are included in the project.

In some cases a dip analysis, made from multi-offset VSP data, can help to analyse difficult structures. A polarization analysis must be made when shear waves are involved, especially when polarization effects have to be considered.

STATE OF ADVANCEMENT :

****

Ongoing investigations concern the onsite data processing and the azimuthal orientation tests. Both are in the construction phase.

RESULTS :

****

The receiver chain has been successfully tested in boreholes of up to 3000 m. In these tests the capabilities for multi-offset VSP could not yet be fully exploited. The gimbled 3-component element, designed for deviated wells, was successfully tested for dip angles up to 55 deg.C against vertical. The application of the Disco-software system was abandoned in favour of an onsite data processing, based on a personal computer system. The software for modeling and migration has undergone final acceptance tests. Dip analysis and polarization analysis were successfully applied to field data.

REFERENCES :

****

KOEHLER, K., KOENING,M., 1986: IMPROVEMENT OF MULTI-OFFSET VSP DATA BY SPECIAL DATA PROCESSING. 48 TH EAEG-MEETING, OSTENDE, BELGIUM
WIERCEYKO,E., NOLTE, E., 1987: ERFASSUNG SEISMICHER DATEN IM BOHRLOCH. 47, JAHRESTAGUNG DER DGG IN CLAUSTHALZELLERFELD

```
*******************************************************************************
* TITLE : DEVELOPMENT OF A TRANSIENT-            *        PROJECT NO          *
*         ELECTROMAGNETIC DEPTH SOUNDING SYSTEM  *                            *
*         FOR HYDROCARBONS AND GEOTHERMAL        *    TH./01059/85/DE/..      *
*         RESOURCES EXPLORATION                  *                            *
*******************************************************************************
* CONTRACTOR :                                   * PROGRAM :                  *
*   UNIVERSITY OF COLOGNE                         *   HYDROCARBONS             *
*   INST FOR GEOPHYSICS AND ME TEL 49/221 4702552 *                           *
*   ALBERTUS-MAGNUS-PLATZ      TLX 8882279        *                           *
*   DE - 5000 KOELN 41                            * SECTOR :                  *
*                                                 * GEOPHYSICS                *
* PERSON TO CONTACT FOR FURTHER INFORMATION :     *                           *
*   DR K.M. STRACK/PROF. F.M.                     *                           *
*                                                 *                           *
*                                                 *                           *
*******************************************************************************
```

AIM OF THE PROJECT :
****
In 'no seismic record' areas transient electromagnetic soundings can provide
information of the subsurface structure. Our aim is the development of a
transient EM field system including a portable receiver, field application,
development of data processing and interpretation techniques. This work is to
be carry out in conjunction with the oil industry in order to directly evaluate
the potential application of the technique.

PROJECT DESCRIPTION :
****
A digital portable data acquisition system has been designed, built and tested.
 Due to the high cultural noise in West Germany new concepts of data processing
techniques have been developed including mainly prestack processing and
statistical or selective stacking. The basic research for the 2.5 and 3D
interpretation has been done and the more promising routes for the theoretical
development outlined. Via the test surveys close contact with the oil industry
has been vital to the success of the research. Two test surveys were carried
out with a total of 180 stations at which depth sounding were done. The data
from the fist test area was used for check out and correction of the
acquisition solftware and interpretation procedure. The second test area
posed a more complicated geologic target of a conductive layer under
crystalline rock. Both test survey required drastic changes in processing
sotfware, the first one due to the high cultural noise level (prestack data
processing), the second one due to the shortness of the signal and resulting
instability of the deconvolution.

STATE OF ADVANCEMENT :
****
The data processing is further refined by using the field data as test data. A
solid state transmitter is being developed as well as a floating point
amplifier. Safety standards with the rebuilding of the successfully tested
first transmitter are implemented. Improved inversion techniques with more
statistics are being developed as well as interpretation procedures for
additional field components which would allow to map also resistive horizons.

RESULTS :

The first test survey (hydrocarbon) show that geologic structures could be successfully mapped underneath basalt layers even in areas with very high cultural noise levels. The second test survey (geothermal) allowed us to map a conductive layer underneath crystalline rock at 10 km depth with only 60 a transmitter current. This was mainly due to the improved data processing procedures and the portable battery operated data acquisition system including a high power computer, In the theoretical development first sucess has been achieved by finding a more stable way the solve the equations numerically. In both test areas survey sites have been found with 3-D effects which - although yet non-interpretable - will be very valuable for the 3-D development.

REFERENCES :

STRACK, K.-M.; HANSTEIN, T.; PETRY, H.; ZIEGON, J.; 4 TALKS AT THE FKPE EM WORKSHOPS 1986
HANSTEIN, T.; EILENZ, H.; STRACK, K.-M.; PETRY, H.; 2 TALKS AT THE DGG MEETING 1986
REIMERS, HANSTEIN, SCHALKOWSKI, STRACK, TALK AT THE EAEG MEETING 1986.

```
*****************************************************************************
* TITLE : BOREHOLE SOURCE                            *      PROJECT NO      *
*                                                    *                      *
*                                                    *    TH./01067/86/FR/..*
*                                                    *                      *
*****************************************************************************
* CONTRACTOR :                                       * PROGRAM :            *
*   GERTH                                            *   HYDROCARBONS       *
*   4, AVENUE DE BOIS PREAU      TEL 1 47.52.61.39   *                      *
*   FR - 92502 RUEIL-MALMAISON TLX 203 050           *                      *
*                                                    * SECTOR :             *
*                                                    *   GEOPHYSICS         *
* PERSON TO CONTACT FOR FURTHER INFORMATION :        *                      *
*   MR. LAURENT                  TEL 1 47.49.02.14   *                      *
*                                                    *                      *
*                                                    *                      *
*****************************************************************************
                                                       VERSION : 01/01/87
```

AIM OF THE PROJECT :
****
The project consists in developing a borehole seismic source able to work down
to oil field depths (around 3000 m). It will allow the use of new borehole
seismic methods (well-to-well measurements, inverse method) and will improve
the knowledge of reservoir structures.

PROJECT DESCRIPTION :
****
The project is to be developed in three phases:
PHASE 1 - Theoretical prototypes
Two theoretical prototypes have been built:
- the first one uses a drill string to lift a weight which is released by the
triggering signal and drops on a striking surface equipped with a suitable
damper. The potential energy can be changed by changing the weight (2000 or
3000 J). The coupling of the source with the wall is provided by a packer.
- the second one uses a conventional logging cable with seven conductors to
transmit energy from the surface to the tool. It uses the mud pressure to
actuate a mobile system and can be adapted to work as an implosion source or a
mechanical impact source. A large-surface hydraulically driven clamping system
has been developed to insure the best coupling conditions. The potential energy
is about 2000 J at a 1000 m depth.
The site (property of Elf Aquitaine) includes four boreholes.
STATE OF ADVANCEMENT :
****
The first prototype has been carefully tested on the surface before a borehole
test performed in October 1986. Recording devices (3-component borehole
geophones and 48 channel borehole streamers) were lowered in two of them. The
source was placed in a third one. A surface spread was also set. The source was
fired between 550 and 450 m depth. More than 700 shots were recorded without
any failure. The second prototype has now been surface-tested.
RESULTS :
****
The borehole and surface data collected in the equipment with the first
prototype are now being processed. The results will be available at the
beginning of 1987.

```
****************************************************************************
*  TITLE : CONTROL OF THE PROPAGATION OF HYDRAULIC   *      PROJECT NO     *
*          FRACTURES                                 *                     *
*                                                    *                     *
*                                                    *   TH./01068/86/FR/.. *
*                                                    *                     *
****************************************************************************
*  CONTRACTOR :                                      *  PROGRAM :          *
*    GERTH                                           *    HYDROCARBONS      *
*    4, AVENUE DE BOIS PREAU      TEL 1 47.52.61.39  *                     *
*    FR - 92502 RUEIL-MALMAISON TLX 203 050          *                     *
*                                                    *  SECTOR :           *
*                                                    *    GEOPHYSICS        *
*  PERSON TO CONTACT FOR FURTHER INFORMATION :       *                     *
*    MR. SARDA                    TEL 1 47.49.02.14  *                     *
*                                                    *                     *
*                                                    *                     *
****************************************************************************
```
VERSION : 01/01/87

AIM OF THE PROJECT :
****
Provide methods and techniques allowing a better evaluation of the technical
risk associated with hydraulic fracturing techniques in difficult practical
cases : reservoirs presenting natural fractures, small productive tickness,
possible extension of the fracture either up into the overburden or down into
the water level. As the values of the horizontal principal stresses in the
reservoir and in the adjacent layers is the principal active parameter, methods
wil be developed for measuring these stresses. In order to evaluate the
efficiency of these methods the dimensions of the fractures will be evaluated
using several techniques.

PROJECT DESCRIPTION :
****
The project is to be developed along three trends :
PHASE 1 : Measurement of the in-situ horizontal stresses and fracture
propagation. In the problem of hydraulic fracturing the measurement of the
horizontal stresses requires a pression which is not usual in the field of
earth sciences. Indeed the minimum in-situ horizontal stress has to be
evaluated with an absolute error inferior to one megapascal in deep or very
deep geological layers. Presently, the adequate methods and/or techniques do
not exist. In order to set-up these methods the following technical
achievements are foreseen on each experimental site :
- minifacturations at several levels in the overburden and in the reservoir
  rock, measurement of the downhole pressure and of other useful parameters
- minifracturation in the openhole under the casing shoe
- prefracturation (or fracturation with a limited amount of fluid and without
propping agents) in the reservoir and with measurement of the bottomhole
pressure.
PHASE 2 : Determination of the dimensions of hydraulic fractures. In order to
evaluate the dimensions of hydraulic fractures three techniques are envisaged :
measurement with inclinometers of the displacements induced at the earth
surface by the hydraulic fracture, detection of the acoustic emissions
accompanying propagation and/or closure of the hydraulic fracture. An attempt
will be made to evaluate the length and direction of the fracture with the help
of the first method, the length direction and height of the fracture using the
second method, the tickness of the fracture using the third method.

PHASE 3 : Synthesis. All the data obtained on several experimental sites will
be interpreted. So will be compared the various techniques and methods
experimented on sites and the existing models which simulate fracture
propagation.
Will be proposed :
- recommended methods for the measurement of stresses and of fracture
  extension
- recommended criteria for the selection of a numerical model of fracture
  propagation depending on the actual mechanical and geological
  conditions of the hydraulic fracturing operation
- feasibility criteria hydraulic fracturing operations
- recommendations for the on-site control of fracture propagation.

STATE OF ADVANCEMENT :
****
Ongoing program. Three main activities : drafting of a technical guide book for
the minifracturation operations, minifracturations on sites and associated
measurements, interpretation of the minifracturation tests.

RESULTS :
****
Technical guide book for minifracturation tests :
A first version of a technical guide for minifracturation tests is being
written. It will consist of three parts : basic consideration on the use of
minifracturation tests for measuring the minimum in-situ stress, list of the
possible wellbore completions and criteria for the choice of the appropriate
completion, summary of the presently available interpretation methods. This
first version will be modified depending on the technical problems encountered
during the research program and taking into account the technical results of
this program.
Minifracturation tests :
Three minifracturation tests have been run on two sites : one test in a deep
gas well in the South West of France, two tests in two neighbour wells in the
South of France.
- First minifracturation test :
   Bottomhole completion : open-hole section at the basis of a casing, packer
   Depth of the open-hole section : 4387 m
   tickness of the open-hole interval : 5 m
   Physical parameters measured : wellhead pressure, annulus pressure,
   bottomhole tubing pressure, flowrate (approximately 0.05 m3 per second).
   Numerical recording of these parameters at a sampling
   rate of one point per second.
   Other available data : a core has been taken after the min HF operation
   and a METC log (experimental multielectrodes resistivity log) has been run.
- Second and third minifracturation tests :
   Bottomhole completion : perforated casing and packer
   Depths of the intervals tested : 2244 m in one well, 2294 m in the other
   well
   Physical parameters measured : tubing pressure, annulus pressure,
   flowrate. Numerical recording at a rate of one sample per second.
Interpretation

During the first test a problem of fluid leak at the packer level occured. This problem could be taken into account and solved by analyzing the numerical recordings of pressures and flowrate. Indeed, at this occasion, these recordings appeared to be far more useful than the analogical recordings due to the possibility of enhancing any part of a curve to any desired scale if an appropriate curve plotting program can be used. This method made possible the interpretation of very small annulus pressure variations and so the actual fluid volume injected in the formation and the corresponding bottomhole pressure could be established. The results have been interpreted in terms of stress perpendicular to the fracture plan, permeability and filtration coefficient of the formation.

```
****************************************************************************
* TITLE : EVALUATION OF LITHOLOGIC PARAMETERS IN    *      PROJECT NO      *
*         HYDROCARBON RESERVOIR BY SIMULTANEOUS     *                      *
*         APPLICATION OF COMPRESSION AND SHEAR      *    TH./01069/86/DE/.. *
*         WAVES                                     *                      *
****************************************************************************
* CONTRACTOR :                                      * PROGRAM :            *
*   D.G.M.K.                                        *   HYDROCARBONS        *
*   STEINSTRASSE 7            TEL 040 326479         *                      *
*   DE - 2000 HAMBURG 1       TLX 211 466            *                      *
*                                                   * SECTOR :             *
*                                                   *   GEOPHYSICS          *
* PERSON TO CONTACT FOR FURTHER INFORMATION :       *                      *
*   DR. M. ALBERTSEN                                *                      *
*                                                   *                      *
*                                                   *                      *
****************************************************************************
```

VERSION : 01/01/87

AIM OF THE PROJECT :
****
Exploration efforts are recently more and more directed to hydrocarbon
reservoirs of the "faziell type". Reservoirs of this type are indicated by a
lateral change in the pay zone lithology, independent of geological structures.
For the exploration of those reservoirs – e.g. in South Germany – it is
necessary to obtain the total information of seismic data and particularly the
amplitudes which bendes the traveltimes are important for this purpose.
At this background, a project LIPS is scheduled to investigate the potential of
comprehensive analysis of seismic data (p and s) in "seismic stratigraphy".

PROJECT DESCRIPTION :
****
The main goal of this project is the analysis, further development and
combination of methods for the evaluation of lithologic parameters of
hydrocarbon reservoirs.
One of the prerequisites for the application of lithoseismic methods and for
the involvation into exploration programs is a comprehensive data base,
especially of those reservoirs with lithologic transitions.
The quality of registrations and the availability of both p- and s- waves are
the main criteria for the selection of test reservoirs.
It is planned to analyse selected seismograms in detail besides the
installation of full sections.
A special objective is the relation between amplitude behaviour of the
reflection and its angle of incidence.
In addition to the seismic registrations, well-based observations by means of
logs and VSP are planned. Wellbore waves like the "Stoneley wave" are to be
investigated on their potential for density and permeability evaluations.
Laboratory investigations on rock materials are directed to obtain correlations
between seismic and petrophysical parameters.

The main technical goals summarized as follows:
- evaluation of reservoir parameters by the combined use of different types of
seismic waves (p-, s- and stoneley waves)
- optimization of the analysing methods by wavelet processing with concern to
the reflection amplitudesand phases
- interrelation of logging- and VSP -parameter with seismic data along surface
related profiles
- further improvement of the generation, application and interpretation of
shear waves.
STATE OF ADVANCEMENT :
****
Ongoing, project is in the design phase.

```
******************************************************************************
*  TITLE : SIMULATION MODEL FOR RESERVOIR          *      PROJECT NO       *
*         HETEROGENEITIES                          *                       *
*                                                  *   TH./01070/86/FR/..  *
*                                                  *                       *
******************************************************************************
*  CONTRACTOR :                                    *  PROGRAM :            *
*   GERTH                                          *   HYDROCARBONS        *
*   4, AVENUE DE BOIS PREAU     TEL 1 47.52.61.39  *                       *
*   FR - 92502 RUEIL-MALMAISON TLX 203 050         *                       *
*                                                  *  SECTOR :             *
*                                                  *   GEOPHYSICS          *
*  PERSON TO CONTACT FOR FURTHER INFORMATION :     *                       *
*   MR. RAVENNE                                    *                       *
*                                                  *                       *
*                                                  *                       *
******************************************************************************
                                                      VERSION : 01/01/87
```

AIM OF THE PROJECT :
****
Elaboration of a geological deposit model shaped by a mathematical model in
order to provide a description of the heterogeneities at metric and hectometric
scale of the physical properties of a sandstone reservoir that can be used by
the reservoir engineer in view of an optimal field exploitation.

PROJECT DESCRIPTION :
****
The project is divided in two successive phases :
PHASE 1 : STUDY OF THE FORMATION
The object of the field acquisition in two and three dimensions was to acquire
geological, geophysical and petrophysical data allowing set-up the model.
These studies will be carried out on the middle Jurasic outcrop of the
Yorkshire cliffs in Great Britain. The selected cliff is a remarkable
representation over 10 kilometers length and a height of 30 to 150 m of a wide
variety fluvial to fluvio-deltaic deposits, hardly altered by the tectonics and
thus providing a good image of their original representation. The sandstonelike
sedimentary elements (eventual reservoirs) show extensions ranging from dozens
of meters (clay dominant) to several kilometers (sandstone dominant). The light
overburden of these elements allows easy access through small core-drills
located at the back of the cliff, and thus makes it possible to obtain physical
properties measurements less altered than those acquired on the outcrops. Core-
drills will also provide us with well calibrated log shapes.
The sedimentologic study of outcrops will include a photogeological study of
the whole cliff, performed with the help of documents obtained by helicopter. A
set of vertical cross-sections will then made, thus providing a detailed
quantitative description of the different sandstone or clay elements in
relation with the deposit mode at a scale below the kilometer.
The execution of core-drills near the outcrops will allow the passage from a 2-
D heterogeneities description to a 3-D description at a kilometric scale and
the comparison of the obtained results with those of the oil drillings thanks
to the recording of logs.
The physical properties will be measured on samples taken from the outcrops and
on plugs sampled from the cores. A geostatistic interpretation of the values
obtained will then be proposed.

High resolution georadar profiles will connect the core-drill with one another and with the outcrops in order to improve the threedimensional image of the studied sedimentary elements.

PHASE 2 : ELABORATION OF THE 3-D GEOSTATISTIC MODEL

This model aims at describing the heterogeneities that can be the object of a direct exploitation by the reservoir engineer in order to optimize the development of fields. The general geological data will thus have to be entered in the numerical model of the field.

The model will be elaborated by integrating the results acquired during the sedimentologic studies and put in shape by the geostatistical processing.

A first step will consist in achieving a 2-D model based, among others, on the study of the images of the cliff.

The second step will consist in achieving a 3-D model deriving from the 2-D model, the 3-D geostatistical modelling and the sedimentologic model.

A third step will involve the validation of the model with real data from the North Sea.

STATE OF ADVANCEMENT :

****

The following operations were carried out in 1986:
- aerial photographs of the cliff and first interpretation.
- seriate sedimentologic cross-sections.
- measure of the physical properties of samples taken from the cliff.
- core-drills (36) and logs on a site where the sandstone phase is very variable.
- geophysical profiles (radar waves) on this same site.

RESULTS :

****

Only partial results concerning phase 1 and namely the twodimensional studies of the formation are available today. Studies relative to the modellings were subject to delay owing to :

1) meteorological problems for the aerial photographs thus delaying the morphomathematical study,

2) problems related to the authorization of running core-drills (detailed examination of request by local authorities). Hence, the core-drill campaign only began end November, delaying the sedimentologic descriptions and the measures of petrophysical properties. The latter will only begin around 15 January 1987.

Last of all, problems of authorization have also delayed the execution of geophysical profiles with radar waves. Also, these were made during a very humid period and thus not favourable : radar waves are indeed strongly absorbed by water and it appears that the results obtained do not correspond to the penetration envisaged.

```
******************************************************************************
* TITLE : DIRECT DETECTION OF HYDROCARBON      *      PROJECT NO          *
*         DEPOSITS USING ADVANCED SEISMIC      *                          *
*         TECHNIQUES                           *      TH./01072/86/UK/..   *
*                                              *                          *
******************************************************************************
* CONTRACTOR :                                 * PROGRAM :                *
*    BRITISH GAS PLC                           *    HYDROCARBONS          *
*    LONDON RESEARCH STATION    TEL 01 736 3344 X 4140*                   *
*    MICHAEL ROAD               TLX 24670      *                          *
*    UK - LONDON SW6 2AD                       * SECTOR :                 *
*                                              *    GEOPHYSICS            *
* PERSON TO CONTACT FOR FURTHER INFORMATION :  *                          *
*    DR. A. MELVIN                             *                          *
*                                              *                          *
*                                              *                          *
******************************************************************************
```

VERSION : 16/03/87

AIM OF THE PROJECT :
****
In the exploration for hydrocarbon deposits, seismic techniques are the methods
of first choice for the initial survey, both onshore and offshore. The aim of a
seismic survey is to delineate the sedimentary structures under the ground or
below the seabed and to use this delineation to identify potential structural
traps for hydrocarbons.
A seismic technique which is specifically sensitive to natural gas deposits and
to just lithology would be highly advantageous. The initial objective of the
British Gas project is to evaluate the feasibility of the developing such a
technique, which would be innovative and would facilitate the reliable location
of gas deposits. If the technique were adopted generally, a saving of some 80%
exploration drilling costs could be possible.

PROJECT DESCRIPTION :
****
Two complementary lines of investigation are being followed, (i) the
development of a seismic source based on new principles, and (ii) the use of
interconversion processes between shear waves and compressional waves which
occur in sedimentary lithology.
The development of the new source must overcome the difficulty that
conventional sources which generate shear waves cannot be used near the surface
of the sea since shear waves cannot propagate through water.
Such sources must therefore be used at seabed level. Onshore, both vibrator and
explosive charge techniques are available for shear wave generation. A design
for a new seismic source based on controlled (supersonic) shock-wave generation
(without the use of explosives) has already been developed. This source can be
operated also for conventional developments as a high efficiency compressional
source of wide frequency bandwidth, but it has features in the mode of
generating shock waves which will allow it to generate shear waves of known
properties in sedimentary strata. The project as envisaged in this application
would involve both new source development and advanced signal processing.

The first stage of the current Project (starting in April 1986) is the optimisation of the repetitive source design based on shock wave valve technology. We have therefore designed and had constructed a 25 mm internal diameter seismic gas gun containing as its main novel design feature a shock wave valve which will permit repetitive operation. The initial task will therefore be to commission the seismic source.

The second stage, with the appropriate choice of fluid in the low pressure chamber, will be to fire the device in the laboratory into water.

The subsequent history of the shock in water will be followed by ultrasomnic transducers located in the water in the path of the shock. Tests will be run in the laboratory on impacting the shock on layered materials submerged on the bottom of a water tank to check the conversion and attenuation of the wave as it transforms to an acoustic wave.

Along with the first and second stages will go the development of wave analysis and signal processing systems.

The third stage will involve the design of a 50 mm internal diameter prototype gas run, based on design information gathered from experience with the 25 mm internal diameter version. After construction and full instrumentation, the gas gun will be used for the main objective of the third stage : the testing of the source in the field appropriate onshore sedimentary structures, which may be gas bearing, oil bearing, coal measures or all three.

The fourth stage will be to mount tests of the gas gun and associated signal processing system offshore on a suitable exploration vessel.

STATE OF ADVANCEMENT :
****
Ongoing. The first prototype seismic device has been designed and constructed. It is currently undergoing commissioning tests in which its performance as a stock wave generator is being evaluated and the shock wave valve design is being optimised.

RESULTS :
****
We are at the commencement of this project and are about to enter the phase of collecting experimental data.

Provided that both onshore and offshore tests of the gas gun prove satisfactory, we shall be in the position of having a fully tested new system for seismic analysis which can then be exploited commercially, within the sphere of exploration operations of British Gas, in Europe in general and elsewhere.

REFERENCES :
****
PATENT APPLICATIONS FOR THE DEVICE HAVE BEEN FILED IN THE U.K., THE UNITED STATES AND THE EEC.

```
***********************************************************************
* TITLE : TWO DIMENSIONAL BALANCED SECTION      *      PROJECT NO      *
*         SOFTWARE                              *                      *
*                                               *    TH./01073/86/UK/.. *
*                                               *                      *
***********************************************************************
* CONTRACTOR :                                  * PROGRAM :            *
*   MIDLAND VALLEY EXPLORATION LTD              *   HYDROCARBONS        *
*   14 PARK CIRCUS           TEL 041 332 2681   *                      *
*   UK - GLASGOW G3 6AX      TLX 8950511        *                      *
*                                               * SECTOR :             *
*                                               *   GEOPHYSICS          *
* PERSON TO CONTACT FOR FURTHER INFORMATION :   *                      *
*   DR. A.D. GIBBS                              *                      *
*                                               *                      *
*                                               *                      *
***********************************************************************
```

VERSION : 01/01/87

AIM OF THE PROJECT :
****
To develop interactive software for use in prospect generation and mapping. The
software will enable the interpreter to validate geometries represented on
cross-sections and to test models for prospect development on a variety of
computer hardware.

PROJECT DESCRIPTION :
****
New advances in balanced section construction and basin development have been
pioneered by MVE. These new advances facilitate the identification of new
exploration prospects and aid their interpretation. The construction of
balanced sections is now a vital step in validating seismically derived maps of
prospects. The same techniques can be used to improve mapping of structurally
complex fields.

The planned program allows generation of new sections from the digitized input
by various operations, such as displacements on fault systems. Many sections
can be held in core at once, with the capability to select, view and operate
upon any one of them. Sections may be written to disk and later retrieved. They
can also be plotted.

The "active" operations which generate new sections from the input include the
simulation of displacement on a fault system. This operation can be used in two
ways: to remove displacements on faults which have offset marker horizons
(inverse modelling), or to simulate the progressive development of faults
(forward modelling). Another "active" operation is the construction of non-
planar fault profiles from the geometries of hanging wall rollovers.

The second class of operations are "passive", that is, they do not alter the
appearance of the section. There are two such operations: a calculation of
total bed length for each stratigraphic horizon, and a calculation for the area
of a specified formation or region of the section. Area calculations are used
to predict the estimated depth to detachment of controlling faults.

STATE OF ADVANCEMENT :
****
Work on developing the technical concepts and designing core program structure
began in July 1986. Design work on the workstation and PC environment was
started in October 1986 and both of these are currently ongoing.

RESULTS :
****
Problems so far encountered have been in designing portability and
compatibility into the system.
Technical trials so far have confirmed that the program core is scientifically
sound.

REFERENCES :
****
PROGRESS IN RESEARCH ON THE PROBLEM OF BALANCED SECTION CONSTRUCTION HAS BEEN
PRESENTED AT 2 UK, 1 NORWEGIAN, 1 CANADIAN CONFERENCES. 3 RELEVANT PAPERS ARE
IN PRESS.

```
*******************************************************************************
* TITLE : RELATIVE PERMEABILITY MEASUREMENTS          *       PROJECT NO       *
*         UNDER SIMULATED RESERVOIR CONDITIONS        *                        *
*                                                     *     TH./01078/86/DE/..  *
*                                                     *                        *
*******************************************************************************
* CONTRACTOR :                                        * PROGRAM :              *
*   PREUSSAG AG                                       *   HYDROCARBONS         *
*   ARNDTSTRASSE 1              TEL 05176-17288        *                        *
*   DE - 3000 HANNOVER 1        TLX 92655              *                        *
*                                                     * SECTOR :               *
*                                                     *   GEOPHYSICS           *
* PERSON TO CONTACT FOR FURTHER INFORMATION :         *                        *
*   DR. R. SOBOTT                                     *                        *
*                                                     *                        *
*                                                     *                        *
*******************************************************************************
```

VERSION : 01/01/87

AIM OF THE PROJECT :
****
The object of this project is to develop suitable methods for the determination
of oil/gas/water relative permeabilities under reservoir conditions in the
laboratory. The relative permeability data is of great importance in the
planning stage as well as the production development of tertiary recovery
projects. The reservoir behaviour is simulated with a mathematical computer
model. The simulator depends heavily on reliable parameters such as
oil/gas/water relative permeability data for the pay zone. It is expected that
measurements under simulated reservoir conditions yield more meaningful results
and will improve the reservoir engineering calculations. Of innovative
character will be the installation of a system which allows the continuous
monitoring of saturation conditions in the core sample during linear water/oil
replacement experiments under reservoir conditions.

PROJECT DESCRIPTION :
****
The project comprises five stages:
1. Design and construction of a small-scale installation for linear flooding
experiments on sanstone plugs (3 cm diameter and 3 cm length)
2. Determination of oil/water relative permeabilities with the small flooding
installation. The suitable procedures will afterwards be employed on PREUSSAG's
high-pressure autoclave.
3. Reconstruction of the high-pressure autoclave for flooding experiments with
reservoir oil.
4. Measurement of oil/water and gas/water relative permeabilities under
reservoir conditions of an oil reservoir from the Osthannover area and a gas
reservoir from the Ostfriesland area.
5. Evaluation and interpretation of the collected data.
During the first stage of the project a small flooding installation is designed
and built in order to screen the suitable methods available for relative
permeability evaluation. This will save time and cost because only successfully
tested methods will be used for PREUSSAG's high-pressure autoclave.
The second stage includes the testing program for the small flooding
installation. Oil/water relative permeability will be evaluated from linear
flooding experiments and gas/water relative permeability measurements carried
out after the Penn State Method. Hereby the average oil/water/gas saturations
in the core will be determined by electrical resistivity measurements.

The third stage will be the implementation of the selected methods on the high-pressure autoclave. This requires the installation of pressure vessels for reservoir oil and a separator at the autoclave exit.
In the fourth stage the high-pressure autoclave is put into operation for oil/gas/water relative permeability measurements under reservoir conditions simulating real reservoir parameters such as temperature, pore pressure, overburden pressure, viscosity of fluids, etc...
Finally, the collected data is evaluated with respect to relative permeability as a function of reservoir parameters. The results are presented in a commented graphical and tabulated form. They will also be published in respective journals and presented at appropriate conferences.

STATE OF ADVANCEMENT :
****
Ongoing with phase 1. Two small flooding installations for steady and unsteady-state (waterflooding) experiments have been assembled and their working condition been proven experimentally.
Also first tests have been carried out for the set-up of the monitoring system for the saturation conditions by AC resistivity measurements.

```
******************************************************************************
* TITLE : DEVELOPMENT OF A GEOMATHEMATICAL MODEL    *      PROJECT NO        *
*          APPLIED TO OIL RESERVOIR                 *                        *
*                                                   *    TH./01083/86/PO/..  *
*                                                   *                        *
******************************************************************************
* CONTRACTOR :                                      * PROGRAM :              *
*   PARTEX-COMPANHIA PORTUGUESA DE SERVICOS SA       *   HYDROCARBONS         *
*   AVENIDA 5 DE OUTUBRO, 160   TEL 73.50.13        *                        *
*   PO - 1000 LISBOA            TLX 14708           *                        *
*                                                   * SECTOR :               *
*                                                   *   GEOPHYSICS           *
* PERSON TO CONTACT FOR FURTHER INFORMATION :       *                        *
*   MR. A. DIOGO PINTO                              *                        *
*                                                   *                        *
*                                                   *                        *
******************************************************************************
```
                                                      VERSION : 16/03/87

AIM OF THE PROJECT :
****
The objective of the project is to build an integrated software package for
modelling oil reservoirs and describing the nature and disposition of the
heterogeneities that inevitably occur in petroliferous formations. This project
aims to develop a new approach based on a geomathematical model resulting from
the Theory of Regionalized Variables and from Kriging Technique, in order to
describe the spatial behaviour of reservoir properties, generate unbiased
profiles for use at different levels, define the error associated with the
estimation taking in account the position and spatial distribution of the
variable, detect structure trends and anisotropies, optimise of new wells
location, determine the best description of reservoir zonation.
The package will be flexible enough to permit the introduction of the practical
experience obtained by field geologists and reservoir engineers and designed to
be used by any petroleum technicien without computer science expertize.

PROJECT DESCRIPTION :
****
The plan of the project is divided in 3 phases and a final integrated stage :
1ST PHASE :
1. Screen review and assemble of the basic data regarding different oil
reservoirs;
2. Construction of the geomathematical model working out a set of computing
programs;
3. Initialization of the model, Kriging known values and checking the resulting
error of calculating various theoretical dispersion variances and comparing
them to the experimental values;
5. Analysis and evaluation of the results and impact on several areas.
a). variographic analysis and detection of the major structural characteristics
and anisotropies of the reservoirs, quantitative evaluation;
b). characterization and description of reservoir heterogeneities;
c). description of the reservoir properties spatial behaviour, generated
profiles and precision of the estimation;
d). contour mapping and comparision with classical techniques, evaluation of
the results;
e). degree of error associated with the estimation, real statistical meaning of
the values; impact on reservoir simulation outputs and reservoir estimation;

f). co-regionlization aiming to study inter-correlated variables simultaneously.
2ND PHASE :
1. Geological modelling using conditional simulation to provide a set of values with the same variability as the studied phenomenon and coinciding with the experimental values at the sample locations, and Minkovsky transforms for characterizing the geometry of the reservoir;
2. Optimization of new wells locations based on the fact that the Kriging variance does not depend on experimental values, but on the structure (variogram model), on the geometric configuration of the domain to be estimated and on the locations of the data points;
3. Subsurface reservoir model and impact on reserves calculation. Application of the geomathematical model and Kriging technique to the study of the reservoir geometry in order to reduce the uncertainty about it simulating the field boundaries in a more accurate way.
3RD PHASE :
1. Automatic method of capturing and storing data. Design of a data-base with specific characteristics of Petroleum information;
2. A new method for reservoir zonation based on the Correspondence analysis, allowing the simultaneous treatment of quantitative and qualitative data;
3. Contribution of variography analysis and Mandelbrot's Theory of fractal objects for the solution of the problems of scale in a reservoir; attempt to relate them quantitatively.
LAST STAGE :
It consists of the integration of the results of all phases in a portable and integrated package to be used by field engineers in order to solve practical oil industry problems in exploration and production.

STATE OF ADVANCEMENT :
****
Ongoing

REFERENCES :
****
- MATHERON, G. "THE THEORY OF REGIONALISED VARIABLES AND ITS APPLICATIONS", C.M. M. FONTAINEBLEAU, 1971.
- JOURNEL, A. AND HUIJBREGTS, CH. "MINING GEOSTATISTICS", WILEY, 1981.
- COSTA E SILVA, A. "A NEW APPROACH TO THE CHARACTERIZATION OF RESERVOIR HETEROGENEITY BASED ON THE GEOMATHEMATICAL MODEL AND KRIGING TECHNIQUE" SPE 60TH TECHNICAL CONFERENCE, LAS VEGAS 1985.
- RIBEIRO, L. AND COSTA E SILVA, A. "A GEOMATHEMATICAL MODEL TESTED ON AN OIL RESERVOIR", APCOM, PENSILVANYA, 1986.

```
***********************************************************************************
* TITLE : GEOCHEMICAL EXPLORATION BY NITROGEN       *      PROJECT NO           *
*         ISOTOPE ANALYSIS OF UNDISTURBED SEA BED    *                           *
*         SAMPLES (GENIUS)                           *      TH./01084/86/DK/..    *
*                                                    *                           *
***********************************************************************************
* CONTRACTOR :                                       * PROGRAM :                 *
*   COWICONSULT CONSULTANT ENGINEERS                 *   HYDROCARBONS            *
*   TEKNIKERBYEN 45              TEL 45 2 85 73 11    *                           *
*   DK - 2830 VIRUM              TLX 37 280           *                           *
*                                                    * SECTOR :                  *
*                                                    *   GEOPHYSICS              *
* PERSON TO CONTACT FOR FURTHER INFORMATION :        *                           *
*   MR. A. BJERRUM                                   *                           *
*                                                    *                           *
*                                                    *                           *
***********************************************************************************
                                                            VERSION : 19/03/87
```

AIM OF THE PROJECT :
****
The project includes a new method to determine if source rock is present in the
subsurface. The method is based on survey of the natural seepage of
hydrocarbons and nitrogen from the subsurface formations. The combination of
seismic survey and of natural seepage survey is considered to give important
information for source rock mapping.
An easy and safe technique shall be developed to recover undisturbed samples
from the sea bed. The samples shall be analyzed for their content of traces of
hydrocarbons and nitrogen which may be trapped in the soil. The purpose is to
determine the real extent and the age of source rock for oil and gas.

PROJECT DESCRIPTION :
****
The project is carried out by Cowiconsult (contractor) and Danpec
(subcontractor).Danpec has tested an existing sampling equipment in Kattegat
and the West Baltic Sea. After the test several improvements in the sampling
technique were suggested. Further it was suggested to include nitrogen in the
analysis to improve the interpretation.
The innovative idea in the proposed project is to improve the sampling
technique to avoid contamination of the samples by nitrogen in the atmosphere.
Further the samples shall be recovered in a depth of say 2 m to avoid the gas
from the bioactive zone at the seabed. Then it should be possible to get
reliable information on any traces of hydrocarbons and nitrogen due to seapage
from the deep formations.
The reason to include nitrogen in the analysis is to improve the interpretation
since most source rock is assumed to generate hydrocarbons as well as nitrogen.
The primary object is to verify the method by testing carboniferous source rock.
 The information on the extent of Carboniferous in the North European region is
poor.
In the test programme the method shall be tested in an area where carboniferous
source rock is known to be present. Later on when the method has been proved
the technique may be extended to Triassic and Jurassic source rock.
The project is divided into 5 phases comprising 12 activities :

PHASE 1 : CONCEPTUAL STUDIES
- Geochemical analyses
- Isotope mesurements
- Core sampling requirements
PHASE 2 : DEVELOPMENT OF A CORE SAMPLER
- Design of equipment
- Construction and test
- Marketing
PHASE 3 : Pilot test
- Selection of test areas
- Field work
- Laboratory work
PHASE 4 : FINAL EVALUATION
- Evaluation of GENIUS results
PHASE 5 : REPORTING AND PROJECT MANAGEMENT
- Reporting
- Project management.

STATE OF ADVANCEMENT :
****
Ongoing. Start of the project has been March 1987.

DRILLING

```
******************************************************************************
* TITLE : DEVELOPMENT OF A DEEP WATER DRILLING      *      PROJECT NO        *
*          UNIT.                                    *                        *
*                                                   *                        *
*                                                   *    TH./02017/83/NL/..  *
*                                                   *                        *
******************************************************************************
* CONTRACTOR :                                      * PROGRAM :              *
*    V.O.F. DESDEC  DEEP SEMI DESIGN CONSULTANTS     *    HYDROCARBONS        *
*    NIJVERHEIDSSTRAAT 54 POSTB TEL 01846 - 6111     *                        *
*    NL - 3370 AC HARDINXVELD-G TLX 25628 MSC NL     *                        *
*                                                   * SECTOR :               *
*                                                   *    DRILLING            *
* PERSON TO CONTACT FOR FURTHER INFORMATION :        *                        *
*    MR. G.J. SCHEPMAN                               *                        *
*                                                   *                        *
*                                                   *                        *
******************************************************************************
```

VERSION : 01/12/85

AIM OF THE PROJECT :
****
Aim of the project is the design of a drilling vessel for severe environment
and deep water.

PROJECT DESCRIPTION :
****
The basic objective is to design a vessel which combines the advantages of a
drill-ship and a semi-submersible. This resulted in the DSS-10.000. This is a
dynamic positioned column stabilized vessel, characterized by the two floaters,
four stability columns, one large central column with an enclosed moonpool and
a box-type upper hull.

Overall length   : 109,5 m
Overall width   : 64 m
Operating draft   : 26,3 m
Operating displacement : 65,000 tons
Accommodation   : 114 persons
Variable deckload   : 4,700 mtons
Total variable deckload : 10,000 mtons
Drilling depth   : 25,000 ft
Water depth   : 10,000 ft

The center column will be used for the storage of the entire marine riser and
creates the space for an enclosed moonpool and a BOP assembling and testing
area. Riser joints (85 ft. length) are stored vertically in carousels. Patents
have been applied for.

STATE OF ADVANCEMENT :
****
Basic design package has been finished. Model testing has been completed and
analyzed. Approval has been obtained from DNV and ABS. Technical evaluation
with oil company engineering departments is ongoing. Building cost has been
determined. Adaptations for specific requirements are being studied.

RESULTS :
****
The design of the DSS-1000 has been completed and documented in drawings,
reports and specifications. Patents have been applied for a specific system on
board. The 5 column principle stands as a feasible and valuable feature for
deep water drilling. The design package is at sufficient level to enable
immediate start of a building project if and when conditions permit.
The design will be valuable for drilling in deep, rough water such as north-
west of the U.K. or in Norwegian waters. It may serve as well as a base case
for different applications, e.g. floating production offshore. No immediate
follow up is expected in view of the present (1987) disarray of the offshore
industry.

```
********************************************************************************
* TITLE : APPLIANCE FOR AUTOMATIC CONTROL OF      *      PROJECT NO         *
*         HOISTING OPERATION IN A DRILLING MAST    *                         *
*         OF DERRICK.                              *   TH./02018/84/FR/..     *
*                                                  *                         *
********************************************************************************
* CONTRACTOR :                                     * PROGRAM :               *
*   P.S.O.                                         *   HYDROCARBONS          *
*   AVENUE ALBERT 1ER 116 BIS   TEL 1 47.49.15.52  *                         *
*   FR - 92500 RUEIL-MALMAISON TLX 203310          *                         *
*                                                  * SECTOR :                *
*                                                  *   DRILLING              *
* PERSON TO CONTACT FOR FURTHER INFORMATION :      *                         *
*   MR. G. GAZEL-ANTHOINE                          *                         *
*                                                  *                         *
*                                                  *                         *
********************************************************************************
```

VERSION : 01/01/87

AIM OF THE PROJECT :
****
The aim of the project is to set an apparatus allowing an accurate stop of the drill string at a predetermined position without waste of time and with full safety precautions.

PROJECT DESCRIPTION :
****
The overall development involves a sensor, which detects rotation or displacement of an element of the cinematic chain. In addition, a sensor detects tool joints passage to permit correction of divergence between actual and calculated position. A tension load sensor gives a signal proportional to the suspended load. Finally a control unit acquires and treats the data and allows displaying of the information.
The project will be carried out with the following phases :
1. completion of a prototype
2. industrial preparation and preliminary simulation
3. comprehensive operational tests.

STATE OF ADVANCEMENT :
****
- all data acquisition and all actuators are running reliably
- drawworks control is working for single movement
- several drawworks movements are chained just by push button validation by operator
- mainly the chain must be extended to several cycles and safety control by software must be implemented
- communication between operator and system must be improved.

```
*************************************************************************
* TITLE : DEVELOPMENT OF A SUBSEA WIRELINE SYSTEM.   *     PROJECT NO    *
*                                                    *                   *
*                                                    *  TH./02019/84/UK/..*
*                                                    *                   *
*************************************************************************
* CONTRACTOR :                                       * PROGRAM :         *
*   ADVANCED PRODUCTION TECHNOLOGY LTD.              *   HYDROCARBONS     *
*   3RD FLOOR, TRAFALGAR HOUSE TEL 01 748 4600       *                   *
*   HAMMERSMITH INTERNATIONAL   TLX 262227           *                   *
*   UK - LONDON W6 8DM                               * SECTOR :          *
*                                                    *   DRILLING         *
* PERSON TO CONTACT FOR FURTHER INFORMATION :        *                   *
*   DR. C.F. BAXTER                                  *                   *
*                                                    *                   *
*                                                    *                   *
*************************************************************************
```
VERSION : 31/12/86

AIM OF THE PROJECT :
****
To demonstrate the feasibility of preforming wireline service from a Monohull
Vessel thus drastically reducing the cost of such operations.
PROJECT DESCRIPTION :
****
Wireline interventions are an accepted method of performing a wide variety of
downhole maintenance, inspection and monitoring taks. These taks are normally
performed either from the Platform Celler Deck or from the Deck of a Semi
Submersible Work-Over Vessel in the case of subsea completions. In both cases a
riser from seabed to deck (either permanent riser or temporarily installed) is
used as the means of access to the borehole.
this project targeted the ability of a high-technology Field Support Vessel to
be used in place of the much more expensive Semi-Submersible Work-Over Vessel.
To achieve the project objectives a design study was undertaken which defined
the necessary parameters that would be required by a system.
Operational availibility; the range of tasks that could be required; the motion
response characteristics and many other factors were considered.
A Basic Offshore Trial Unit was engineered and built during 1985. This basic
unit was aimed at an early demonstration that the system could be successfully
operated in the North Sea. The system was deployed on the British Argyll in
june 1986 and successfully performed a trial operation on a Duncan well under
the operatorship of Hamilton Brothers Oil + Gas Ltd.
The system comprised a Derrick and Heave compensated load and wireline units
together with a subsea lubricator. The whole system was mounted on a
Dynamically Positioned; Roll-Stabilised; Twin-Moonpol support vessel the
BRITISH ARGYLL which is owned and operated by APT's parent company British
Underwater Engineering Ltd.
Since successful demonstration of the basic unit the project has progressed.
Its current status is that a commercially viable unit is now being constructed
which embodies the lessons from the Basic Offshore Trial. In particular APT are
totally redesigning the Subsea Lubricator element so as to provide access for
both slick line and electric braided cable applications. This technology
continues to be supported by oil company operators. In particular the support
of Amerada Hess is to be acknowledged in the forthcoming build phase.

The commercial unit will be modular and will therefore be able to be depolyed from a range of suitable vessels and to be operable in water depths of +/- 350 meter over most well configurations.
Proving of this system will take place in the weather window of 1987.

STATE OF ADVANCEMENT :
****
Ongoing. System for future commercial operations is currently being built.

RESULTS :
****
The Basic Offshore Trial unit demonstrated the feasibility of this mode of operation.
Numerous improvements were identified as a results of the trial. These particularly concerned the subsea lubricator which displayed several undesirable characteristics of a troublesome and time-consuming nature. A more robust Derrick and compensation unit was also indicated to be desirable when operating over more complex xmas trees.
Further results are commercially sensitive but interested parties may register specific queries with APT at the above address.

REFERENCES :
****
- "THE WHOLE OF THE FIELD SUPPORT VESSEL IN SUBSEA FIELD DEVELOPMENTS" - BAXTER AND EDE, MARGINAL + DEEPWATER OILFIELD DEVELOPMENT CONFERENCE, LONDON APRIL 86.
- THE BRITISH ARGYLL DSV WIRELINING SYSTEM - HUBER (HAMILTON BROS), OFFSHORE EUROPE EXHIBITION 1985.
- "MARGINAL FIELDS" SPECIAL FEATURE - OCEAN INDUSTRY MAGAZINE, NOVEMBER 86.

```
*******************************************************************************
* TITLE : STUDY OF REALIZATION OF A SYSTEM      *      PROJECT NO         *
*         ENABLING THE DISPLAY OF DRILLING      *                         *
*         PARAMETERS AND THE OPTIMIZATION OF    *   TH./02020/84/FR/..    *
*         COSTS.                                *                         *
*******************************************************************************
* CONTRACTOR :                                  * PROGRAM :               *
*   SYMINEX                                     *   HYDROCARBONS          *
*   BOULEVARD DE L'OCEAN 2      TEL 91 739003   *                         *
*   FR - 13275 MARSEILLE CEDEX TLX 400563       *                         *
*                                               * SECTOR :                *
*                                               *   DRILLING              *
* PERSON TO CONTACT FOR FURTHER INFORMATION :   *                         *
*   MR. A.J. KERMABON                           *                         *
*                                               *                         *
*                                               *                         *
*******************************************************************************
                                                     VERSION : 31/12/86
```

AIM OF THE PROJECT :
****

To develop oil drilling rigs equipment which helps the operator to take
decisions during drilling operations by informing him in a reliable manner and
in real time about the tendancies of the present or of past situations. This is
achieved by means of theoretical computations and forecasted statistical
reports to guarantee a reliable display and recording of all the parameters
necessary to enable the rational exploitation of the rig.

PROJECT DESCRIPTION :
****

The project will start from the "VISUFORA" system, developed by SYMINEX to
acquire and graphically display drilling and deviation parameters and to record
them.
This system will be developed along 2 directions :
- improvement and extension of existing programmes using only information from
the well being drilled.
- exploitation of the records obtained from the wells previously drilled on the
field.

STATE OF ADVANCEMENT :
****
Ongoing

RESULTS :
****
Stage 1 : extension of existing programs
Drilling parameter computations and their digital and analogical displays have
been completed. Round tripping detection is possible automatically and the
stock control of drill pipes is operational. Regarding casing, the casing phase
parameters and stock control of the casing can be monitored. Definition of the
cementing programme is achieved. Drilling cost software has been implemented.
Additional software (hydraulics, alarms, tables edition, editions (print-out)
have been included.
Stage 2 : programs for exploitation for statistical analysis of the results on
previous wells of the same field is operational. Directional well path analysis
software is completed.

```
************************************************************************
* TITLE : NEW LINING TECHNOLOGIES FOR DRILLING    *      PROJECT NO      *
*         AND PRODUCTION EQUIPMENT.               *                      *
*                                                *                      *
*                                                *   TH./02021/84/FR/..  *
*                                                *                      *
************************************************************************
* CONTRACTOR :                                   * PROGRAM :            *
*   COATING DEVELOPMENT S.A.                      *   HYDROCARBONS       *
*   S/C S.I.T.T.              TEL 75 41.40.38     *                      *
*   RUE DU BAC                TLX 345079          *                      *
*   FR - 07500 GRANGES LES VAL                    * SECTOR :             *
*                                                *   DRILLING           *
* PERSON TO CONTACT FOR FURTHER INFORMATION :     *                      *
*   MR. LE COAT               TEL 75 41.40.38     *                      *
*                             TLX 345079          *                      *
*                                                *                      *
************************************************************************
```

VERSION : 09/06/87

AIM OF THE PROJECT :
****
This project consists in developing a method allowing to deposit on drilling
and production equipment, a product presenting mechanical characteristics
(resistance to abrasion and shocks) superior to those of tungsten carbide and
slightly inferior to those of diamond, a synthetic polycristalline of
equivalent cost or even less expensive than tungsten carbide.

PROJECT DESCRIPTION :
****
The project consists of building a pilot unit allowing to process a great
number of elements, within conditions representative of an industrial
manufacture.
The number of parts to be treated and their diversity require an important
number of tests within various domains involved in the oil industry: drilling
(progress test, wear test...) and production (erosion and abrasion test,
corrosion tests...)
Parallel to the pilot assembling and tests surveys, the project also includes a
fundamental study covering the physical and chemical analysis of the
constituents of the lining, the texture of the deposit, and the interaction
between the deposit and the different substrata studied.
The project comprises 4 phases:
PHASE 1 - Fundamental study:
Systematic study of deposit parameters (temperature, pressure, gas type,
flowrates, proportions of mixtures), by characterising the structural
properties of the products (nature of phases, interaction with substratum,
deposit, balance diagram...), by determining the mechanical properties and the
correlations between the structural properties and the mechanical properties.
PHASE 2 - fabrication of pilot unit:
These tasks will consist, depending on the database deriving from the
fundamental study, in determining the adequate instrumentation, in computing
the thermal capacities of the kilns and in optimising working parameters so to
elaborate the equipment which will be tested later on a real site.
PHASE 3 - Manufacture of prototype:

The pilot unit allows application of hard lining on real size prototypes specially made for the cell and in-situ tests: drill bits, stabilizers, tool-joints, swiveling equipment, nozzles, cutting tools, cutting plates.
PHASE 4 - On-site tests
Prototype elements are first tested on cells or test benches, so to minimise the risk of interrupting a drill by presenting an equipment, whose reliability would not have been previously tested.
STATE OF ADVANCEMENT :
****
Ongoing
RESULTS :
****
PHASE 1 - Fundamental study
It consisted in developing the manufacture of a ceramic deposition on substrata based on tungsten carbide, while using the CVD technique (Chemical Vapor Deposition). Deposits obtained in various configuations present very interesting characteristics concerning the hardness (hardness KNOOP varying from 5000 to 6000 kg/mm2) to be compared with the hardness of the tungsten carbide (1500 kg/mm2) and that of the diamond (8000 kg/mm2), and the wear resistance through friction (tribologic tests) and through abrasion.
PHASE 2 - Construction of a pilot unit
A unit formed by a quartz tube in vertical position (of internal diameter of 135 mm and a length of 1500 mm, electrically heated by the wall, has been manufactured with five levels allowing to line at full load 90 small plates (inserts of 20 mm diameter and 5 mm thickness).
Tests were run while using different feed configurations and different partial pressures.
These tests provided very different results depending on the chosen feed configuration, and it was very difficult to reproduce the excellent results of the fundamental study. These tests proved that an industrial pilot was necessary (400 mm diameter, 800 mm height) and also that is was indispensable to find the best gas feed configuration in order to control the characteristics of the deposition.
PHASE 3 - Manufacture of prototypes
Deposits were made on substrata formed of tungsten carbide (owing to resilience)
3.1 - Rock drilling : inserts have been lined
3.2 - Wear : drilling nozzles 10/32", graphite casting, sanding nozzle, cable-guide, bearings.
3.3 - Steel machining : rotary plates
PHASE 4 - In-cell and on-site tests
4.1 - Rock drilling
Tests were run in granite. It seemed difficult to obtain results similar to those provided by the polycrystalline diamond. Nevertheless, new drilling tests with important diameter tools are being envisaged in soft rocks
4.2 - Wear
Bearings : obtain very interesting dry friction coefficients.
Nozzles : very satisfactory results
Cable-guide : very satisfactory results.
Steel machining: qualitative tests on throw-away plates in tungstene carbide showed a lifespan two to threee times longer than that of plates lined with titanium carbide and titanium nitride.
REFERENCES :
****
A PATENT HAS BEEN APPLIED FOR BY DIAMANT BOART IN FRANCE ON 27 JUNE 1984 UNDER THE NUMBER 84-10290

```
********************************************************************************
* TITLE : OPTIMIZATION OF DRILLING OPERATIONS    *        PROJECT NO          *
*                                                *                            *
*                                                *    TH./02024/84/FR/..      *
*                                                *                            *
********************************************************************************
* CONTRACTOR :                                   * PROGRAM :                  *
*   GERTH                                         *   HYDROCARBONS            *
*   AVENUE DE BOIS PREAU 4       TEL 1 47 52 61 39 *                          *
*   FR - 92502 RUEIL-MALMAISON TLX 203050         *                          *
*                                                * SECTOR :                   *
*                                                *   DRILLING                 *
* PERSON TO CONTACT FOR FURTHER INFORMATION :    *                            *
*   MR TRAONMILIN                 TEL 59 83 40 00 *                          *
*                                 TLX 560 804     *                          *
*                                                *                            *
********************************************************************************
                                                         VERSION : 01/01/87
```

AIM OF THE PROJECT :
****
The object of this project is to develop new methods allowing the improvement
of the immediate drilling procedures, in view of a better control of drilling
operations and reduction of their costs.

PROJECT DESCRIPTION :
****
The main trend of the project is to exploit and integrate, in real time, the
information linked to the major physical phenomena which intervene in the
optimization of drilling procedures. These phenomena concern mainly the
following three aspects :
1. directional behaviour of drill strings
2. hold of walls through friction analysis
3. behaviour of drill bits while cutting formation.
The present project consists in preparing, not only the in-well experiments to
be ran during drilling operations, but also tests on the appropriate test bench.

STATE OF ADVANCEMENT :
****
The project will be completed within the scope of this contract by 31 December
1986 and pursued within the scope of contract TH./02028/85 with further tests
in wells and on bench scale.

RESULTS :
****
The 4 phases of the project were :
- Phenomenological studies : bibliographical study on each item,
- Physical modelling : design of a test bench for down scale modelling of
bottom hole assembly,
- Numerical modelling in three dimensional manner of the directional behaviour
of the bottom hole assembly,
- Methodology for test programmes and data acquisition systems.

PHENOMENOLOGICAL STUDIES
The major part of the work consisted in bibliographical studies relative to
above mentioned aspects. These studies allowed to specify the aspects which
were to be the object of a more thorough study, namely :
- the importance of friction in the behaviour of drill strings
- the utilisation of MWD tolls to limit wedging against the walls and the
detection of abnormal pressure areas
- the mechanical aspect of the tool progress at the cutting front,
particurlarly modelling of the relation between the weight on the bit and its
torque so to obtain characterization of the bit or of the rock and information
on the state of wear of the bit while drilling
- the importance of hydraulic phenomena linked to :
-   the flowrate and the mud pressure
-   the characteristics of the drilled rock versus filtration
-   the geometry of the drilling tool.

NUMERICAL MODELLING
The composition of drill strings and the choice of parameters during
directional drilling are generally based on acquired know-how. The purpose is
to create computing aids to measure the distorsions and the contact force of
the drill strings. Studies thus consisted in choosing mathematical modelling
methods for the drill strings and coding them to enable computer calculation.
These models predict the behaviour of the inclination and the azimuth of drill
strings.

PHYSICAL MODELLING
The purpose was to prepare bench tests so to complete the in-well tests
scheduled during the next test phase. The studies covered both the detailed
design and engineering of a drill string simulation bench at reduced scale and
the possibility of modelling drilling turbines compatible with the simulation
algorithms of the drill strings. It appears necessary to mesure the linear
weight and the inertia of turbine. Negotiations were carried out with various
suppliers of turbines for the adaptation of their benches and experimentation
procedures have been set-up.

TEST METHODOLOGY
The purpose was to define the appropriate methodology for each set of
measurements in-well and studies allowed to specify :
- the tests programme
- the nature of parameters to be measured
- the frequency of recordings
- the equipment
- choice of wells
- choice of data acquisition methods.
Two preliminary sets have been run to adapt methodology to operational
realities : quality of the measurements, stresses on normal drilling. These
tests were run on wells L4A4 in Holland then on HAA443 and TM25 in Indonesia.
During these tests the data acquisition method has been adapted to suit the
field conditions. Data analysis software packages have also been elaborated.

```
********************************************************************************
* TITLE : COMPLETION OF HORIZONTAL DRAINS.         *       PROJECT NO          *
*                                                  *                           *
*                                                  *    TH./02025/84/FR/..     *
*                                                  *                           *
********************************************************************************
* CONTRACTOR :                                     * PROGRAM :                 *
*    GERTH                                         *    HYDROCARBONS           *
*    AVENUE DE BOIS PREAU 4      TEL 1 47 52 61 39 *                           *
*    FR - 92500 RUEIL-MALMAISON TLX 203050         *                           *
*                                                  * SECTOR :                  *
*                                                  *    DRILLING               *
* PERSON TO CONTACT FOR FURTHER INFORMATION :      *                           *
*    MR. SPREUX               TEL 59 83 40 00      *                           *
*                             TLX 560 804          *                           *
********************************************************************************
                                                      VERSION : 01/01/87
```

AIM OF THE PROJECT :
****

   The object of this project is to determine the techniques, equipment and
procedures specific to horizontal wells regarding both the acquisition and
interpretation of field data and the means of achieving the appropriate well
completion. Considering the field heterogeneties encountered by the drain, its
optimized exploitation may require the implementation of a selective completion
providing the possibility of chosing the production areas.

PROJECT DESCRIPTION :
****

The project was divided in two main phases, a study or engineering phase and a
completion phase.
The object of the engineering phase was to determine the means regarding the
acquisition and interpretation of field data around a well. These are
characterized by two main elements, production measurements or logs and
interpretation methods.
Regarding the first element, an attempt was made to obtain a number of
measurements and their interpretation while considering the particular feature
of the horizontal flow. As for the second element, an attempt was made, first,
to schematize the horizontal variations of the geological characteristics of
the reservoir and to interprete and model the well tests and production with or
without interface.
The completion phase is more technological and its object consisted in making a
selective completion.
Depending on the nature of the reservoir, the liner can be either cemented and
slotted or separated in several preslotted sections by means of special
equipment. Both systems constitute the first two studies of this phase. In
order to improve production conditions, an attempt was also to have been made
to find the means of simulating and treating such a drain. Last of all, and in
as much as possible, all these results would have been finalized or confirmed
by real in-well tests.

STATE OF ADVANCEMENT :
****
Abandoned. Studies within the engineering phase have made possible to finalize
a set of tools allowing, to assess the opportunity of exploiting a reservoir by
horizontal drains, and also second, to interprete the data collected during the
development and exploitation of a field. Considering the progress achieved with
development of sufficient tools, the project has been interrupted.
RESULTS :
****
a) ENGINEERING HORIZONTAL DRAINS
Measurements
-   Determination of the section of a horizontal drain. Tests run on a test
bench showed that a measuring equipment of the SHDT type provided diameters by
defect within an error rate of 8%, which is rather important and needs to be
taken into account when defining the shape and the volume of the drain.
-   The study on sensor transport means in a well fitted with its tubing has led
to the solution of two methods "PUMP OUR STINGER" and "COILED TUBING CONVEYED".
-   A test bench resulted in testing of measurements made by bottomhole
flowmeters under horizontal diphasic flow conditions. It revealed, that to be
properly interpreted, these measures required additional data on parameters
such as the water content of the fluid and the nature of the flow.
Interpretation methods
-   The interpretation of well tests (analytical and numerical) provides data on
the vertical and horizontal permeabilities near the well and on the residual
skin.
-   A study on interface evolution provides easy analytical methods for the
preliminary studies of specific cases and numerical models have also been
studied. Regarding the reservoir, this allows to predict the production before
and after the breakthrough, the breakthrough duration, the critical flowrate,
etc...
-   Regarding lateral changes of reservoir geological characteristics , studies
enabled to evaluate the influence of the horizontally on the logs and provided
specifications for recalibration.
b) COMPLETION
Cementing : to perform a hole-layer interface taking into account the variation
of the characteristics of the reservoir, the liner has been cemented then
perforated according to chosen zones. Laboratory studies and tests showed the
possibility of achieving a correct liner cementation while modifying the
formulation of the slurries, the rheological stresses with the spacer, the
positioning scheme.
Sectioning the drain by use of external packers has proved feasible. Regarding
the well equipment, studies have also been carried out on the internal means
allowing to complete the selectivity made by the liner.
For the well treatment, methods and tools have been chosen allowing
interventions before and after tubing. Before tubing, locating can be done with
gell, after tubing cup tools which are necessary to force the treating fluid to
enter in contact with the reservoir.
Tests on horizontal wells confirmed the validity of our studies on production
logging and on external packers and completed our know-how deriving from
previous experiments carried out within the scope of the horizontal drilling
project.

REFERENCES :
****
- "ECOULEMENT DES FLUIDES AUTOUR D'UN PUITS HORIZONTAL-ESSAIS DE PUITS"
SPE-ROME-FEBRUARY 1985
- "HORIZONTAL WELLS PRODUCTION TECHNIQUES IN HETEROGENEOUS RESERVOIRS"

SPE-BAHRAIN-MARCH 1985
- "HORIZONTAL WELLS PRODUCTION AFTER FIVE YEARS"
SPE-LAS VEGAS-SEPTEMBER 1985
- "PRESSURE ANALYSIS FOR HORIZONTAL WELLS"
SPE-LAS VEGAS-SEPTEMBER 1985
- "NEW PRODUCTION LOGGING TECHNIQUES FOR HORIZONTAL WELLS"
SPE-LAS VEGAS-SEPTEMBER 1985
- "ANALYTIC 2D MODELS OF WATER CRESTING BEFORE BREAKTHROUGH FOR HORIZONTAL
WELLS"
SPE-NEW ORLEANS-OCTOBER 1986
- "SOME PRACTICAL FORMULAS TO PREDICT HORIZONTAL WELL BEHAVIOUR"
SPE-NEW ORLEANS-OCTOBER 1986
- "MUD AND CEMENT FOR HORIZONTAL WELLS"
SPE-NEW ORLEANS-OCTOBER 1986

```
******************************************************************************
* TITLE : DRILLING AUTOMATION (PART 1)              *       PROJECT NO       *
*                                                   *                        *
*                                                   *     TH./02026/84/FR/.. *
*                                                   *                        *
******************************************************************************
* CONTRACTOR :                                      * PROGRAM :              *
*   GERTH                                           *   HYDROCARBONS         *
*   AVENUE DE BOIS PREAU 4      TEL 47 52 61 39     *                        *
*   FR - 92502 RUEIL-MALMAISON TLX 203050           *                        *
*                                                   * SECTOR :               *
*                                                   *   DRILLING             *
* PERSON TO CONTACT FOR FURTHER INFORMATION :       *                        *
*   MR. R. LE ROC'H             TEL 40 47 31 32     *                        *
*                                                   *                        *
*                                                   *                        *
******************************************************************************
                                                         VERSION : 01/01/87
```

AIM OF THE PROJECT :
****
The final purpose of this project is the full automation of the drilling
operation. The aim of the part 1 is to obtain the individual automation of main
components such as screwing and unscrewing machine for the drilling string,
handling of triple, and power-swivel.

PROJECT DESCRIPTION :
****
The project is divided into three phases :
Phase 1 - Automation of screwing and racking in the derrick.

For the screwing and unscrewing machine new sensors (centering over of the
rotating table, positionning over the tool joint, torque measurement, rotation
accounting, ...) are to be developped and connected to a programmable automat.
In additions the automatic racking of triple at the levels of the drilling
floor and the upper racking platform, is obtained with two remote-controlled
automatic machines.
Phase 2 - Semi-automatic drilling.

This phase is directed to a better knowledge of three important parameters
directly applied on the drilling tool : the weight, the torque and the rotating
speed. Specific sensor are to be developped.
Phase 3

Electrical drive of the power swivel including the development of a new
reducting gear box, refrigeration, ... with study and tests of components.

STATE OF ADVANCEMENT :
****
The project was completed by end of June 1986.

RESULTS :

****

Phase 1 was completed by the end of 1985 on a drilling site in the south of
France with simultaneous test runs on the following equipment :
- an automated drill pipe spinner (BIL-K). This equipment is ready to be
commercialized with automatic control directly on the machine.
- an automatic slips system (BIL-PACK). This equipment is already
commercialized.
- a machine for the automatic hooking and racking at the derrick hooking
platform (UPPER BIL-STAB). This equipment has to be automated within the
contract TH 02.29/85.
- a machine for the racking of thribbles at the feet of the derrick (LOWER BIL-
STAB). This equipment has to be automated within the contract TH 02.29/85
Phase 2 comprised the study of an automation allowing semi-automatic drilling
and the study of a band-brake servo system. However, the latter cannot be
adapted to the various band brakes existing on the market.
Phase 3 consisted in fabricating and testing at the workshop an electric power-
swivel prototype. The resistance of this equipment proved to be very
satisfactory. This equipment will be completed with handling systems and
overall tested within the scope of contract TH 02.29/85.

REFERENCES :

****

OSEA 1986 (SINGAPOOR)
OTC 1986 (HOUSTON)
ONS 1986 (STAVANGER)
TECHNICAL BULLETIN FOR COMMERCIALIZATION.

```
*******************************************************************************
* TITLE : OPTIMIZATION OF DRILLING OPERATIONS     *        PROJECT NO        *
*          (PHASE II)                             *                          *
*                                                 *     TH./02028/85/FR/..   *
*                                                 *                          *
*******************************************************************************
* CONTRACTOR :                                    * PROGRAM :                *
*   GERTH                                         *   HYDROCARBONS           *
*   4, AVENUE DE BOIS PREAU     TEL 1 47.52.61.39 *                          *
*   FR - 92502 RUEIL-MALMAISON TLX 203 050        *                          *
*                                                 * SECTOR :                 *
*                                                 *   DRILLING               *
* PERSON TO CONTACT FOR FURTHER INFORMATION :     *                          *
*   MR. TRAONMILIN              TEL 59.83.40.00   *                          *
*                               TLX 560 804       *                          *
*                                                 *                          *
*******************************************************************************
                                                    VERSION : 01/01/87
```

AIM OF THE PROJECT :
****
The aim of the project is to develop new methods to improve the conduct of
drilling operations, while using the physical data related to the drilling
process. The topics selected to conduct this research are directly related to
the making of new hole. This activity is given priority here, as it uses up to
40% of the time spent on a well. The other basis for this research is in
assuming that progress relies on an improved knowledge of downhole physics,
until now mostly approached through statistical models. The necessary basis for
the expected improvements is the acquisition of models. The area concerned are
for this project:
- the behaviour of the borehole walls as seen through the transmission
  losses of torque and weight on bit from surface to the drill bit,
- the directional behaviour of drill strings,
- the behaviour of the drill bit cutting face while drilling.

PROJECT DESCRIPTION :
****
The main activities of the project are:
- well data acquisition,
- test bench data acquisition,
- interpretation of data,
- integration of results in the conduct of drilling operations.
These steps have been prepared by the project nr TH 2024/84 during which
bibliographies, physical and numerical modelling and preliminary tests where
performed.
These activities are carried out in a specific manner for each of the topics
mentioned, except for the well data acquisition which is common to all of them.
WELL DATA ACQUISITION
Drilling data are acquired essentially by specialized teams using data
recorders brought to and used on the well site under their direct supervision.
Both downhole and surface data are recorded. A complex treatment is afterwards
required to put all the needed data in files adapted to the interpretation work
pertaining to the other activities.

TEST BENCH DATA ACQUISITION

Such data are acquired in a laboratory environment much less subject to environmental and process noises.

The test benches involved are:

- a drilling bench for bits up to 6" (152 mm) in diameter able to drill 50 mm long rock samples,
- a reduced scale bottom hole assembly (BHA) model, representing about 50 m length of BHA,
- turbine test benches for static or dynamic measurements.

INTERPRETATION OF DATA

The comparison between well and/or bench data and the output of the theoretical models attempting to simulate downhole physics is made systematically. The models are thereafter completed or modified to fit better with the processes observed.

INTEGRATION OF RESULTS IN THE CONDUCT OF DRILLING OPERATIONS

The integration of the results in the daily drilling procedure requires various approaches depending on the topic studied. It will often require the use of computing means and their adaptation to the relatively rough environment of the drilling site. The results will either be integrated in automation processes, or at a lower frequency of occurence, participate in decision making and equipment selection.

STATE OF ADVANCEMENT :
****

Ongoing: most of the test bench and well data acquisition has been performed. The major part of interpretation of the acquired data has been done. The integration of the results in the drilling procedures has been started. The main part of the remaining work is completing the study of integration.

RESULTS :
****

The main results are:

- the discovery of practical laws for the transmission of torque and weight on bit,
- computer codes for the simulation in three dimensions, statically and dynamically, of the bottom hole assemblies directional and vibratory behaviour,
- a better knowledge of the torque-weight relationships at the bit cutting face.

REFERENCES :
****

PETROLE ET TECHNIQUE NR 322 - JANVIER 1986
OPTIMISATION DE LA CONDUITE DU FORAGE
SPE 15466 (SOCIETY OF PETROLEUM ENGINEERS MEETING - NEW ORLEANS - OCT 86)
OTC 5510 (OCT 87)
FIELD DATA ANALYSIS OF WEIGHT AND TORQUE TRANSMISSION TO THE DRILL BIT (PLANNED)

```
********************************************************************************
* TITLE : DRILLING AUTOMATION (PART 2)              *      PROJECT NO        *
*                                                    *                        *
*                                                    *   TH./02029/85/FR/..   *
*                                                    *                        *
********************************************************************************
* CONTRACTOR :                                       * PROGRAM :              *
*   GERTH                                            *   HYDROCARBONS         *
*   4 AVENUE DE BOIS PREAU       TEL 1 47.52.61.39   *                        *
*   FR - 92502 RUEIL-MALMAISON TLX 203 050           *                        *
*                                                    * SECTOR :               *
*                                                    *   DRILLING             *
* PERSON TO CONTACT FOR FURTHER INFORMATION :        *                        *
*   MR. R. LE ROC'H              TEL 40 47.31.32     *                        *
*                                TLX 710 960         *                        *
*                                                    *                        *
********************************************************************************
                                                     VERSION : 01/01/87
```

AIM OF THE PROJECT :
****
The final purpose of this project is the full automation of the drilling
operation. The aim of Part 2 is to obtain a complete automation of screwing and
racking devices in the derrick. These devices have already been automated
individually within the scope of project TH 02024/84 Drilling automation Part 1.
  The last phase of this project is to complete automation of the power swivel
previously developed within part 1 thanks to the handling equipment depending
directly on the power swivel allowing overdrill and automatic connection and
disconnection operations at any height inside the derrick. The new design of
this second generation Power Swivel handling equipment is modular, so future
improvements could be made separately for each module and also the drilling
process will not be stopped because of troubles of only one module.

PROJECT DESCRIPTION :
****
1. Improve the automation of hooking systems namely by increasing the handling
capacity of the lower racking machine for the triples up to 9 tons so to handle
the drill collars.
2. Realise a single automation linking the following operations :
 - screwing-unscrewing (BIL-K)
 - transfer of BIL-K to bottom racking
 - racking on lower position
 - racking upper position.
3. Improve handling beyond the drilling floor area to replace pneumatic
wirelines and winches presently used by safer systems based on jacks, winches
and robots.
4. Manufacture of an actuator for the brake band
5. Realisation of power swivel automation equipment, namely :
 - connection module (or tool joint breaker): this module is able to make
   the connection between the power swivel and the drill stem at any height
   inside the derrick, and also to dismantle or to connect lower
   parts of the power swivel shaft for easy maintenance
   (i.e. : Kelly cocks, Saver sub).

- handling module : this module improves handling operations during tripping phases by tilting, to convenient position, the elevator which catches the pipe standing out the racking board,
- threading compensation module : this module avoids the damage of the tool joint thread during connection of power swivel to drill stem. Without this module the total weight of the power swivel (19 tons) should be supported by the power swivel saver sub pin threads and by the upper tool joint box threads of the drill stem.

STATE OF ADVANCEMENT :
****
Ongoing. Phases 1 and 3 are presently in their study state. Equipment inherent to Phase 5 have been manufactured and mounted onto the power swivel for a first test run. Some modifications are presently being made connection module and handling module after this first test run.

RESULTS :
****
The only results available today concern Phase 5. All the equipment planned within the contract have been manufactured and operate satisfactorily. Their functioning as an assembly is scheduled on January 1987 at the workshop. The complete assembly could be despatched to a rig for test under real service conditions at the end of February 1987.

```
********************************************************************************
* TITLE : CONTROLLED DRILLING FOR LAYING OF       *      PROJECT NO        *
*         PETROLEUM PIPELINES IN ROCKY GROUND     *                        *
*                                                 *    TH./02036/86/FR/..  *
*                                                 *                        *
********************************************************************************
* CONTRACTOR :                                    * PROGRAM :              *
*   GERTH                                         *   HYDROCARBONS         *
*   4, AVENUE DE BOIS PREAU     TEL 1 47.52.61.39 *                        *
*   FR - 92502 RUEIL-MALMAISON TLX 203 050        *                        *
*                                                 * SECTOR :               *
*                                                 *   DRILLING             *
* PERSON TO CONTACT FOR FURTHER INFORMATION :     *                        *
*   MR. Y. CEZARD               TEL 1 46.87.31.45 *                        *
*                               TLX 202 010       *                        *
*                                                 *                        *
********************************************************************************
                                                     VERSION : 01/01/87
```

AIM OF THE PROJECT :
****
This project consists of studying, designing and manufacturing a complete
lowdepth, controlled drilling equipment able to drill in hard soils where
petroleum pipelines are to be laid.
This method allows the crossing of natural or artificial obstacles (rivers,
shores, urban sites...) found on the laying paths.
For many years controlled drilling has been employed in industrial
constructions all over the world but for technical reasons its use at the
present time is restricted to loose and soft ground only.
The aim of this project is to extend the use of horizontal drilling to sites
where consolidated or rocky facies are found. The technical objective is to lay
petroleum pipelines with diameters of less than 42 inches and lengths of
several hundred meters.
Controlled drilling is extremely competitive with conventional techniques such
as dredging and bridge building from both economical and safety points of view.
Environmental protection is also ensured by this drilling method.

PROJECT DESCRIPTION :
****
Controlled drilling is generally carried out in three stages.
The first stage consists in drilling a small diameter pilot hole, beneath the
obstacle. The drilling device is controlled by means of a steering tool which
controls the direction of the drilling tool.
The second stage consists in reaming the hole to the desired diameter in
successive passes if required.
The third stage is the pipeline insertion into the hole.
The different technical parts which will be studied are the following :
- drilling tool
- reamer
- device providing thrust on drilling string
- deviation of the drilling
- removal of cuttings
- stabilization of the walls
- bottom/surface links
- surface equipments.

The project programme, planned for a period of three years, comprises five phase :

Phase 1- Study of technological transfers

The aim here is to obtain full knowledge of both equipment specifications and adoptable technical transfers by means of a market study and technical research of the different existing methods which could be used

Phase 2- Manufacturing and testing of a drilling equipment prototype

The selection, resulting from the previous phase, would enable a prototype to be manufactured with existing and proved materials.

This prototype would have to be able to carry out the first stage of the controlled drilling : the drilling of a small diameter pilot hole (less than 12 inches).

It would be tested in order to determine its capacities and any eventual improvements to be made.

Phase 3- Detailed preliminary project and construction of a pilot tool

The object of this phase is to define and design a complete drilling pilot tool, from the results of the previous phase.

The different parts of the pilot tool are :

- the drilling tool
- the loads transmission device
- the drilling deviation equipment
- the device for consolidating the walls
- the bottom/surface link system.

Phase 4-Detailed preliminary project and construction of a reamer

This phase is identical to the previous one but the equipment to be made here is th reamer which enlarges the diameter of the hole.

Phase 5- General tests, optimization and qualification of the drilling rig

The drilling equipment will be subjected to a series of tests, and particularly to on-site tests in order to debug and qualify the system.

STATE OF ADVANCEMENT :
****
Ongoing.

Beginning of the project.

Phases 1 and 2 are planned to be completed in the first year of the study. The first step, which is ongoing consists in studying the different existing drilling in both the petroleum sector, mining and geothermal sectors, enabling rock drilling at sufficient speed.

PRODUCTION SYSTEMS

```
*******************************************************************************
* TITLE : DESIGN AND DEVELOPMENT OF A FIXED STEEL    *      PROJECT NO        *
*         PLATFORM FOR A DEPTH OF 650 M.             *                        *
*                                                    *    TH./03103/81/IT/..  *
*                                                    *                        *
*******************************************************************************
* CONTRACTOR :                                       * PROGRAM :              *
*    SUB SEA OIL SERVICES SPA                        *    HYDROCARBONS        *
*    VIA DELLA SCAFA, 19         TEL 601321          *                        *
*    IT - 00054 FIUMICINO        TLX 611156          *                        *
*                                                    * SECTOR :               *
*                                                    *    PRODUCTION SYSTEMS  *
* PERSON TO CONTACT FOR FURTHER INFORMATION :        *                        *
*    MR. GASPERINI                                   *                        *
*                                                    *                        *
*                                                    *                        *
*******************************************************************************
```

VERSION : 26/03/87

AIM OF THE PROJECT :
****
Design and development of a fixed steel platform for 650 m water depth.

PROJECT DESCRIPTION :
****
The present fixed platforms can work up to 300 m water depth: feasibility
studies to reach oil fields in deeper waters were conducted by several
engineering companies and were oriented towards cable tension platforms.
S.S.O.S. Project tends on the contrary towards a classic fixed steel platform.
The platform has been designed as follows:
- fixed structure on four legs (their bases are anchored on the bottom at 650 m.
  Water depth)
- the legs support jacket and deck structures.

STATE OF ADVANCEMENT :
****
The project has been abandoned.

```
********************************************************************************
* TITLE : SUBSEA PRODUCTION SYSTEM FOR          *       PROJECT NO        *
*         HYDROCARBON FIELD EXPLOITATION.        *                        *
*                                                *     TH./03108/81/IT/..  *
*                                                *                        *
********************************************************************************
* CONTRACTOR :                                   * PROGRAM :               *
*   TECNOMARE                                    *   HYDROCARBONS          *
*   SAN MARCO 2091           TEL 041 796711      *                        *
*   IT - 30124 VENEZIA       TLX 410 484         *                        *
*                                                * SECTOR :                *
*                                                *   PRODUCTION SYSTEMS    *
* PERSON TO CONTACT FOR FURTHER INFORMATION :    *                        *
*   MR. A. RODIGHIERO                            *                        *
*                                                *                        *
*                                                *                        *
********************************************************************************
```

VERSION : 01/01/87

AIM OF THE PROJECT :
****
To study and design an underwater production system for the exploitation of
hydrocarbon fields in 1,000 m water depth.

PROJECT DESCRIPTION :
****
The project is subdivided in the following phases:
- definition of the system configuration
- design of the system components, definition of the installation and operative
procedures, design of the interfaces with other subsystems necessary for the
field exploitation
- design, construction and test of critical component models.

STATE OF ADVANCEMENT :
****
Completed

RESULTS :

****

The underwater production system has been designed for 100,000 bbl/day and is composed of:
- Two large, parallel cylindrical underwater modules held by their weight on the sea bottom. The modules are sandwich structure (steel, concrete steel, overall thickness 1 m, diameter 14 m, lenght 170 m) which provide housing to the full remote control process plant, manifold, etc.
- A monopile, a slender tensioned structure providing a permanent structural link to surface.
- The topsides where power generation and 40 people living quarters and flare are located.
The production scheme foresees remote wellheads, oil treatment inside modules and sealines to shore. No loading point at surface.
Process plant operates inside modules in atmospheric pressure, inert gas environment and without human presence except for maintenance (every 2 months on average). Maintenance crew operate using masks. Link to surface is provided by a shuttle module running along a rail placed on the monopile. An emergency module is provided on bottom, independent from the sandwich vessels, which ensures 7 days life support for maintenance crew.
Construction and installation procedure have been defined and costs estimated. The system can be built in existing shipyards and is transported to site in three parts: modules, momopile, topsides.
Extensive model test have been carried out to validate sandwich vessel computer programs.
The configuration is easily adaptable to different water depths and production rates.
The concept has been precertified by DNV.

REFERENCES :

****

C. PENNINO, P. VIELMO
"TSPS, A NEW CONCEPT OF SUBSEA PRODUCTION SYSTEM FOR VERY DEEP WATER", ROYAL INST. OF NAVAL ARCHITECTS (RINA) SYMPOSIUM ON "DEVELOPMENTS IN SUBSEA ENGINEERING SYSTEMS",
LONDON, MAY 14-16, 1985
A. TATTO, C. BROGI
"DESIGN METHOD FOR DEEP WATER LARGE SANDWICH VESSELS", AND C. PENNINO, W. CAMPANA, "TSPS, COMPLIANT STRUCTURE FOR EXPLOITATION OF OIL FIELD IN DEEP WATER", 1ST AIOM CONGRESS, VENICE, JUNE 3-6, 1986.
G. SEBASTIANI, R. BRANDI, T. NAESS, P. SCHAMAUN,
"DESIGN OF A NEW CONCEPT OF SINGLE-POINT MOORING SYSTEM FOR VERY DEEP WATERS", PAPER OTC 4350, HOUSTON, MAY 3-6, 1982.

```
******************************************************************************
* TITLE : DESIGN AND DEVELOPMENT OF A HOMING-IN    *      PROJECT NO         *
*         DEVICE FOR BLOW-OUT CONTROL              *                         *
*                                                  *      TH./03112/81/NL/.. *
*                                                  *                         *
******************************************************************************
* CONTRACTOR :                                     * PROGRAM :               *
*   SHELL INTERNATIONALE PETROLEUM MAATSCHAPPIJ B.V. * HYDROCARBONS          *
*   CAREL VAN BYLANDTLAAN 30    TEL 070-773867     *                         *
*   NL - 2501 AN DEN HAAG       TLX SHELL NL 31005/KSE*                      *
*                               PL NL 31527        * SECTOR :                *
*                                                  *   PRODUCTION SYSTEMS    *
* PERSON TO CONTACT FOR FURTHER INFORMATION :      *                         *
*   MESSRS T.N. WARREN/F. DRIE                     *                         *
*                                                  *                         *
*                                                  *                         *
******************************************************************************
                                                       VERSION : 01/01/87
```

AIM OF THE PROJECT :
****
Design and development of an acoustic homing-in device for use in the oil
industry to determine both the relative distance and direction from a relief
well to a blow-out well by monitoring, in the relief well, the noise generated
by the blowing well.

PROJECT DESCRIPTION :
****
During the feasibility study, suitable methods have been established for the
measurement and analysis of acoustic signals in a borehole. Then engineering-
model logging tools have been designed and constructed to evaluate these
acoustic principles under simulated blow-out conditions. During third stage, a
prototype homing-in tool is to be constructed, which can be used under actual
field conditions.

STATE OF ADVANCEMENT :
****
A field prototype of the direction determination part of the tool (HIT-A) has
been developed in cooperation with the service company Schlumberger. This tool
has been successfully tested in shallow test holes. A test under actual field
conditions in a deep and deviated well is being prepared. A field prototype
tool of the distance determination part of the tool will be developed after
successful testing of the HIT-A.

RESULTS :
****
Passive acoustic methods have been developed. The validity of basic assumptions
concerning the sound generated by a blowing well, has been confirmed by noise
logs recorded in actual relief wells. Engineering model logging tools have been
developed and tested in a configuration of three shallow test holes which were
specially drilled for this purpose. Results obtained demonstrated the validity
of the underlying theoretical principles and also proved that both direction to
and distance from an artificial sound source can be determined with the help of
the engineering model tools. In cooperation with Schlumberger, a homing-in
prototype tool (HIT) is being developed which can be used under actual field
conditions with regard to pressure and temperature. The first test with
Schlumberger's HIT-A (direction determining sonde) in the test holes was
successful. The acquired data compared very well with those obtained with the
engineering model logging tool under identical conditions, and the direction
towards a remote noise source was properly indicated. It can be concluded from
this test that the HIT-A is suitable for passive acoustic direction
measurements.

```
********************************************************************************
* TITLE : DEEP WATER PLATFORM                    *         PROJECT NO          *
*                                                *                             *
*                                                *      TH./03113/81/FR/..     *
*                                                *                             *
********************************************************************************
* CONTRACTOR :                                   * PROGRAM :                   *
*   SEA TANK CO.                                 *   HYDROCARBONS              *
*   RUE DU DESSOUS DES BERGES  TEL (1) 584 11 64 *                             *
*   F - 75013 PARIS            TLX 270263 F      *                             *
*                                                * SECTOR :                    *
*                                                *   PRODUCTION SYSTEMS        *
* PERSON TO CONTACT FOR FURTHER INFORMATION :    *                             *
*   M. VACHE                                      *                             *
*                                                *                             *
*                                                *                             *
********************************************************************************
                                                       VERSION : 09/06/87
```

AIM OF THE PROJECT :
****
The aim of the project is to develop a concept for oil and gas production in
400 m of water depth based on a concrete platform with modular construction and
specific assembly methods in order to cope with existing construction site.

PROJECT DESCRIPTION :
****
The resulting structure looks like a stack of several elementary conventional
concrete platforms having 3 or 4 columns each. The elementary structures are
built in a conventional way. The major innovation consists of the assembly
method between the different elementary structures. Each structure is tilted
from vertical to horizontal by differential ballasting. When two elementary
structures are horizontal, they are approached and connected. Specific devices
and procedures have been developed during the first phase of this study. When
the complete structure is achieved, it is towed to side in either the vertical
or the horizontal position. Norwegian companies took part in the activities as
subcontractors.

STATE OF ADVANCEMENT :
****
The first phase of the project, the conceptual design phase, was completed in
June 1983. The second phase relating to detail studies was completed at the end
of 1985.

RESULTS :
****
The conceptual design phase has demonstrated the feasibility of the concept
from structural, construction-installation and operational view points.
Between the two phases, an optimization study was conducted to reduce concrete
quantities thanks to lighter designs (hexagonal basement, 3 concrete legs only
in a dock and suppression of the assembly procedure in horizontal floating
position)
Detail study of the second phase has been forcusing on structural analysis by
finite elements for sensitive parts and calculation under seismic conditions
according to Norwegian standards.

```
********************************************************************************
* TITLE : DOWN HOLE PUMPING DEVELOPMENT            *      PROJECT NO         *
*                                                  *                         *
*                                                  *    TH./03120/82/UK/..   *
*                                                  *                         *
********************************************************************************
* CONTRACTOR :                                     * PROGRAM :               *
*   PEEBLES ELECTRICAL MACHINES                    *   HYDROCARBONS          *
*   EAST PILTON              TEL 031-552 6261       *                         *
*   EDINBURGH               TLX 72125 (PP EDING)   *                         *
*   SCOTLAND EH52XT                                * SECTOR :                *
*                                                  *   PRODUCTION SYSTEMS    *
* PERSON TO CONTACT FOR FURTHER INFORMATION :      *                         *
*   MR. CAMPBELL                                   *                         *
*                                                  *                         *
*                                                  *                         *
********************************************************************************
```

VERSION : 09/06/87

AIM OF THE PROJECT :
****
To develop and test an electric motor driven modular pumping unit, with a
maintenance-free life of at least two years at depths of 3,000 m in deep oil
wells with a deviation of up to 6 deg. C./30 m., at oil temperatures of up to
120 deg. C. and pressures of 210-350 kg/cm2 even when the motor is full of sour
crude oil.

PROJECT DESCRIPTION :
****
Phase 1.
Determination of mechanical and electrical properties of materials, development
of pressure-temperature cycling equipment for well-bottom simulation tests:
design and manufacture of pressure balance unit: development of winding
encapsulation and core corrosion protection systems: investigation of bearing
coating to resist abrasive and chemical effects of sour crude oil containing
sand: measurement of dynamic thermal and vibration response of full scale
simulated shaft-bearing system: design, manufacture and test of prototype
motor to establish dynamic characteristics at full load.
Phase 2.
Manufacture of second prototype motor emboying the solutions from
phase 1: life test of motor-pump unit, supplied by static converter equipment,
 and pumping brine solution contaminated with sharp sand.
Phase 3: manufacture of a module for testing in a north sea production well, to
be provided by amoco. Life-testing in the well environment.

STATE OF ADVANCEMENT :
****
Abandoned after completion of phase I.

RESULTS :
****
The dynamic thermal and vibration responses of the shaft-bearing system in the
test apparatus have been established with "soft" bearing surfaces. Results so
far confirm that bearing temperature rises and shaft dynamic responses are
satisfactory. Stator core ceramic coating for corrosion protection has been
demonstrated to be effective in sea water over the above temperature and
pressure range. The main technical problem encountered during the project was
concerning the development of a suitable encapsulation resin. Several resin
samples have been tested. After extensive tests, particularly with crude
oil/gas mixture at a temperature of 200 deg.C and pressure of about 4000 psi, a
suitable material was selected.
But those tests took considerably larger time than anticipated. In addition to
use the selected material in downhole prototype requires long term testing of
this important component. Considering increased project expenses and further
delay required for solving technical problems it has been decided to interupt
the project after phase I completion.

```
*******************************************************************************
* TITLE : FLOATING PRODUCTION SYSTEM FOR DEEP      *      PROJECT NO         *
*         MEDITERRANEAN WATERS                     *                         *
*                                                  *      TH./03121/82/IT/.. *
*                                                  *                         *
*******************************************************************************
* CONTRACTOR :                                     * PROGRAM :               *
*   AGIP SPA                                        *   HYDROCARBONS          *
*   TEIN                      TEL 39 2 5205969      *                         *
*   P.O. BOX 12069            TLX 310246            *                         *
*   IT - 20120 MILAN                                * SECTOR :                *
*                                                  *   PRODUCTION SYSTEMS    *
* PERSON TO CONTACT FOR FURTHER INFORMATION :       *                         *
*   MR. P. TASSINI                                  *                         *
*                                                  *                         *
*                                                  *                         *
*******************************************************************************
```

VERSION : 01/01/87

AIM OF THE PROJECT :
****
Development of a floating production system for exploitation of oil fields in
very deep waters in the Mediterranean sea, reference case the Aquila field.

PROJECT DESCRIPTION :
****
Analysis and selection of the production system configuration with definition
of its main components (platform, risers, tendons, etc.). Development of
computer procedure for hydrodynamic and structural analysis. Preliminary design
of the floating platform and of the other main subsystems. Definition of
construction and installation procedures. Model testing of the platform in
ocean basin. Cost of schedule evaluation.

STATE OF ADVANCEMENT :
****
Completed. Preliminary design of the platform and of the main subsystems has
been carried out and basic construction and installation procedures have been
defined. Critical components needing detailed design and functional tests have
been pointed out. A precertification review performed by Det Norske Veritas
supports the project overall conclusions.

RESULTS :

****

The project has demonstrated that the concept of the Tension Leg Platform
proposed is feasible and it can be used as floating production system for the
production of deep water Mediterranean oil fields.
The TLP is basically composed of a floating hull which tensions a system of
vertical tethers connected to the base.
The concept may be subdivised into four main subsystems :
- Platform consisting of a double symetric semisubmersible unit having four
cylindrical columns connected by pontoons and an integrated deck designed to
house process plants and auxiliaries.
- Anchoring system made up of 16 tendons (4 per column) that are composed of
steel pipes welded together during the platform installation phase.
- Foundation made up of 4 piled bases connected 2 by 2 by means of a lattice
structure provided in order to facilitate and reduce operations.
- Completion consisting of : eight production risers, four water injection
risers and one export riser connected to the well template which has 16 slots.
- Specific criteria were developed to analyse hydrodynamic aspects, in order to
have a correct assessment of tether forces and overall motion of the platform.
It was found that tether loading was essentially driven by the platform
response to the waves at their own frequencies, whilst a large contribution to
the overall (unconstrained) motions of the platform came from long drift
effects.
As a consequence, the analysis procedure was built up in such a way as to have
correct characterization of the platform motions in both mentioned aspects.
Fatigue of the tethers was checked by means of a stochastic approach to take
properly into account the contributions of the sea states occuring during
lifetime.
Model tests have been also carried out in a pool and in a wind tunnel.

REFERENCES :

****

SEVERAL (7) TECHNICAL PAPERS HAVE BEEN PRESENTED AT THE FOLLOWING SEMINARS AND
CONFERENCES :
- "NEW TECHNOLOGIES FOR THE EXPLORATION AND EXPLOITATION OF OIL AND GAS
RESOURCES" LUXEMBOURG, 5-7 DECEMBER 1984.
- "3RD DEEP OFFSHORE TECHNOLOGY CONFERENCE" SORRENTO, 21-23 OCTOBER 1985.
- "1ST OMAE SPECIALITY SYMPOSIUM" NEW ORLEANS, FEBRUARY 1986.
- "1ST AIOM CONGRESS" VENEZIA, JUNE 1986.

```
********************************************************************************
* TITLE : CONCRETE PLATFORMS FOR ARCTIC          *      PROJECT NO          *
*         CONDITIONS.                            *                          *
*                                                *   TH./03123/82/UK/..     *
*                                                *                          *
********************************************************************************
* CONTRACTOR :                                   * PROGRAM :                *
*   MC ALPINE                                    *   HYDROCARBONS           *
*   P.O. BOX 74              TEL (01)8373377     *                          *
*   40, BERNARD STREET       TLX 22308           *                          *
*   UK - LONDON WC1N 1LG                         * SECTOR :                 *
*                                                *   PRODUCTION SYSTEMS     *
* PERSON TO CONTACT FOR FURTHER INFORMATION :    *                          *
*   MR. CLARE                                    *                          *
*                                                *                          *
*                                                *                          *
********************************************************************************
                                                      VERSION : 03/03/87
```

AIM OF THE PROJECT :
****
To develop concrete gravity designs for production platforms in arctic regions,
based on the experience already gained in the north sea.

PROJECT DESCRIPTION :
****
The project is in two phases.
Phase 1 :
 - state of the art study - has been to investigate the precise requirements
for platforms in the arctic regions and how to extend the existing design
concepts by the application of concrete technology already proven in the north
sea. Phase 2 :
 - will then prepare a detailed design proposal and economic evaluation of one
or maybe more particular applications.

STATE OF ADVANCEMENT :
****
Phase 1 has been completed.

RESULTS :
****
Study of the exploration programme and forecasts, environmental conditions and
ice properties followed by discussions with operators and government agencies
to determine requirements. Design experience has been used to formulate outline
designs for platforms for oil or gas production in the beaufort sea and the
scotia shelf and grand banks area.
Due to the evaluation of commercial demand, further detailed work as proposed
for phase 2 has been deferred.

```
*********************************************************************************
* TITLE : DEVELOPMENT OF A DEEP WATER BARGE          *       PROJECT NO         *
*                                                    *                          *
*                                                    *     TH./03124/82/UK/..   *
*                                                    *                          *
*********************************************************************************
* CONTRACTOR :                                       * PROGRAM :                *
*   TAYLOR WOODROW ENERGY LTD.                       *   HYDROCARBONS           *
*   TAYWOOD HOUSE, 345 RUISLIP TEL (01)5782366       *                          *
*   SOUTHALL, MIDDLESEX UB1 2Q TLX 24428             *                          *
*   ENGLAND                                          * SECTOR :                 *
*                                                    *   PRODUCTION SYSTEMS     *
* PERSON TO CONTACT FOR FURTHER INFORMATION :        *                          *
*   MR.J.P. GIBSON                                   *                          *
*                                                    *                          *
*                                                    *                          *
*********************************************************************************
                                                        VERSION : 09/06/87
```

AIM OF THE PROJECT :
****
To carry out, and report on, the development of a barge mounted production and
storage system (BPSS), in particular to prove that the system is a commercial
viable contender for the exploitation of oil fields in the north sea, with
water depths from 100 m to 300 m and greater.

PROJECT DESCRIPTION :
****
The project has been carried out in three phases.
Phase 1 :
The computer model of the barge hydrodynamics was confirmed by model tests.
Additionally an outline design for a BPSS in 300 m of water was produced and
costed.
Phase 2 :
A BPSS for 100 m of water has been designed and detailed costing and economic
analyses were being carried out. Lloyds "approval in principle" has been
obtained.
 Phase 3 :
The areas of uncertainty disclosed by phase 2 were studied. The phase 2 design
was modified to incorporate a drilling facility and gas injection.

STATE OF ADVANCEMENT :
****
Completed

RESULTS :
****
The BPSS will have better heave motions than existing semi-submersibles and
will cost less than a semi-submersible production system with in-line storage.
With storage, it is similar in cost to a semi-submersible based system without
in-line storage. Equipment used for BPSS can all be of field proved design.
It is commercially and technically attractive in water depths greater than 200
m for oil fields with large quantities of associated gas. It lowers the
threshold of field viability and provides substantially greater deck load
capacity than a semi-submersible based system.

```
********************************************************************************
* TITLE : DEEP WATER GRAVITY TOWER                *      PROJECT NO         *
*                                                 *                         *
*                                                 *   TH./03127/82/FR/..    *
*                                                 *                         *
********************************************************************************
* CONTRACTOR :                                    * PROGRAM :               *
*   C.G. DORIS                                    *   HYDROCARBONS          *
*   58 RUE DU DESSOUS DES BERG TEL (1) 5841164    *                         *
*   F - 75013 PARIS            TLX 270263F        *                         *
*                                                 * SECTOR :                *
*                                                 *   PRODUCTION SYSTEMS    *
* PERSON TO CONTACT FOR FURTHER INFORMATION :     *                         *
*   F. SEDILLOT / L. DES DESER                    *                         *
*                                                 *                         *
*                                                 *                         *
********************************************************************************
```
VERSION : 09/06/87

AIM OF THE PROJECT :
****
Within a previous separate phase, the concept of the deep water gravity tower
has been developed. The purpose of this present project is now to further
investigate some particular points related to this articulated structure, in
order to reach a sufficient level of technological development.

PROJECT DESCRIPTION :
****
The main items studied are the following:
- detailed dynamic response, fatigue and seismic analysis. This study
  Includes the assessment of stress history corresponding to wave and
  Wind spectra, and steel mode finite element analysis to establish
  Exact stress concentration factors
- study of drilling and conductor systems, and risers
- tests on the laminated rubber articulated joint, including compression -
shear fatigue tests, laboratory and deep sea environmental tests
- detailed construction and installation procedures.

STATE OF ADVANCEMENT :
****
Completed

RESULTS :
****
A new methodology for fatigue analysis has been established. It is based on the
particular geometry of the structure which allowed a constant relationship
between forces in members arriving at a node, for a given wave heading. This
methodology permitted and optimisation of steel quantities.
Detailed fatigue analyses have been made on conductors and risers, and their
installation and connection procedure has been developed.
Extensive laminated rubber joint tests were conducted under hydrostatic
pressure. They proved to be fully positive. Corrosion tests which will last
several years are initiated.
Detailed contruction and installation procedures have been prepared, including
marine operations and equipment specification.

```
*******************************************************************************
* TITLE : OFFSHORE TRIALS OF COMPOSITE TUBES          *      PROJECT NO       *
*                                                     *                       *
*                                                     *    TH./03128/82/FR/.. *
*                                                     *                       *
*******************************************************************************
* CONTRACTOR :                                        * PROGRAM :             *
*   GERTH/AEROSPATIALE                                *   HYDROCARBONS         *
*   4, AV.DE BOIS PREAU        TEL (1) 47.52.61.39    *                       *
*   F - 92502 RUEIL-MALMAISON  TLX (1) 47.52.69.27    *                       *
*                                                     * SECTOR :              *
*                                                     *   PRODUCTION SYSTEMS  *
* PERSON TO CONTACT FOR FURTHER INFORMATION :         *                       *
*   MR. GUESNON                TEL (1) 47.52.63.57    *                       *
*                              TLX 203050             *                       *
*                                                     *                       *
*******************************************************************************
                                                         VERSION : 01/05/87
```

AIM OF THE PROJECT :
****
The analysis of the static and dynamic behaviour of drilling extension tubes in
deep water clearly evidences the interest of reducing the weight of the kill
and choke lines and that of the booster line by making them in composite
materials rather than in steel. Such tubes have been developed by IFP and the
AEROSPATIALE (SNIAS) while using the wirewinding technique with fibre glass,
kevlar or carbon impregnated in epoxy resin. After study and computation phases
that allowed the optimization of the dimensioning of the tubes and to qualify
the prototypes for the envisaged application, it seemed preferable to check the
in-situ behaviour of the scale 1 tubes, submitted to real tests following their
utilisation during offshore drilling operations for a significant period.

PROJECT DESCRIPTION :
****
The offshore trial project of composite tubes being used as peripheral lines
for the drilling extension tubes under real utilisation conditions has been
carried out in three major phases :
2.1 Test preparation
- Definition of the experimental tube specifications according to two types :

Designation	Kill & Choke	Booster line
Internal diameter	4" - 101,6 mm	4" - 101,6 mm
Service pressure	70 MPa - 10000 psi	35 MPa - 5000 psi
Effective length	15,24 m - 50 ft	15,24 m - 50 ft

- Definition of practical test conditions :
  - parallel assembling of each type of tube on the two upper elements of the
    riser of the semi-submerisble PENTAGONE 84 operating for TOTAL OIL MARINE
    in the North Sea
  - periodical raise at service pressure of the two additional lines thus
    formed
  - holding of assembly during several drilling operations
  - return to laboratory for appraisal of tubes.
  - Study of tubes dimensioning of their integration in the riser elements, of
    ancillary equipment required to run the tests and to define the procedures.

2.2 Manufacture and testing of tubes and ancillary equipment

- Construction and development of a machine for the fabrication of long tubes (up to 24 m) in composite materials
- fabrication of 5 tubes of each type according to selected definition
- verification of performance obtained by bursting the first tubes of each series.

Designation	Kill & Choke	Booster line
bursting pressure	176 MPa	81,2 MPa
service/bursting ratio	2,5	2,3

- verification of axial bursting resistance under internal pressure of each type of tube
- fabrication of ancillary equipment and testing of tubes pressurization flexibles

2.3 Tests

- stamping of all tubes at 1,5 times their service pressure
- truck transport of tubes and their integration with other elements of the riser
- transfer on the P84 platform and implementation during three drilling campaigns held between May and December 1983. Total immersion period of most solicited tube : 133 days, including 22 cycles under pressure service (534 h)
- dismounting of tubes and return in January 1984
- Tubes appraisal after tests : bursting of a kill and chokee tube at 133,5 MPa,

showing a limited damage of the tube comapred to initial rupture pressure.

STATE OF ADVANCEMENT :
****
Completed. The composite tubes have been separated from the other riser elements and brought back to the plant.

RESULTS :
****
The first phase of the project, described above, has allowed the definition and development of a methodology to test on-site the components of the offshore drilling and production systems (peripheral tubes in composite materials) without disturbing other operations (development drilling operations), nor even interfering with them. This "uncoupling" has allowed to simplify considerably the organization of the test and to reduce its cost. This method, the validity of which was later established (phase 3), could be applied in other circumstances and for other components. From this point of view this phase is a success.

The second phase of the project, described above, has allowed to fabricate a small series of 2 x 5 tubes of a length of 15 m in composite materials by means of a preindustrial machine designed, manufactured and developed outside the scope of this project. These tubes were used, either for destructive qualification tests, or for the third phase of the project. The fabrication technique used for the first time at this scale has proved satisfactory and it was possible to demonstrate the implementation aptitude under industrial conditions of composite tubes developed in laboratory by SNIAS and IFP. This phase also represents a success.

The third phase of the project, described above, has allowed the characterization of the behaviour on-site of composite tubes made in the previous phase, used as peripheral lines for a drilling riser. Booster line tubes, the 350 bar service pressure of which only requires a relatively low wall thickness, present an excessive sensitivity versus handling and transport constraints (pinching and shock effect). They need to be redesigned to be adapted to operational service. On the other hand, the kill and choke lines tubes, significantly thicker due to their service pressure of 1050 bar, supported similar tests without showing any failure after the trials. These tubes are now qualified for the envisaged application. However, additional studies would be necessary for their homologation, specially concerning their external protection and the development of control and in-situ rehabilitation processes. Lastly, it will be necessary to significantly reduce production costs of these tubes before they can be proposed for commercialization. This third phase of the project is thus only half a success.

REFERENCES :
****
DEEP OIL TECHNOLOGY SYMPOSIUM - SORRENTE (ITALY) OCTOBER 1985
"DRILLING RISERS FOR GREAT WATER DEPTHS : ADVANïAGE OF MASS REDUCTION BY MEANS OF COMPOSITE MATERIALS" BY P. ODRU (IFP) AND J.C. GUICHARD (AEROSPATIALE)
OMAE SYMPOSIUM - NEW ORLEANS (LOUISIANE) - FEBRUARY 23-27 1986
"HIGH PERFORMANCE COMPOSITE TUBES FOR DRILLING OR PRODUCTION SYSTEMS"
BY P. ODRU (IFP) AND J.C. GUICHARD (AEROPSATIALE)
"TECHNOLOGIES FOR THE EXPLOITATION AND EXPLOITATION OF OIL AND GAS RESOURCES"
PROCEEEDINGS OF THE 2ND EEC SYMPOSIUM HELD IN LUXEMBOURG (5-7 DECEMBER 84)
"RISER ET TUBES EN MATERIAUX COMPOSITES" (03.63/78, 03.115/81 AND 03.128/82)
VOLUME I P 470-482 BY M. PEINADO (IFP) AND J.C. GUICHARD (AEROPSATIALE).

```
********************************************************************************
* TITLE : DESIGN OF A FLOATING PRODUCTION          *      PROJECT NO          *
*         FACILITY FOR USE ON MARGINAL FIELDS      *                          *
*         (PART 1 AND 2)                           *      TH./03131/82/UK/..   *
*                                                  *                          *
********************************************************************************
* CONTRACTOR :                                     * PROGRAM :                *
*   BRITOIL PLC                                    *   HYDROCARBONS           *
*   301 ST VINCENT STREET      TEL 041 204 2525    *                          *
*   UK - GLASGOW G2 5DD        TLX 777 633         *                          *
*                                                  * SECTOR :                 *
*                                                  *   PRODUCTION SYSTEMS     *
* PERSON TO CONTACT FOR FURTHER INFORMATION :      *                          *
*   MR. K.J. WELLS                                 *                          *
*                                                  *                          *
*                                                  *                          *
********************************************************************************
                                                      VERSION : 01/01/87
```

AIM OF THE PROJECT :
****
To develop a completely integrated floating production system, designed for
optimal development of small North Sea fields.

PROJECT DESCRIPTION :
****
The project set out to develop a design of a semi-submersible to be purpose
orientated for production to overcome the incompatibilities of converted drill
rigs and to investigate the possibility of reducing downtime due to
environmental conditions by optimising the column stabilising characteristics
in association with innovative technology in riser and mooring systems for semi-
permanent location without the need for transit considerations.
The study includes the investigation of subsea templates and manifolds with the
associated control and instrumentation systems, clustering of satellite wells
and the problems of commingling of production flow. Various configurations for
the fpf have been considered, e.g. Producer only, driller and producer,
producer with simultaneous work-over, producer with consecutive workover all
with or without storage and a comparison made of the economic viability of
these various alternatives.

STATE OF ADVANCEMENT :
****
Completed

RESULTS :
****
This study developed a design for a low cost, lightweight, unsophisticated semi-
submersible production only vessel without drilling capability. The vessel
study addressed vessel size reduction, limiting motions, on-board storage,
seabed loading, buoy loading downtime, rigid and flexible riser design and
analysis, mooring systems with and without winches, and optimisation of deck
layout.

Studies were carried out on the subsea equipment, templates, clusters, test loops, drilling programme, wellheads and trees, field configuration development, detailed field layout planning and maintenance and protection philosophy. Three basic field layouts were studied, with variations in location of production and injection wells commensurate with the mooring line pattern determined by the riser restraints. The various manifold and small cluster systems were assessed both from the technical and early flow aspects. A performance simulation programme was developed to aid in the quantification of the production efficiency of the subsea systems and to optimise their design.

```
***************************************************************************
* TITLE : NEW CONCEPT FOR FIXED OFFSHORE          *      PROJECT NO       *
*         PLATFORMS.                              *                       *
*                                                 *   TH./03132/83/NL/..  *
*                                                 *                       *
***************************************************************************
* CONTRACTOR :                                    * PROGRAM :             *
*   CONSULTING ENGINEERS H. VETH BV               *   HYDROCARBONS        *
*   POSTBUS 274            TEL 078 131 944         *                       *
*   NL - 3300 AG DORDRECHT  TLX 29452             *                       *
*                                                 * SECTOR :              *
*                                                 *   PRODUCTION SYSTEMS  *
* PERSON TO CONTACT FOR FURTHER INFORMATION :     *                       *
*   MR. IR. R. VAN DE WAAL                         *                       *
*                                                 *                       *
*                                                 *                       *
***************************************************************************
                                                     VERSION : 31/12/86
```

AIM OF THE PROJECT :
****
The development of a monotower production platform of steel. Preliminary study
shows a cost advantage of a steel monotower of 20 percent compared with
conventional jacket structures. Moreover the innerside of the column is
suitable for use of oil-storage of 500.000 barrels oil. This project must prove
in more detail the advantages of this concept and must beacon its boundaries.

PROJECT DESCRIPTION :
****
In close cooperation with one or more oil companies case-studies have to be
done for some locations. Therefore next items must be examined:
- foundation;
- structural analysis;
- process requirements;
- transport and installation;
- construction method;
- drifting ice problems;
- IMR aspects.

STATE OF ADVANCEMENT :
****
After finishing the basic designs, studies has been done after special
applications of the F.S.P. The application of the F.S.P. in artic regions is
very promising. Oil-companies operating in this regions have shown their
interest in this concept. To examine are the possibilities of further
cooperation between these oil-companies and Veth.

RESULTS :
****
New interest has been shown by Petrobras in Brasil. The environment conditions
there make the FSP a very good and competitive alternative compared with other
solutions. The chances to be shortlisted for new projects in near future are
reasonable.

```
**************************************************************************
* TITLE : GRAVITY PLATFORM FOR 300-400 M. IN POOR   *     PROJECT NO      *
*         ENVIRONMENTS.                             *                     *
*                                                   *    TH./03135/83/NL/.. *
*                                                   *                     *
**************************************************************************
* CONTRACTOR :                                      * PROGRAM :           *
*   ANDOC                                           *   HYDROCARBONS      *
*   H.J. NEDERHORSTSTRAAT      TEL 01820 102 22      *                     *
*   NL - 2801 SC GOUDA        TLX 20925             *                     *
*                                                   * SECTOR :            *
*                                                   *   PRODUCTION SYSTEMS *
* PERSON TO CONTACT FOR FURTHER INFORMATION :       *                     *
*   IR. B.J.G. VAN DER POT                          *                     *
*                                                   *                     *
*                                                   *                     *
**************************************************************************
                                                    VERSION : 03/03/87
```

AIM OF THE PROJECT :
****
To develop a concrete gravity type platform which can be used as a support
structure for offshore oil and gas production facilities in waterdepths between
300 and 400 m, in areas with severe environmental conditions and poor soil
quality.

PROJECT DESCRIPTION :
****
A platform concept has been developed with the following characteristics :
- minimal forces on the foundation for obtaining sufficient foundation
stability even on very weak soils.
- conventional construction methods.
- the possibility to tow the completed platform with installed top-sides in
vertical position to its location.
The project consists of the following phases:
1. conceptual design (dimensioning of the substructure)
2. optimisation (computer analyses and modification)
3. final design (structural analyses, detailing)
4. construction, planning/budgetting.

STATE OF ADVANCEMENT :
****
The project is completed.

RESULTS :
****
The results are positive; the expectations have been confirmed. Applications
are presently discussed with oil companies who are operating in the Norwegian
trench (Troll field, Haltenbanken area).

```
******************************************************************************
* TITLE : USE OF HYDRAULICALLY DRIVEN SUBSEA        *       PROJECT NO       *
*         PUMPS FOR ARTIFICIAL LIFT AND REMOTE      *                        *
*         FIELD PRODUCTION.                         *    TH./03136/83/UK/..   *
*                                                   *                        *
******************************************************************************
* CONTRACTOR :                                      * PROGRAM :              *
*   BRITOIL PLC                                     *   HYDROCARBONS         *
*   ST VINCENT STREET 150      TEL 01 409 25 25     *                        *
*   UK - GLASGOW G2 5LJ        TLX 881 20 71        *                        *
*                                                   * SECTOR :               *
*                                                   *   PRODUCTION SYSTEMS   *
* PERSON TO CONTACT FOR FURTHER INFORMATION :       *                        *
*   MR. J. ANDERSON                                 *                        *
*                                                   *                        *
*                                                   *                        *
******************************************************************************
                                                          VERSION : 18/03/86
```

AIM OF THE PROJECT :
****
To appraise the detailed engineering and economic aspects of alternative
downhole and subsea pumping systems both proven and novel, including an
assessment of proven systems based on operator experience in the North Sea
where natural drive is weak or absent.

PROJECT DESCRIPTION :
****
The study envisaged the study of the historical record of the problems
experienced in the artificial lift pumping systems as applied to North Sea
reservoirs. A review of the philosophies adopted by the various alternatives
and the resulting consequences which affect the downhole completion details,
topside power supply and interface connections with satellite wells. Sufficient
detailed engineering to be carried out to give a firm basis for costing a
complete integrated system to allow a cost/benefit analysis to be made.

STATE OF ADVANCEMENT :
****
Because of Britoil's involvement in a North Sea Field where a hydraulically
driven downhole pump was being tested, the study was delayed to enable an
assessment to be made on its performance which could have altered the direction
and necessity for the study. The rapid advance in subsea pump technology in the
intervening period negated the major part of the study and it was decided to
limit the study to considering the reliability of the electrically driven
submersible pumps on Beatrice.

```
********************************************************************************
* TITLE : THE DEVELOPMENT OF A DIVERLESS AND          *      PROJECT NO        *
*         GUIDELINELESS SUB-SEA CHRISTMAS TREE TO     *                        *
*         BE USED IN DEEP WATER.                      *    TH./03138/83/IT/..  *
*                                                     *                        *
********************************************************************************
* CONTRACTOR :                                        * PROGRAM :              *
*   AGIP SPA                                          *   HYDROCARBONS         *
*   C.P.120 69            TEL 02/5201                 *                        *
*   IT - 20120 MILANO     TLX 310 246 - ENI           *                        *
*                                                     * SECTOR :               *
*                                                     *   PRODUCTION SYSTEMS   *
* PERSON TO CONTACT FOR FURTHER INFORMATION :         *                        *
*   ING. TASSINI-TEIN                                 *                        *
*                                                     *                        *
*                                                     *                        *
********************************************************************************
                                                         VERSION : 01/01/87
```

AIM OF THE PROJECT :
****
Scope of this project is the development of a new generation of subsea system,
specifically conceived for hydrocarbon production in deep (200-600 m w.d.) and
very deep (beyond 600 m) waters, including a dedicated maintenance device.

PROJECT DESCRIPTION :
****
The object of the research in the evaluation of the feasibility of sub-sea
production systems which do not require divers("DIVERLESS" systems) or
guidelines ("GUIDELINELESS" systems) for installation, operation and
maintenance. Therefore, the concepts of maintenance studied in previous
projects will be transferred into a purpose designed operating system and the
design a subsea christmas tree suitable for use in deep and very deep waters
(including flowline connection system and maintenance device) will be produced.
The design of the maintenance device will be made simultaneously with the
system to be serviced, in order ot obtain an intervention tool integrated in
the system, with obvious advantages from the operating point of view, and
consequently improved reliability.
The prototype construction and long-term tests (scope of the phase II and III
of the project) will allow the evaluation of the system's functioning and
reliability.

STATE OF ADVANCEMENT :
****
The detailed design of the systems is ongoing. The basic design of the
production system and of the maintenance device is complete. Different subsea
power generation systems have been evaluated.

RESULTS :
****
A prototype will be installed and tested on a live well. This constitutes the
phase II and III of the project.

```
****************************************************************************
* TITLE : TRIPOD TOWER PLATFORM;PHASE 2.          *      PROJECT NO       *
*                                                 *                       *
*                                                 *   TH./03139/83/NL/..  *
*                                                 *                       *
****************************************************************************
* CONTRACTOR :                                    * PROGRAM :             *
*                                                 *   HYDROCARBONS        *
* HEEREMA ENGINEERING SERVICE   TEL  (071)351535  *                       *
* 47 VONDELLAAN                  TLX   32483       *                       *
* NL-2332 AA LEIDEN                               * SECTOR :              *
*                                                 *   PRODUCTION          *
* PERSON TO CONTACT FOR FURTHER INFORMATION :     *   SYSTEMS             *
* MR. HEEREMA/MR MEEK                             *                       *
*                                                 *                       *
*                                                 *                       *
****************************************************************************
                                                  VERSION : 12*03*85
```

AIM OF THE PROJECT :
****
To provide the offshore industry with a fixed steel structure, economically and
technically attractive for application in deep and shallow waters as support
structure for production, wellhead, living quarter platforms a.o.

PROJECT DESCRIPTION :
****
Extensive feasibility studies were conducted to prove the Tripod solution for
design conditions of the Troll field (340 m) Norwegian Sector, North Sea. These
studies included dynamic and fatigue analyses, material and welding tests on
200 mm thick steel plates, tests on grouted joints, hydrodynamic tests,
foundation designs and more.

STATE OF ADVANCEMENT :
****
Technical feasibility is fully proven for the Troll application. Minor items
are still being optimized, sensitivity and reliability studies will be carried
out in future.

RESULTS :
****
Main results are : a proven novel platform concept, know how in grouted joints
and welding and fabrication of thick walled (200 mm) tubulars, new computer
programmes for spectral dynamic and fatigue analysis. Application in a wide
range of offshore construction areas, not at least in shallow waters (from 25
metres), where structural steel weight savings of up to 50% are attainable.

```
*****************************************************************************
* TITLE : CONTRUCTION AND ASSEMBLY TECHNIQUES FOR    *      PROJECT NO      *
*         STEEL PLATFORMS IN DEEP WATER.             *                      *
*                                                    *    TH./03140/83/IT/..*
*                                                    *                      *
*****************************************************************************
* CONTRACTOR :                                       * PROGRAM :            *
*    TECNOMARE SPA                                   *    HYDROCARBONS       *
*    SAN MARCO 2091          TEL 041 796711          *                      *
*    IT - VENEZIA            TLX 410484 MAREVE       *                      *
*                                                    * SECTOR :             *
*                                                    *    PRODUCTION SYSTEMS *
* PERSON TO CONTACT FOR FURTHER INFORMATION :        *                      *
*    MR. L. BEGHETTO                                 *                      *
*                                                    *                      *
*                                                    *                      *
*****************************************************************************
                                                      VERSION : 01/01/87
```

AIM OF THE PROJECT :
****
Purpose of the present research project is the sudy of both the basic problems
and the procedures concerning the construction, assembling, transportation
phases of the fixed offshore platform in 350 m W.D.

PROJECT DESCRIPTION :
****
The research project is divided into two different stages : the first studies
the basic problems regarding the construction and safety of fixed structures
in 350 m W.D. while the second stage examines the importance and consequences
of the know-how on the structural configuration and on the check procedures of
such types of platforms in order to determine investement costs and time.
First phase : problems affecting the offshore structures in deep water.
Status : opened.
Percentage of qualitative progress : 90%
Second phase : basic problems of construction and assembling.
Status : opened
Percentage of qualitative progress : 60%.
Third phase : final analyses and procedures.
Status : not opened.
Fourth phase : economical evaluation.
Status : not opened.
Identification of preliminary platform configurations and techniques for
construction and assembling.

STATE OF ADVANCEMENT :
****
Ongoing

RESULTS :
****
The obtained results are in accordance with the progress of the project, and at
present are not available to third parties.

```
****************************************************************************
* TITLE : OIL PRODUCTION FROM EXPOSED,DEEP WATER,    *      PROJECT NO      *
*         MULTI-WELL MARGINAL OFFSHORE OIL FIELDS    *                      *
*         USING A DYNAMICALLY POSITIONED SHUTTLE     *   TH./03142/83/IR/..  *
*         TANKER                                     *                      *
****************************************************************************
* CONTRACTOR :                                       * PROGRAM :            *
*   ARAN ENERGY PLC                                  *   HYDROCARBONS       *
*   CLANWILLIAM COURT          TEL 760 696           *                      *
*   LOWER MOUNT STREET         TLX 304 88            *                      *
*   IR - DUBLIN 2                                    * SECTOR :             *
*                                                    *                      *
* PERSON TO CONTACT FOR FURTHER INFORMATION :        *                      *
*   MR. PETER D. GORMAN                              *                      *
*                                                    *                      *
*                                                    *                      *
****************************************************************************
                                                        VERSION : 30/06/85
```

AIM OF THE PROJECT :
****
The aim of the project was the examination of the application of the "multiwell
offshore shuttle tanker" (MOST) concept for oil production in deep water (400
metres) and severe marine environments such as those of the Eastern Atlantic
Ocean.

PROJECT DESCRIPTION :
****
The project was an examination of the application of the MOST concept,
consisting of :
- a number of subsea-completed satellite wells connected to a manifold;
- a dynamically-positioned ship for use as a production tanker vessel
  containing all field process and storage facilities;
- deployable riser connecting the production tanker to the subsea manifold;
- use of the production vessel as a periodic shuttle tanker to a shore terminal.
- system to be capable to operation in 400 metres water depth;
- system to give commercially-acceptable production regularity in
  North Atlantic weather conditions, though not necessarily with year-round
operation.
Technical solutions in many of the problem areas involved in the development of
this concept were examined and assessed, particularly in the areas of :
- subsea completion systems;
- riser systems;
- vessel motion performance and station-keeping;
  estimation of production system regularity and downtime.
Certain economic data, particularly as regards capital and operating costs,
were also produced.
STATE OF ADVANCEMENT :
****
The project has concluded; not all the work originally envisaged was actually
conducted, as possible technical solutions which did not appear to be
economically acceptable were not pursued. However a basis has been laid for
further technical investigation if warranted for a future project.
RESULTS :
****
The project was only a preliminary study; for principally economic reasons it
has not proved justifiable to proceed with further work at this time.

```
*******************************************************************************
* TITLE : DEVELOPMENT OF HYDRAULICALLY POWERED        *      PROJECT NO        *
*           DOWNHOLE PUMPS.                           *                        *
*                                                     *    TH./03143/83/UK/..  *
*                                                     *                        *
*******************************************************************************
* CONTRACTOR :                                        * PROGRAM :              *
*   WEIR PUMPS LTD                                    *   HYDROCARBONS         *
*   CATHCART WORKS              TEL 041 637 7141      *                        *
*   UK - GLASGOW G44 4EX        TLX 771 612           *                        *
*                                                     * SECTOR :               *
*                                                     *                        *
* PERSON TO CONTACT FOR FURTHER INFORMATION :         *                        *
*   MR. M.L. RYALL                                    *                        *
*                                                     *                        *
*                                                     *                        *
*******************************************************************************
                                                        VERSION : 01/03/86
```

AIM OF THE PROJECT :
****
To develop a range of hydraulically powered downhole pumps, exploiting the
potential of the reliability product concepts for a wide range of market
applications in as short a time as possible.

PROJECT DESCRIPTION :
****
To prepare detail designs for each of a range of downhole pumps and turbines,
manufacture hydraulic stage components to confirm manufacturing methods and
stage performance through laboratory testing, and to design and manufacture for
field trials full size prototype pumpsets and systems.

STATE OF ADVANCEMENT :
****
The project is now complete and prototype pumpsets are, through field trials,
demonstrating performance in different applications.

RESULTS :
****
A range of hydraulically powered downhole pumps for a flow range between 1,000
BPD abd 100,000 BPD and for heads up to 7,000 feet are now available and are
suitable for both oil and water lift applications primarily related to enhanced
oil recovery operations. Product performance and the potential for high
reliability has been demonstrated.

```
********************************************************************************
* TITLE : SUB-SEA PRODUCTION WITHOUT ONSITE      *        PROJECT NO         *
*         SEPARATION.                            *                           *
*                                                *                           *
*                                                *    TH./03144/83/FR/..     *
*                                                *                           *
********************************************************************************
* CONTRACTOR :                                   * PROGRAM :                 *
*   GERTH                                         *   HYDROCARBONS           *
*   4, AVENUE DE BOIS PREAU    TEL 1/47 52 61 39  *                          *
*   FR - 92502 RUEIL-MALMAISON TLX A 203 050      *                          *
*                                                 * SECTOR :                 *
*                                                 *   PRODUCTION SYSTEMS     *
* PERSON TO CONTACT FOR FURTHER INFORMATION :     *                          *
*   MR. GILBERT BLU                               *                          *
*                                                 *                          *
*                                                 *                          *
********************************************************************************
```

VERSION : 01/12/85

AIM OF THE PROJECT :
****
To develop techniques for exploiting offshore hydrocarbon reservoirs by a fully
underwater system and to specify a production system.

PROJECT DESCRIPTION :
****
The work consisted of :
- a parametric study to evaluate the economics of the project by comparing the
operating costs of a resevoir with an "all sub-sea" configuration and those
obtained with a "conventional" configuration;
- the definition of a study case in the preferential field defined by the
economic evavaluation study; the technical contraints set for this study case
were selectedto cover all the problems liable to occur on this type of
application;
- a study of all subassemblies forming the system in order to reveal the
technological node points;
- elaboration of mainguide plans for the subsea station and modules fulfilling
all the required functions (valves, remote control, electric power supply,
grouping of production, pumping...);
- examination of pipeline operating problems. In particular, general solutions
have been proposed for a sorting remote-controlled high capacity yard of
scrapers for the pipeline connection system to the subsea station and in order
to combat the internal corrosion of the pipeline.
STATE OF ADVANCEMENT :
****
The project started in January 1983 and ended April 1984.
RESULTS :
****
This study has enabled the privileged field of application of an "all
underwater" system to be defined : distance from the coast less than 200 km,
depth of water above 200 m, in zones where the cost of surface installations is
high; in addition, the advantage of such a system is all the more evident, the
more severe the conditions of environmemt, the smaller the relative size of the
reservoirs and the higher the proportion of gas in the effluent.
The combination of knowledge and essential results thus acquired has enabled
the development programme of an all underwater hydrocarbon deposits operating
system known as POSEIDON to be defined (contract no. TH 03164/84)

```
********************************************************************************
* TITLE : REDUCTION OF DEVELOPMENT COSTS OF        *        PROJECT NO        *
*         OFFSHORE PROJECTS.                        *                          *
*                                                   *     TH./03145/83/FR/..   *
*                                                   *                          *
********************************************************************************
* CONTRACTOR :                                      * PROGRAM :                *
*   GERTH                                           *   HYDROCARBONS           *
*   AVENUE DE BOIS PREAU 4      TEL 1 47.52.61.39   *                          *
*   FR - 92502 RUEIL-MALMAISON TLX 203 050          *                          *
*                                                   * SECTOR :                 *
*                                                   *   PRODUCTION SYSTEMS     *
* PERSON TO CONTACT FOR FURTHER INFORMATION :       *                          *
*   MR. P. FABIANI              TEL 1 42.91.40.00   *                          *
*                               TLX 615 700         *                          *
*                                                   *                          *
********************************************************************************
```
VERSION : 01/01/87

AIM OF THE PROJECT :
****
This project aims to reduce in a significant manner offshore development costs,
either by reducing equipment weight and consequently reducing the weight of the
overall structure, or by reducing equipment costs.

PROJECT DESCRIPTION :
****
The project is divided in three phases :
1. Evaluation and supervision of offshore projects.
Using the simulation modelling, this phase aims to develop software which will
be applied in real cases in the North Sea. This software will be extended to
other areas.
2. Weight reduction of metallic structures.
This phase will improve the present calculation methods on two points : fragile
breaking-down and flexibility of the tubular nodes.
3. Weight reduction of processing equipment.
The study concerns the processing equipment for injected waters and the flaring
equipment.
The arrangements described below have been approved by Lloyds Register adn DNV.

STATE OF ADVANCEMENT :
****
Completed. The work started in December 1982 and ended in December 1985.

RESULTS :
****
Phase 1 - Evaluation and control of execution of offshore projects.
A computing tool named SPOT capable of simulating a conventional offshore
project has been developed.
On the basis of parameters accessible to the operator, this simualtion enables
the elements of the project to be evaluated in terms of the following factors :
- technical description
- project planning schedule and manning requirements
- cost (in the different countries of implementation)
- investment programme
SPOT comprises the following modules :
- superstructures (surface installations),
- metal structure (jacket) (water depth from 50 to 200 m)
- offshore installation operations
- hook-up
- metal structure for water depths of less than 50 m
- subsea lines
- concrete gravity structures.
These modules are completed by a data bank set up on the basis of offshore
oilfied developments, particularly those located in the North Sea.
Many tests have been carried out on a real cases embodied or not in the initial
data base; the results have invariably proved highly satisfactory. The overall
results very rarely stray beyond +/- 10%.
The SPOT model predicts the final result in terms of planning schedule, cost,
investment programme and manning requirements. Not only does it enable the
technical scheme of a development project to be optimized when the decisions
are taken, but also the development itself to be controlled during the
construction phases and subsequently the operating phases.
SPOT is par excellence a comparison tool and consequently an aid for decision-
making, but does not claim to replace the studies usually carried out, once the
development decision has been taken.
Phase 2 - Lightening of the metal structures.
A program using integral equations enables the behaviour of a metal part with a
fault in it to be computed and the corresponding risks of brittle fracture to
be evaluated. However, the specificity and complexity of the computing
operations does not enable it to be built into a general application program of
the STRUDL type, as had originally been envisaged.
In addition, a model for introducing a rigidity junction node matrix into the
structural design program has been established.
Phase 3 - Lightening of the treatment equipment.
With regard to study of tangential filtration of the injection waters,
promising results were obtained with ceramic membranes, though a clogging
phenomenon was noted when using reservoir waters, which may render use of these
membranes doubtful.
Elsewhere, study of new degassing systems of the injection waters enabled the
processes proposed by NEYRTEC(F), B.H.R.A. (UK) and NORSK HYDRO (Norway) to be
evaluated.
Lastly, following an initial technico-economic evaluation, it turns out that
buffer storage of the gas for flaring represents only a limited interest. This
has led to studying a method of protection against overpressures enabling the
importance of the flaring systems to be reduced.

```
****************************************************************************
* TITLE : PRODUCTION RISER FOR 1000/3000 M WATER   *      PROJECT NO      *
*         DEPTH                                     *                      *
*                                                   *    TH./03146/83/FR/..*
*                                                   *                      *
****************************************************************************
* CONTRACTOR :                                      * PROGRAM :            *
*   GERTH  S.N.I.A.S.                               *   HYDROCARBONS       *
*   AV. DE BOIS PREAU 4          TEL 1 47.52.61.39  *                      *
*   FR - 92502 RUEIL-MALMAISON TLX 203 050 F        *                      *
*                                                   * SECTOR :             *
*                                                   *   PRODUCTION SYSTEMS *
* PERSON TO CONTACT FOR FURTHER INFORMATION :       *                      *
*   MR. P. JOUBERT              TEL 1 47.52.61.46   *                      *
*                                                   *                      *
*                                                   *                      *
****************************************************************************
```

VERSION : 01/01/87

AIM OF THE PROJECT :
****
The object of the project is to develop the design of a production riser
linking the riser base on the sea bottom to the floating production facility
(which could be either a tanker or a semisubmersible platform)in water depths
ranging from 1000 to 3000 meters. Since the weight of the riser is a major
parameter in such water depths, the riser is made of lightweight tubulars
called "composite pipes"using newly developed technology of carbon and glass
fibers. The composite pipes are suspended from a subsurface buoy, which is
linked by flexible pipes to the surface facility to accommodate its motions.

PROJECT DESCRIPTION :
****
The riser is divided into two sections :
- The lower vertical rigid section is made of tubes and installed using typical
drilling methods. This section includes a central core and some 11 peripheral
lines made of composite pipes. The central core is the structural element of
the riser. It is made of steel,the present degree of development of composite
pipes being not compatible with the required specifications. All lines are
hydraulically connected to a riser base and held in tension by a subsurface
buoy. Tubular transition pieces couple the pperipher lines to the flexible
pipes of the upper section.
- The upper flexible section consists of a linear arrangement of flexible pipes
terminating at the floating production facility. For a shipshape vessel, this
termination is into a buoy plug mechanically latched in the ship turret.
The project consisted in three phases:
- an overall study to determine the main parameters governing the behaviour of
the riser and establish the specifications for all its elements. This study
included the development of new numerical tools and adaptation of existingones.
- the preliminary design of the riser elements and their installation
procedures:bottom connectors..
- the design of the composite pipes for all functions:gathering, water
injection, gas injection, gas lift. etc.. and tests of representative
prototypes.

STATE OF ADVANCEMENT :
****
Completed.
- the overall study produced an optimization of the design for the cases
examined, as well as numerical tools required for a site specific application.
- the study of components and their installation procedures is almost completed,
 as well as prototype testing.
- design and test of composite pipesdesigned for this application are completed.

RESULTS :
****
The results of the design effort is a compliant production riser for water
depths ranging fro 1000 to 3000 meters.
The main features of the riser are :
- the lower section includes carbon/glass fiber composite pipes
- installation and maintenance do not require deep diving
- the riser is designed to stay connected in the most severe environment
- each tubular and flexible are individually retrievable.
The riser has been fully engineered by a detailed analysis and by prototype
testing of newly developed components.

REFERENCES :
****
"DEVELOPMENT OF A COMPLIANT PRODUCTION RISER FOR DEEPWATER APPLICATION"(PAPER
PRESENTED AT THE DEEP OFFSHORE TECHNOLOGY CONFERENCE
SORRENTO,21,23 OCTOBER 1985)
"HIGH PERFORMANCE COMPOSITE PIPES FOR DEEPWATER MULTILINE PRODUCTION
RISERS"(PAPER PRESENTED AT THE OFFSHORE MECHANICS AND ARCTIC ENGINEERING
CONFERENCE
HOUSTON, 1-6 MARCH 1987.)

```
***************************************************************************
* TITLE : FLEXIBLE DRILLING TOWER FOR PRODUCTION    *      PROJECT NO       *
*         OF HYDROCARBONS IN GREAT WATER DEPTH (>   *                       *
*         300 M)                                    *     TH./03147/83/FR/.. *
*                                                   *                       *
***************************************************************************
* CONTRACTOR :                                      * PROGRAM :             *
*   E.T.P.M.                                        *   HYDROCARBONS        *
*   33,35 RUE D'ALSACE          TEL (1) 47.59.60.00 *                       *
*   FR - 92531 LEVALLOIS PERRE TLX ETPM612021       *                       *
*                                                   * SECTOR :              *
*                                                   *   PRODUCTION SYSTEMS  *
* PERSON TO CONTACT FOR FURTHER INFORMATION :       *                       *
*   MR. B. ANDRIER              TEL (1) 47.59.60.00 *                       *
*                               TLX ETPM612021      *                       *
*                                                   *                       *
***************************************************************************
                                                        VERSION : 01/04/87
```

AIM OF THE PROJECT :
****
The object of this project was to prove the feasibility of steel structures
supporting hydrocarbon production and drilling equipment and presenting the
same display as structures presently installed in water depths ranging from 100
to 300 m, but which could be installed at much more important depths.

PROJECT DESCRIPTION :
****
The ROSEAU column is designed to support drilling and production equipment in
deep waters. Contrary to the very rigid structures of the jacket type, the
functioning of the ROSEAU is based on the flexibility of the structure and the
resulting dynamic effect due to a first natural period very much superior to
the wave periods. This allows a relatively light structure. Compared to other
flexibles structures, ROSEAU presents numerous advantages, the main being is
its resemblance with conventional jackets and the fact that it requires neither
buoyancy tanks, guylines nor articulation.
Studies relative to the ROSEAU platform concerned mainly its feasibility. The
purpose was to provide a better understanding of the phenomena determining the
behaviour of the structure, in order to define the analysis methods, find the
possible architectural designs and the appropriate installation methods.

STATE OF ADVANCEMENT :
****
The project was completed by the 31st of March 1986.

RESULTS :
****
The following ROSEAU platforms have been studied:
- in various water depths ranging from 450 to 600 m for the
oceanometeorological conditions of the Gulf of Mexico;
- for North Sea conditions (water depth of 600 m);
- for the Gaspian Sea conditions (300 m);
- for the Californian Coast conditions (600 m);
- for the Mediterranean Sea (825 m);
- for the West African Coast (400 m);
The following aspects were covered for each platform:
- definition of the database, particularly the oil exploitation conditions
(topload, number of conductor pipes, etc..) and the oceanometeorological data;
- platform dimensioning by means of an analysis method used on a single-wine
model;
- a parametric study which allowed to determine optimal values and critical
applications for this project;
- a study of the main installation methods;
- a study on specific subjects : fatigue, seismic, torsion, foundation effect,
low frequency phenomena;
- a study on the main architectural options possible.
In view of a detailed inspection concerning the solidity of the platform,
program has been developed to analyse the behaviour of the ROSEAU structure.
This "step by step" calculation program operates within a temporal domain, that
is to stay it simulates the response of the structure versus the excitation of
a random surge. The validity of this program has been basin tested on a model
in June and July 1984. A comparison of the recordings of certain motion
parameters of the model with the results of a simulation calculated with the
same surge as that in the test basin has proved that the program makes a good
analysis of the phenomena at low and high frequency and takes into account the
mean value of forces applied. The program also determines the extreme values
and the fatigue damages according to several distinct methods corresponding to
different mathematical theories.
Futhermore, basin tests were also run on the behaviour of the stabilizer, so to
provide a more accurate definition of its hydrodynamic properties.
Last of all, problems concerning the launching of long and flexible structure
from a barge have also been the object of basin tests which allowed to specify
the behaviour of the struture during its launching, particularly concerning the
efforts on the stabilizer.
To conclude, these studies showed that the ROSEAU concept allowed the extension
this jacket concept into deeper waters, without increasing the manufacture,
installation and exploration costs.
This ROSEAU concept has been particularly well received and is now in a good
position within the list of options chosen by oil companies concerned by deep
offshore oil production. This has been proven by the success obtained by the
"ticket" proposed for the detailed study of a platform in a 600 m water depth
in the Gulf of Mexico to 10 different oil companies: Chevron, Elf, Marathon,
Occidental, Petrobras, Philips, Shell, Sohio, Texaco, Union.

REFERENCES :
****
THE ROSEAU CONCEPT WAS PRESENTED TO :
- D.O.T. MALTA OCTOBER 1983
- D.O.T. SORRENTE OCTOBER 1985
- O.T.C. HOUSTON MAY 1986.

```
********************************************************************************
* TITLE : DEVELOPMENT OF A MOBILE PRODUCTION      *       PROJECT NO        *
*         SYSTEM INCORPORATING A COMPLIANT        *                         *
*         PRODUCTION RISER.                       *   TH./03149/84/UK/..    *
*                                                 *                         *
********************************************************************************
* CONTRACTOR :                                    * PROGRAM :               *
*   FOSTER WHEELER PETROLEUM DEV. LTD             *   HYDROCARBONS          *
*   SHAFTSESBURY AVENUE 125    TEL 01 836 80 30   *                         *
*   UK - LONDON WC2H 8AD       TLX 296523         *                         *
*                                                 * SECTOR :                *
*                                                 *   PRODUCTION SYSTEMS    *
* PERSON TO CONTACT FOR FURTHER INFORMATION :     *                         *
*   MR. H. PASS                                   *                         *
*                                                 *                         *
*                                                 *                         *
********************************************************************************
```

AIM OF THE PROJECT :   VERSION : 21/01/86
****
Recognising that half the world's future oil discoveries will be made offshore,
the three participating companies formed an association with the aim of
developing a floating production system tailored for use in marginal fields.
PROJECT DESCRIPTION :
****
The project was executed in two phases.
In the first phase the group selected two floating production schemes for
design development. One of these schemes is based on the use of a semi-
submersible as a production vessel whilst the other is based on a ship-shaped
vessel. A market census and preliminary economic analysis indicated that the
ship-shaped system, known as COMPASS, was the more likely to find application
in the next few years.
The second phase of the project principally concentrated on developing the
engineering design of COMPASS in the areas of the floating vessel, the Turret
system and moorings, process and utility systems, the Riser system and the
Subsea Production systems. In parallel, a more limited series of studies were
undertaken on the semi-submersible option, both for oil field development and
for a gas-condensate field. The COMPASS system, including the riser tower, was
modelled at 1:75 scale and tank tested under various environmental conditions.
Finally a detailed economic appraisal was carried out for the various floating
schemes and compared with a base-case jacket structure.
STATE OF ADVANCEMENT :
****
The project started in December 1983 and was completed in August 1985. The
results were presented in a final report in four volumes.
RESULTS :
****
An economic evaluation of the base case using COMPASS indicates that a Real
Rate of Return on investment of 17.4% is possible. This compares favourably
with a target rate of 15% and with figures of 12.6% and 5.4% for the semi-
submersible and a conventional steel jacket respectively. Sensitivity studies
indicate areas for further improvement on these systems, and the related cases
for lower throughput and for gas-condensate fields. The overall project
objective of establishing a development concept suitable for deployment on
marginal fields and offering significant advantages over current technology has
therefore been achieved. To date no direct application has been found, though
the large potential market, and active marketing of the system, will no doubt
yield results in time.

```
*************************************************************************
* TITLE : TENSION LEG PLATFORM WITH HIGH PAYLOAD     *      PROJECT NO       *
*         FOR A NATURAL GAS LIQUEFACTION PLANT IN    *                       *
*         WATER-DEPTHS OF 500 TO 1,000 METERS.       *   TH./03150/84/DE/..  *
*                                                    *                       *
*************************************************************************
* CONTRACTOR :                                       * PROGRAM :             *
*   SALZGITTER AG                                     *   HYDROCARBONS        *
*   POSTFACH 15 06 27          TEL 030/88.42.97-15    *                       *
*   DE - 1000 BERLIN 15        TLX 308 611            *                       *
*                                                    * SECTOR :              *
*                                                    *   PRODUCTION SYSTEMS  *
* PERSON TO CONTACT FOR FURTHER INFORMATION :         *                       *
*   DR-ING. PIETSCH                                   *                       *
*                                                    *                       *
*                                                    *                       *
*************************************************************************
                                                          VERSION : 03/03/87
```

AIM OF THE PROJECT :
****
The project will concentrate on the development of a tension leg platform for
high deck loads as a supporting structure for a natural gas liquefaction plant
in water up to 1,000 m deep, in order to economically exploit natural gas
reservoirs in deep-water locations.

PROJECT DESCRIPTION :
****
The proposed development work will examine from a technical standpoint the
possibility of using a TLP with a buffer store as a support structure for a
natural gas liquefaction plant in water up to 1,000 m deep. The overall system
comprises the foundation components, tensioning system, the buoyancy body, the
jack-up platform and the transfer system for the product.

STATE OF ADVANCEMENT :
****
 Ongoing the project is in the design and construction phase.

RESULTS :
****

The intention of a further developed anchoring of the tension legs both on the buoyancy body and on the foundation body was continued. Instead of the cardan joints, work was begun using flexible link joints.

The basis of this is the mooring system of the English company Vickers, which has already been used in the Hutton TLP of CONOCO.

The calculations for the applications of the TLP system in water depths between 500, and 1,000 m were evaluated and namely for the buoyancy body being positioned 40, 50 and 60 m below the water surface and for the position of the platform at 0 deg.C and 45 deg.C to the wave. These comprehensive works could not yet be completed.

The evaluation of the calculations to date reveals that a relatively good agreement with the results of the model tests is present.

The FUGRO Ltd. company was able to supply an evaluation of the soil relationships in the deep water areas of the North Sea. According to FUGRO no geotechnical boreholes have yet been sunk north-west of Scotland. However the general geological composition can be judged on the basis of the available surface samples taken using dropped probes in this area.

With this data and the stated literature, design parameters could be determined for the calculation of the designed foundation. The trend is probably tending towards pile foundations, as design up until now gravity foundations in these areas demand. Caisson-type foundations with deep penetrating aprons.

The integration of the integrated intermediate storage in the platform could be solved satisfactorily.

Already at this point it is evident that this variant of the LNG intermediate storage can be dealt with more easily than with Variant II, the storage on the buoyancy body.

```
********************************************************************************
* TITLE : COMPLIANT MONOTOWER FOR OIL/GAS        *        PROJECT NO          *
*         PRODUCTION IN DEEP WATER MARGINAL       *                            *
*         FIELDS.                                 *      TH./03151/84/IT/..    *
*                                                 *                            *
********************************************************************************
* CONTRACTOR :                                    * PROGRAM :                  *
*   TECNOMARE SPA                                 *   HYDROCARBONS             *
*   SAN MARCO 2091            TEL 041 796 711      *                            *
*   IT - 30124 VENEZIA        TLX 410484           *                            *
*                                                 * SECTOR :                   *
*                                                 *   PRODUCTION SYSTEMS       *
* PERSON TO CONTACT FOR FURTHER INFORMATION :     *                            *
*   MR. C. FERRETTI                               *                            *
*                                                 *                            *
*                                                 *                            *
********************************************************************************
                                                    VERSION : 03/03/87
```

AIM OF THE PROJECT :
****
Development of a system which should render economically attractive the
exploitation of numerous oil and gas marginal fields in deep water up to 1,000
m w.d.

PROJECT DESCRIPTION :
****
Study, design and model testing of a new concept of structure, the COMPLIANT
MONOTOWER, proposed for oil and gas production from marginal fields in very
deep waters up to 1,000 m w.d.
The concept is a very slender and flexible structure, with complete structural
continuity from the seabed to the sea surface, put in tension by a buoyancy
chamber at its top. Such structural continuity allows a certain number of
conductors to run along the structures as in a fixed platform and to put the
wellheads on the deck surface, with strong benefit of the operation maintenance
process.

STATE OF ADVANCEMENT :
****
Ongoing, at the study phase.

RESULTS :
****
Different structure configurations have been identified and are going to be
compared. This phase will lead to the choice of one configuration for gas and
one for oil that will be later designed in detail.

```
****************************************************************************
* TITLE : OIL FIELD TESTING OF A NEW COMPACT      *      PROJECT NO        *
*         SEPARATOR OF 25,000 BBL/DAY CAPACITY.   *                        *
*                                                 *    TH./03153/84/FR/..  *
*                                                 *                        *
****************************************************************************
* CONTRACTOR :                                    * PROGRAM :              *
*   BERTIN & CIE                                  *   HYDROCARBONS         *
*   B.P. 3                      TEL 34.81.85.00   *                        *
*    FR - 78373 PLAISIR CEDEX   TLX 696231        *                        *
*                                                 * SECTOR :               *
*                                                 *   PRODUCTION SYSTEMS   *
* PERSON TO CONTACT FOR FURTHER INFORMATION :     *                        *
*   MM. J.Y. DEYSSON/M. REYBILLET                 *                        *
*                                                 *                        *
*                                                 *                        *
****************************************************************************
                                                      VERSION : 03/03/87
```

AIM OF THE PROJECT :
****
Test a new separator concept for off-shore oil production, incorporating novel
features for improved gas, oil and water separation, resulting in reduced
weight for both vessel and structure.

PROJECT DESCRIPTION :
****
The current project is the test of a 25 000 BBL/day separator under true oil
field operating conditions. The program is comprised of the following phases :
1. Preparation of a 25 000 BBL/day compact separator for the tests. Design,
procurement and setting of the necessary control and safety devices.
2. Shipment of the skid and erection of same on the Obagi Site.
3. Test of performance under various flow rate and GOR configurations : oil in
gas, gas in oil, water in oil, oil in water.

STATE OF ADVANCEMENT :
****
Completed. The whole program has been completed on March 1986 with good results.
In order to make some operational improvements and to carry out a long duration
testing of the separator on an off-shore platform, a community technological
development (Hydrocarbons) project was registered at the EEC (DG 17) in April
1986.

RESULTS :
****

The testing of the BERTIN three-phase compact separator on the OBAGI oilfield (ELF NIGERIA) occured according to the programme laid down. The tests made it possible to set the figure for the separator practical operating capacity in its present state at 15,000 bbl/d. This figure corresponds to that of separators currently used in off-shore production.

Under such conditions, the gain in volume and in ground surface area in relation to a conventional horizontal separator with equivalent performance is very considerable.

The BERTIN separator has a tank volume of 11 m3 as against 26 m3 for an equivalent classical horizontal separator and a ground surface area of 2 m2 as against 18 m2.

Naturally, these advantages are slightly attenuated if one considers the entire skidmounted separator rather than the tank, but they are still highly appreciable when considered from the point of view of a policy of price reduction involving a reduction of the load at the head of the structures.

Also, all the improvements proposed should make it possible to increase the separator capacity from 15 to 20 or 25,000 bbl/d (which is the practical capacity limit found on a platform) and, by a long-duration campaign on an off-shore production platform, to prepare the marketing of the apparatus.

Even now, it is already possible to anticipate the following off-shore applications for the BERTIN three-phase compact separator :

. 1st stage HP separator
. 2nd stage MP separator
. test separator

```
*********************************************************************************
* TITLE : RECOVERY FROM VERY SMALL FIELDS        *        PROJECT NO        *
*                                                *                          *
*                                                *    TH./03154/84/UK/..    *
*                                                *                          *
*********************************************************************************
* CONTRACTOR :                                   * PROGRAM :                *
*   BRITOIL PLC                                  *   HYDROCARBONS           *
*   ST VINCENT STREET 301      TEL 041 2042525   *                          *
*   UK - GLASGOW G2 5DD        TLX 777633        *                          *
*                                                * SECTOR :                 *
*                                                *   PRODUCTION SYSTEMS     *
* PERSON TO CONTACT FOR FURTHER INFORMATION :    *                          *
*   MR. K.J. WELLS                               *                          *
*                                                *                          *
*                                                *                          *
*********************************************************************************
                                                     VERSION : 03/03/87
```

AIM OF THE PROJECT :
****
The study objective was to consider methods by which a small reservoir of
8/10,000 barrels per day throughput and having a 5 year life could be developed
at minimum capital expenditure.

PROJECT DESCRIPTION :
****
The study is exploring the hypothesis that the use of a buoy/tanker system
could offer significant advantages for the development of small reservoirs and
could reduce the time span between project start and first oil export. The
study is being carried out in two phases. The first phase studied a number of
design options including a Britoil developed Flexible Riser and Mooring System
(FRAMS). This system embodies a flexible riser conveying well fluid and
injection water, and envisages utilising standard elastomeric flexible joints.
On the vessel bow a simple deck-mounted swivel unit mates with the flexible
riser.

STATE OF ADVANCEMENT :
****
Ongoing. The first phase of the project has been completed and the economic
evaluation of several options for the typical synthesized field parameters used
in the study has indicated that it may be possible to achieve a viable concept.
A scope of work is being compiled for the second phase of the study, which will
enhance the basic design concept to achieve a reduction in the technical
complexity of the novel untried components. A parametric study on field size
and throughput will be undertaken.

```
******************************************************************************
* TITLE : PLATFORM CONCEPT FOR MARGINAL FIELDS    *      PROJECT NO         *
*                                                 *                         *
*                                                 *    TH./03157/84/DK/..   *
*                                                 *                         *
******************************************************************************
* CONTRACTOR :                                    * PROGRAM :               *
*    AALBORG VAERFT                               *    HYDROCARBONS         *
*    P.O. BOX 661              TEL 45 8155000      *                         *
*    DK - 9100 AALBORG         TLX 69664          *                         *
*                                                 * SECTOR :                *
*                                                 *    PRODUCTION SYSTEMS   *
* PERSON TO CONTACT FOR FURTHER INFORMATION :     *                         *
*    MR. A. TOFT                                  *                         *
*                                                 *                         *
*                                                 *                         *
******************************************************************************
                                                   VERSION : 20/01/86
```

AIM OF THE PROJECT :
****
The purpose of the project is to improve the economy of production concerning
hydrocarbons from marginal fields in the North Sea shallow waters and to
decrease the field size limit for economically viable field developments.

PROJECT DESCRIPTION :
****
The project is carried out in 2 phases :
- Conceptual phase evaluating the economical and technical feasibility of
different concepts of free standing conductors and innovative chemical process
plant associated with light weight jackets distinguishing between oil and gas
field applications.
- Preliminary design phase developing preferred concepts of the free standing
conductor and the light weight jacket within certain ranges of the flow rates,
gas to oil ratios, water depths etc... .

STATE OF ADVANCEMENT :
****
Completed

RESULTS :
****
A number of options has been considered resulting in the identification of the concepts of a light weight jacket and a monotower remotely controlled from a central production platform.

In order to enhance the oil recovery as much as possible, several production features have been considered :

1. Conventional process with oil/gas separation.
2. Hydraulical/electrical downhole pumps.
3. Multiphase pumps.

Each of the three features have been studied in connection with the two different support structure options.

A parameter study of the six options has been carried out in order to achieve a knowledge of the susceptibility due to oil reservoir characteristics, distance to central production platform, crude composition, environmental and soil conditions.

For each of the six options a cost estimate has been performed and used as a point of view in an overall economical evaluation with special emphasis on the Danish taxation model.

The most economical viable options on the basis of the chosen design criteron are :

- a light weight jacket
- a hydraulic downhole pump
- a multiphase export pipeline
- power supply via subsea cable.

```
*********************************************************************************
* TITLE : CONCRETE PLATFORMS FOR DEEP WATER          *        PROJECT NO        *
*                                                    *                          *
*                                                    *    TH./03159/84/UK/..     *
*                                                    *                          *
*********************************************************************************
* CONTRACTOR :                                       * PROGRAM :                *
*   MC ALPINE OFFSHORE LTD                           *   HYDROCARBONS           *
*   BERNARD STREET 40            TEL 01 837 33 77     *                          *
*   UK - LONDON WC1N 1LG         TLX 22308           *                          *
*                                                    * SECTOR :                 *
*                                                    *   PRODUCTION SYSTEMS     *
* PERSON TO CONTACT FOR FURTHER INFORMATION :        *                          *
*   MR. M.J. COLLARD                                 *                          *
*                                                    *                          *
*                                                    *                          *
*********************************************************************************
                                                    VERSION : 03/03/87
```

AIM OF THE PROJECT :
****
To extend the economic water-depth limit for concrete production platforms by
further developing designs of gravity based and compliant platforms for water
depths to 1,500 metres.

PROJECT DESCRIPTION :
****
To select and develop a range of design concepts for a range of water depth and
production requirements. The contract will be performed in two phases. The
first phase will allow to define the most attractive structure type.
Two options will be examined,namely a tension leg platform and a catenary
moored platform. Dynamic analysis will be completed to assess hull forces and
mooring forces to determine concept feasibility.
Construction and installation will be assessed and cost estimates made.

STATE OF ADVANCEMENT :
****
Ongoing. Investigation presently establishing base case criteria, selecting
candidate concepts and specifying analysis.

```
******************************************************************************
*  TITLE : A PROGRAMME FOR DEVELOPMENT OF          *       PROJECT NO        *
*          STANDARDIZED PRODUCTION FACILITIES FOR   *                         *
*          THE EXPLOITATION OF MARGINAL             *    TH./03160/84/UK/..   *
*          HYDROCARBON RESERVES                     *                         *
******************************************************************************
*  CONTRACTOR :                                     * PROGRAM :               *
*     WIMPEY OFFSHORE ENGINEERS & CONSTRUCTORS LTD  *    HYDROCARBONS          *
*     FLYOVER HOUSE, GREAT WEST   TEL 01 560 31 00  *                         *
*     BRENTFORD                   TLX 933861        *                         *
*     UK - MIDDLESEX TW8 9AR                        * SECTOR :                *
*                                                   *    PRODUCTION SYSTEMS   *
*  PERSON TO CONTACT FOR FURTHER INFORMATION :      *                         *
*     DR. I.E. TEBBETT                              *                         *
*                                                   *                         *
*                                                   *                         *
******************************************************************************
                                                    VERSION : 25/02/87
```

AIM OF THE PROJECT :
****
To demonstrate feasibility of the concept of standardised production facilities
for certain field types and location, achieve cost reduction in engineering
design, construction and hook-up phases and effectively increase the potential
development of recoverable hydrocarbon reserves within the EEC geographic
sphere of interest. (European waters).

PROJECT DESCRIPTION :
****
The project will utilise a data base prepared specifically for this study in
order to identify potential location and establish design premise for the SPF.
A fully optimised conceptual design for a standard production facility will be
developed with details of standardised equipment, systems and modules.
The final design package will be selected to satisfy the technical, economic
and commercial (marketing) requirements. The execution of the project is
divided into four distinct phases of differing duration and extends over a
period of 65 weeks (total).

STATE OF ADVANCEMENT :
****
Ongoing. Phase I of the project - Data gathering and assessment of potential -
has been completed. (Updating of the information database on European Fields
will, however, continue for the duration of the project).
Phase 2, mainly dedicated to formulating a design premise, is almost complete.
However, a recent survey of the market for Standardised Developments has
indicated a possible need to modify the project direction. A review of the
target sector has therefore been scheduled for early 1987.

RESULTS :
****
Awaiting further development of the study, and outcome of project target review
scheduled for early 1987.

```
*******************************************************************************
* TITLE : DEVELOPMENT OF A COST EFFECTIVE SUBSEA    *       PROJECT NO        *
*         PRODUCTION SYSTEM FOR MARGINAL AND DEEP   *                         *
*         WATER HYDROCARBON FIELDS.                 *    TH./03161/84/UK/..    *
*                                                   *                         *
*******************************************************************************
* CONTRACTOR :                                      * PROGRAM :               *
*   KONGSBERG SUBSEA DEVELOPMENTS LTD               *   HYDROCARBONS          *
*   KINGS COURT, CHURCH STREET TEL 04862 27592      *                         *
*   WOKING                      TLX 859429          *                         *
*   UK - SURREY GU21 1HA                            * SECTOR :                *
*                                                   *   PRODUCTION SYSTEMS    *
* PERSON TO CONTACT FOR FURTHER INFORMATION :       *                         *
*   DR. T.S. LUNN                                   *                         *
*                                                   *                         *
*                                                   *                         *
*******************************************************************************
```

VERSION : 20/02/87

AIM OF THE PROJECT :
****
To develop computer based simulation and design methods to analyse various
field development options in order to produce optimum subsea production systems.
PROJECT DESCRIPTION :
****
A new approach is required to design and assess alternate production systems
such that safety, performance and economics are integral part of overall system
development. Such an approach would provide a more complete appreciation of the
"risks" involved and will also permit design optimisation to be conducted. This
should embrace all practical engineering aspects pertaining to technical
categories of field development and, to be effective, must be formalised and
"computerised". The work programme is designed to provide the facility to meet
this need. The essential components of this facility are :
* A computerised database which contains basic data on reliability, repair,
maintenance, operating procedures, weather, repair vessels and costs.
* A method of generating a computerised representation of the various
subsystems.
* A means of generating an analysis model based upon subsystem representation
and the database.
* Rapid analysis of the subsystem performance over field life taking into
account failure, repair and maintenance of all operating modes which effect
production and safety.
* "Designer Friendly" interface for data input, program control and information
output.
STATE OF ADVANCEMENT :
****
Program scope and systems analysis outline specification is completed. Detailed
program specification and coding is in progress together with generation of
equipment data base. Of the three program units (Failure, Repair, and
Availability Models) the Failure Model is nearing completion.
The remaining models are approximately two thirds coded.
RESULTS :
****
Initial Results from one program module give excellent agreement with
alternative method of calculation. Detailed 'proving' will not be possible
until all program units are complete.

```
*********************************************************************************
* TITLE : HYDROCARBON PRODUCTION-SYSTEMS IN ARTIC    *       PROJECT NO        *
*         WATER.                                     *                         *
*                                                    *    TH./03162/84/IT/..   *
*                                                    *                         *
*********************************************************************************
* CONTRACTOR :                                       * PROGRAM :               *
*   TECNOMARE SPA                                    *   HYDROCARBONS          *
*   SAN MARCO 2091            TEL 041 796 711        *                         *
*   IT - 30124 VENEZIA        TLX 410484             *                         *
*                                                    * SECTOR :                *
*                                                    *   PRODUCTION SYSTEMS    *
* PERSON TO CONTACT FOR FURTHER INFORMATION :        *                         *
*   MR. M. BERTA                                     *                         *
*                                                    *                         *
*                                                    *                         *
*********************************************************************************
                                                          VERSION : 01/01/87
```

AIM OF THE PROJECT :
****
Study of hydrocarbon production systems for Artic Seas and design, with model
testing, of two specific solutions. Different production systems will be
analyzed and their optimal ranges of applicability will be defined. Two types
of gravity platform with fast removal capability will be designed: the former,
composed of a concrete base and of an upper conical steel structure, for water
depths in the range of 25-80 meters and for areas with severe ice conditions;
the later, composed of three bases and of an upper tripod steel structure, for
water depths up to 200 meters and for areas with a large number of icebergs. In
both cases, a sealing suction system at the interface with the base allows a
very fast disconnection of the upper part.

PROJECT DESCRIPTION :
****
The main technical aspects of the project are:
- the systematic approach to the problems posed by the exploitation of arctic
fields allows to define the optimal ranges of applicability for the various
proposed solutions.
- the capability to remove the upper structure is assured by a quick
disconnection system.
- from economic point of view gravity structures, which can be removed in short
time and towed to a safe area in case of hazard in extreme conditions, imply
weights and costs lower than those relevant to fixed structures.

STATE OF ADVANCEMENT :
****
Ongoing at the study phase.

RESULTS :
****
An assessment of the design criteria following a probabilistic approach is in
progress.

```
********************************************************************************
* TITLE : POSEIDON (PHASE 1)                        *     PROJECT NO          *
*                                                   *                         *
*                                                   *  TH./03164/84/FR/..     *
*                                                   *                         *
********************************************************************************
* CONTRACTOR :                                      * PROGRAM :               *
*   GERTH                                           *  HYDROCARBONS           *
*   AVENUE DE BOIS PREAU 4      TEL 1 47 52 61 39   *                         *
*   FR - 92502 RUEIL-MALMAISON TLX 203050           *                         *
*                                                   * SECTOR :                *
*                                                   *  PRODUCTION SYSTEMS     *
* PERSON TO CONTACT FOR FURTHER INFORMATION :       *                         *
*   MR A. CASTELA/MR. B. DARDE TEL 1 47 49 02 14    *                         *
*                                                   *                         *
*                                                   *                         *
********************************************************************************
```

VERSION : 01/01/87

AIM OF THE PROJECT :
****
The aim of this project is to develop new techniques allowing a reduction of
the costs associated with offshore hydrocarbon production. In the near future,
the trend will go towards development of smaller hydrocarbon deposits situated
under an ever increasing water depth. In order to stay cost effective, the "all
subsea" scheme appears very promising : it will permit already proven subsea
production stations to be implemented at a longer of two phase pumping, subsea
motors, diverless subsea production, two phase flow transportation techniques.

PROJECT DESCRIPTION :
****
The all subsea production scheme is based on two phase transport technology and
is characterized by :
- Elimination of the petroleum effluent separation units, thus also
  eliminating platforms and all surface installation at the production site.
- Installation of subsea equipment of modular design highly flexible operation
  and ease of maintenance.
- Use of a pumping system and a polyphasic pipeline.
- Reduction to a minimum of the on-site processing installations, which will
  be strictly limited to the requirements of the polyphasic pumping and
  transportation.
The present project will essentially consist of research and development in the
most critical areas of the concept.
Accordingly, the project is divided into two main activities :
MULTIPHASE PUMPING
Basically this activity consists of the study of the basic components of the
polyphasic production pumping unit, together with production of models and
testing of these models in a test bench.
- Homogenizing-regulation cell : regulates the flow at the inlet of the pump,
  and also acts as a choke, when needed.
- Compression cells : compresses the effluent.
- Separation-recycling cell : isolates a portion of the liquid phase,
  enabling lubrication of the bearing with gas-free liquid,
  and also recycling the liquid to the homogenizer,
  in order to lower the gas content, when needed.

126

- Mechanical components ; this activity aims selecting the proper materials
  for the industrial pump.
SUBSEA STATION
- Subsea wellhead compatible with a down hole pump
- Subsea matable electrical connector : design and full scale development of
  a 1 MW prototype.
- Subsea motors for multiphase pumps : design and development of a full scale
  1 MW subsea motor.
- Laying of heavy equipment on sea bottom.
- Connexion between the export pipeline and the subsea production station.
- Subsea pig launcher.

STATE OF ADVANCEMENT :
****
ONGOING:
- Multiphase pumping
ABANDONED:
- Subsea wellhead compatible with a down hole pump
- Heavy equipment laying
COMPLETED:
- Subsea matable electrical connector (full scale development).
- Subsea motors for multiphase pumps (full scale development).
- Subsea pig launcher : detailed engineering available.
- Pipeline connexion : basic engineering available.

RESULTS :
****
ITEM 1 : MULTIPHASE PUMPING
All four subactivities are in a design phase. They should be completed during
1987.
The main results concern the compression cells. several pump protypes were
developed and extensively tested on a test bench. The performance of those
prototypes demonstrated the capability of a larger size pump to cope with
industrial requirements, in the future.
A low pressure prototype of the homogenizing unit has been tested, and improved.
 Results are positive. A high pressure prototype is currently under development.
The recycling cell prototype is still in an early development stage.
Improvements are still required, as performances are too low. This prototype is
of the centrifugal type.
A mechanical components bench test is under development, as well as a short
pump prototype, for test purpose.
ITEM 2 : SUBSEA STATION
Subsea wellhead compatible with a down hole pump : after preliminary
investigations, thsi project was discarded it was not deemed justified, being
too speculative : the reliability of a down hole pump is still low, when
considering the duration and cost of a replacement after failure, below a
subsea wellhead.
Subsea matable electrical connector : the full scale, 1 MW, prototype has been
designed developed and tested in a pressure caisson. It was connected and
discinnected 16 times in sea water, under pressure, without incident."Subsea
motors : a full scale, 1 MW, prototype has been designed, developed and is
currently being integrated into a test structure. It will undergo long term
subsea trials in 1987 out of the scope of the present contract.
Heavy equipment laying : afeasibility study was performed, which did not
identify any major potential problem, whilst in the mean time, heavy lift

vessels were made available; those should further the problem.
Pipeline connection : a basic study was performed, which estblished that two
methods can be contemplated for this duty : either welding connection or a
mechanical connection. The last one should be more readily available, for deep
ater application.
Pig launching station : detailed engineering studies ara available. No major
problem was encountered.

REFERENCES :
****
TECHNICAL MEETING : "ENVIRONMENTAL EXTREMES"
SPONSOR : CESTA
DATE, PLACE : FEBRUARY 25TH, 1986-MARSEILLE
TITLE : ADVANCED TECHNIQUES FOR A HOSTILE ENVIRONMENT
AUTHOR : B. DARDE

```
********************************************************************************
* TITLE : NEW TREATMENT PROCESS FOR OIL AND WATER   *      PROJECT NO         *
*         EMULSION ON OFFSHORE PLATFORMS            *                         *
*                                                   *                         *
*                                                   *    TH./03165/84/FR/..   *
*                                                   *                         *
********************************************************************************
* CONTRACTOR :                                      * PROGRAM :               *
*   GERTH                                           *   HYDROCARBONS          *
*   AVENUE DE BOIS PREAU 4      TEL 1 47 52 61 39    *                         *
*   FR - 92502 RUEIL-MALMAISON TLX 203050           *                         *
*                                                   * SECTOR :                *
*                                                   *   PRODUCTION SYSTEMS    *
* PERSON TO CONTACT FOR FURTHER INFORMATION :       *                         *
*   MR. J.C. GAY & MR. C. SCHR TEL 1 42 91 40 00     *                         *
*                              TLX 615 700          *                         *
*                                                   *                         *
********************************************************************************
```

VERSION : 01/01/87

AIM OF THE PROJECT :
****
The quantity of water associated to the crude generally represents a
significant part of the well effluents. This water must be separated as early
as possible from the commercial part of the output. Specifications of the
maximum water and salt concentration in the crude are stipulated by the
refiners. In addition, the waste waters are subjected to environmental
contraints covering their hydrocarbon concentrations.
Since the efficiency of primary separation process is not sufficient, heavy and
cumbersome equipment are generally set up in order to achieve the required
specifications.
The objective of this study are both :
- to find out means of reducing the volumes and the fineness of the oil and
water emulsions to be treated on offshore platforms,
- to develop smaller and lighter equipments for the settling of these emulsions.

PROJECT DESCRIPTION :
****
Three types of equipment capable of bringing about significant gains, were
initialy selected :
a) HYPO-EMULSIFYING CHOKES AND VALVES
The formulas defining the droplet size distribution of a liquid dispersed in
another liquid by turbulent dissipation show that it is possible to increase
the diameter of the droplets dispersed :
- by reducing the pressure drop, which is achieved by dividing this into
several stages instead of a single stage,
- by increasing the energy dissipation volume, which corresponds to increasing
the frictional surface areas that generate the turbulence,
- by increasing the concentration of the phase dispersed, resulting in partial
local separation and in subsequently applying the pressure drop to each of the
roughly separated phases.

Models applying in the above processes :
- a multi-port stage valve,
- a rotating flow diode,
- a long pipe type valve.
were developed for parallel testing with a single diaphragm representing valves
that already exist, thus providing a reference for comparison with other
systems.

b) DE-OILING CYCLONE

Second a de-oiling cyclone (emulsion of oil in water) capable of accommodating
to wide variations in flow (1 to 4) and oil concentrations whilst at the same
time reducing in a proportion of 1 to 10 the overall dimensions and weights of
existing equipments fulfilling the same functions.
This device, working on a segregation principle, will multiply the difference
in specific gravity between the oil and the water phases by the huge
centrifugal acceleration developed in a vortex flow.
On the basis of a mathematical model initialy set up, a plexiglass scale of a
conventional cyclone was built first.
After extensive testing of this model a new concept of hydrocyclone has emerged
and a modified scale model was developed. Finally an industrial prototype was
built and has been tested in an offshore field.

c) ELECTROSTATIC DEHYDRATOR/DESALTER

Design improvements and increases in efficiency of the electrostatic dehydrator
was found possible, which should result in significant reduction in the size
and volume of the existing equipments. the initial analysis of the problem has
focussed on the following axis of research :
- the design and the development of a sampling/blistering procedure enabling to
clot the droplet size distribution of the emulsion for future characterization,
- the analysis by means of radioacive tracers of the flow of the different
phases in refinery desalter,
- the pilot testing of the new development available on the market like the
three electrode technology (Bielectric from Petrolite) or of the horizontal
flow type (HTI from Hydrotech).

STATE OF ADVANCEMENT :
****
a) Hypo-emulsifying chokes and values. Laboratory work is completed but
industrial development is abandoned due to lack of commercial interest.
b) De-oiling cyclone. Industrial prototype field test being siccessful the
industrialization will be implemented.
c) Electrostatic dehydrator/desalter. A new technology scale pilot is
undergoing trials on a test rig.

RESULTS :
****
a) HYPO-EMULSIFYING CHOKES AND VALVES
The three types of valves were initially tested on low concentration oil in
water emulsions. The efficiency of each type of valve was quantified and the
validity of the theoretical experiments initially set up was confirmed. The
results were checked in the case of oil in brine emulsions, higher
concentration emulsions an finally in the presence of gas. The relationship
between the parameters being defined a conceptual design of two types of
valves/chokes was completed. Although no potential problem was identified in
the technical development of these equipments, the difficulties anticipated in
the marketing approach of the customers in conjunction with the uncertainties
of the oil industry linked to the low oil price level, have made necessary to
cancel this project.

b) DE-OILING CYCLONE

Testing of the conventional type scale model has enabled first to optimize the general set up of the cyclone. Then an adjustable injection head was defined in order to improve the flexibility in flow rate. The efficiency of the cyclone was then improved furthermore by the addition of internal devices increasing the hydraulic stability of the stream. Two patent applications were filed. The cyclone performances were then comparable to those of a tilted plate separator but a much smaller weight and volume. However the results were much lower than what was anticipated from the theoretical calculations.Therefore the laboratory studies were resumed in order identify the origin of this problem. Based on this study a new concept was defined leading to a different type of hydrocyclone. An industrial prototype has been built and tested on an offshore field. Excellent results were obtained but some minor improvements are still required in order to reduced the pressure drop requirement.

c) ELECTROSTATIC DEHYDRATOR/DESALTER

The sampling/blistering procedure studies lead to the development of a method which has been patented. However, this system can only be used on laboratory fluids and attempts to extend it to the field conditions have failed.

Testing of newly available process have shown some possibilities of reduction in the sizes of the equipments. However the work on these pilots was stopped when studies on a new concept of equipment have revealed that larger progress could be achieved. A scale pilot is being tested resulting in significant improvement in performance and residence time. Testing will continue in 1987.

```
******************************************************************************
* TITLE : CONAT-DEEPWATER PRODUCTION TOWER.        *        PROJECT NO      *
*                                                  *                        *
*                                                  *    TH./03167/84/DE/..  *
*                                                  *                        *
******************************************************************************
* CONTRACTOR :                                     * PROGRAM :              *
*   BILFINGER & BERGER BAUAKTIENGESELLSCHAFT       *   HYDROCARBONS         *
*   POSTFACH 76 0240           TEL 040 229 23 124  *                        *
*   DE - 2000 HAMBURG 76       TLX 21 11 86         *                        *
*                                                  * SECTOR :               *
*                                                  *   PRODUCTION SYSTEMS   *
* PERSON TO CONTACT FOR FURTHER INFORMATION :      *                        *
*   MR. H.G. BUTT                                  *                        *
*                                                  *                        *
*                                                  *                        *
******************************************************************************
                                                        VERSION : 01/01/87
```

AIM OF THE PROJECT :
****
Development of an articulated tower used in waterdepths between 330 m and 430 m
in the North Sea as alternative to fixed platforms or as satellite platform to
them.

PROJECT DESCRIPTION :
****
A parameter study has to show the deck weights that can be carried by towers in
the waterdepth range and deflections these towers undergo. A structure that is
optimum with regard to operational condition and the design of which is guided
by practice, safety and cost is subject to further R+D work. A component of
special importance of the R+D works is the design of a joint for the
articulation which is able to transfer high loads from the tower into the
foundation and which is, at the same time, maintenance free and resistant to
wear and tear. Construction methods and configuration of the tower have to be
developed and worked on to ensure a vertical construction of the tower and a
partial installation of the topsides close to shore in sheltered areas.
Description of the components:
A. Deck and topsides have to be developed for each special application.
B. Column
The cylindrical column is made of concrete and prestressed in its vertical axis.
 It is surrounded by a multi-cell buoyancy body, closed at its upper and lower
ends by domes. This buoyancy body is monolithically concreted together with the
column.
C. Joint articulation
Consits of :
- the upper semisphere with the support pipes for the pins and sliding rings
for the torque-lock which are arranged in the equator of the ball joint.
- The lower semisphere with an annulus of slide elements and the slide chair
for the transfer of the moments around the vertical axis.
D. The foundation can be piled or gravity-founded according to soil conditions.
Whether or not the foundation has storage capacity depends on the field
conditions.

STATE OF ADVANCEMENT :
****
Ongoing. A detailed design for an articulated tower in 335 m waterdepth with a deck load of 20.000 to. is in progress considering second order wave loads, earthquake with 10.000 years probability and dynamic wind effects. Special safety measures have been developed. A detailed construction procedure inclusive tow out and installation has been worked out. Conceptual appraisal by Det Norske Veritas is pending. A conceptual design of an articulated tower for 400 m waterdepth has commenced.

RESULTS :
****
For the CONAT used in 335 m waterdepth the following results have been found : THE BASE STRUCTURE is an arrangement of vertical cylinders which are closed on their upper and lower ends by domes and made of prestressed concrete. It bears 16 m long sheet pile skirts which transfer the vertical and horizontal forces as well as torsional moments into the seabed even with existing poor soil conditions.
The ball joint is made of stainless steel X 5 Cr Ni Mo 16.5 and exposed to environmental loads of H = 82 MN V = +/- 62 MN.
M torque = 108 MN. In addition to environmental forces a static vertical load acts on the joint which results from a weight surplus of 200 MN over the buoyancy of the column. This ensures that lifting forces cannot arise at all within the joint.
The joint is designed for a maximum inclination of 8 degrees. All slide parts are made of a metallic matrix which can be compared with a tin-lead-bronze alloy casting. In this metallic matrix solid lubricants are embedded. In addition the sliding pairs are lubricated by a lithium saponified lubricant.
A conservative wearing calculation based on the corresponding long term response distribution resulted in a rate of wear of 5 mm for the slide elements.
 This does not effect at all the operation of the joint. Therefore no maintenance for the ball joint is required.
The inside of the joint is separated by a gas-tight and pressure resistant bulkhead from the column and is permantly filled with compressed air at a sufficient pressure to withstand the outside water. Mechanical sealing elements for the ball joint are therefore not required.
THE CONCRETE COLUMN is bottle shaped with circular cross section. The lower part has an outer diameter of 35 m, the upper part of 20 m. The risers are installed inside the column.
The inclinaison of the column amounts to 4.5 degrees under extreme conditions. 3 degrees is static inclination due to wind and current and 1.5 degrees oscillation due to wind forces.
An oscillation of 0,20 degrees will not be exceeded during 99,59% of the lifetime.

REFERENCES :
****
- " DESIGN AND SYSTEMATIC CONCEPT DEVELOPMENT OF ARTICULATED OFFSHORE TOWERS" OFFSHORE BRASIL 1978
- "TEST CONAT. LARGE-SCALE TEST IN THE VICINITY OF THE RESEARCH PLATFORM 'NORDSEE'" EUROPE LONDON 1980
- "CONAT - THE ARTICULATED OFFSHORE SYSTEM"
  DEEP OFFSHORE TECHNOLOGY CONFERENCE 1981
- ARTICULATED PRODUCTION TOWER FOR DEEPWATER
  OTC 1984
- INTERNATIONAL WORKSHOP ON CONCRETE FOR OFFSHORE STRUCTURES, ST. JOHN'S 1986

```
*******************************************************************************
* TITLE : EXTRA-DEVELOPMENT OF EXPERT SYSTEM        *        PROJECT NO        *
*          TECHNOLOGY FOR RISER ANALYSIS AND        *                          *
*          DESIGN                                   *     TH./03171/84/IR/..    *
*                                                   *                          *
*******************************************************************************
* CONTRACTOR :                                      * PROGRAM :                *
*   STELLAR INT. LTD                                *   HYDROCARBONS           *
*   MCS SCIENCE PARK, UPPER NE TEL 091 24373        *                          *
*   IR - GALWAY              TLX 50094              *                          *
*                                                   * SECTOR :                 *
*                                                   *   PRODUCTION SYSTEMS     *
* PERSON TO CONTACT FOR FURTHER INFORMATION :       *                          *
*   PROF. SEAN F. MCNAMARA                          *                          *
*                                                   *                          *
*                                                   *                          *
*******************************************************************************
```

VERSION : 03/03/87

AIM OF THE PROJECT :
****
The objective of the EXTRA project is the synthesis of expert systems
technology and advanced riser analysis and design techniques into an integrated
marketable software package.

PROJECT DESCRIPTION :
****
The project includes an intelligent pre-processor that assists in setting up
the analysis of a riser system, refined solution techniques for steel and
flexible risers and an intelligent post-processor that incorporates design
checks and code certification.

STATE OF ADVANCEMENT :
****
Ongoing. The EXTRA package had been implemented using the OPS5 production
system language to control the Expert System and to call-out to the various
analysis packages via local BASIC, FORTRAN and system language modules.

RESULTS :
****
The package acts as an expert assistant in the design of exploration and
production, rigid and flexible risers and articulated loading columns and
towers. A preprocessor aids the user in the definition of the model in a form
suitable for the analysis routines; a program strategy controller decides the
most apprppriate analysis techniques to be followed. Finally, a postprocessor
performs design code checks and certification, before advising the user of the
suitability of his model. Graphical output is available to the user at all
stages to provide a clear, understandable representation of results.
Results include the satisfactory design of two API test case risers, namely API-
500-21.5-2-R and API-1500-21.5-2-R, using frequency domain and time domain
analyses respectively. The 2-D flexible riser section is also operational and
satisfactory results have been obtained from the design of a number of flexible
riser production systems.

REFERENCES :
****
THE DEVELOPMENT OF AN EXPERT SYSTEM FOR MARINE RISER ANALYSIS.
D. BOYLE, J.F. MC NAMARA, R. O' SULLIVAN, OFFSHORE MECHANICS AND ARTIC
ENGINEERING, SIXTH INTERNATIONAL SYMPOSIUM, HOUSTON, TEXAS, 1-6 MARCH, 1987.
EXPERT SYSTEM BLACKBOARD CONCEPT APPLIED TO MECHANICAL ANALYSIS AND DESIGN.
J.F. MC NAMARA,R. O' SULLIVAN, TO BE PRESENTED AT THE IRISH MANUFACTURING
CONFERENCE, NIHE, LIMERICK, SEPTEMBER 1987.

```
*******************************************************************************
* TITLE : POSEIDON (PHASE II)                    *      PROJECT NO        *
*                                                *                        *
*                                                *   TH./03172/85/FR/..   *
*                                                *                        *
*******************************************************************************
* CONTRACTOR :                                   * PROGRAM :              *
*   GERTH                                        *   HYDROCARBONS         *
*   4 AVENUE DE BOIS PREAU     TEL 1 47.52.61.39 *                        *
*   FR - 92502 RUEIL-MALMAISON TLX 203 050       *                        *
*                                                * SECTOR :               *
*                                                *   PRODUCTION SYSTEMS   *
* PERSON TO CONTACT FOR FURTHER INFORMATION :    *                        *
*   MR. CASTELA               TEL 1 47.49.02.14  *                        *
*                             TLX 615 700        *                        *
*                                                *                        *
*******************************************************************************
                                                      VERSION : 01/01/87
```

AIM OF THE PROJECT :
****
The aim of the Poseidon project is to develop new techniques allowing a
reduction of the costs associated with offshore hydrocarbon production.
Poseidon (phase I) project focused on the design studies of the main
subcomponents of the "all subsea" scheme. Poseidon (phase II) project is
basically dedicated to the trials of the most critical amongst those
subcomponents, namely the multiphase pumping system and the subsea drive system.

PROJECT DESCRIPTION :
****
The present Poseidon Phase II project comprises :
1) Development of a reduced scale, complete and marine multiphase pumping
system, derived from the results obtained during the Poseidon Phase I project.
This development is divided into four subsequent phases :
- evaluation of the results acquired from the Poseidon Phase I project,
- basic engineering and definition of the architecture of the system,
- detailed engineering of the reduced scale prototype,
- fabrication and bench tests of the prototype.
2) Long term development subsea trials of the full scale drive assembly
developed within Poseidon Phase I project.
The main activities are :
- fabrication of the test platform,
- integration of the drive assembly into the test platform,
- shore tests,
- deep water tests.
3) Theoretical study of the electrical network supplying power to a future
subsea station through the means of long distance subsea cables.

STATE OF ADVANCEMENT :
****
ITEM 1 : MULTIPHASE PUMPING SYSTEM
Ongoing
Simulation of the pumping system behaviour: a call for tender was issued in
order to select the industrial entity to which the development of the marine
multiphase pumping system will be entrusted.

ITEM 2 : SUBSEA STATION AND WELLHEAD
Ongoing
Subsea power connector and motorization assembly are being integrated into a
subsea test platform.

RESULTS :
****
ITEM 1 : MULTIPHASE PUMPING SYSTEM
This activity has just started and no results are yet available.
ITEM 2 : SUBSEA STATION AND WELLHEAD
Fabrication of the subsea test platform and template is completed. Integration
of the motor assembly within this test platform is underway.
Fabrication of the remote control system for the subsea motor is completed.
Basic engineering of the remote control system of a future subsea station is
completed.
Theoretical studies have been performed, concerning the electrical behaviour of
the long distance power supply system. Both steady state and transient states
(start up, short circuits, ...) have been studied. This activity will be
completed when the electronic variable frequency converter power supply case
will have been covered.

```
********************************************************************************
* TITLE : PLATINE (PHASE 1)                        *        PROJECT NO        *
*                                                  *                          *
*                                                  *     TH./03173/85/FR/..   *
*                                                  *                          *
********************************************************************************
* CONTRACTOR :                                     * PROGRAM :                *
*   GERTH                                          *   HYDROCARBONS           *
*   AVENUE DE BOIS PREAU 4       TEL 1 47.52.61.39 *                          *
*   FR - 92502 RUEIL-MALMAISON TLX 560 804         *                          *
*                                                  * SECTOR :                 *
*                                                  *   PRODUCTION SYSTEMS     *
* PERSON TO CONTACT FOR FURTHER INFORMATION :      *                          *
*   MR. P. LEFEVRE               TEL 1 59.83.62.38 *                          *
*                                TLX 560 804       *                          *
*                                                  *                          *
********************************************************************************
                                                       VERSION : 01/01/87
```

AIM OF THE PROJECT :
****
The aim of the PLATINE project is to specify a central production platform,
unmanned, remotely operated from a control room located onshore, with a yearly
(or half-yearly) visit for inspection and overhauling.
The intermediate aims consist of working out and testing pilot units related to
sub-systems and testing new equipments.
The project has been divided into 3 phases : PLATINE - Phase 1 (1986-1988);
PLATINE - Phase 2 (1987-1989); PLATINE - Phase 3 (1988-1990).

PROJECT DESCRIPTION :
****
This project consists of several items :
- methodology studies;
- test and adaptation of equipment;
- development or adaptation of maintenance and production softwares.
The adopted policy will aim at simplifying the general architecture and
examining the existing processes or methods critically. Use of new techniques
or materials will be systematically assessed.
The facilities have been divided into 7 main sub-systems : 4 production subsystems (gas-lift, sea water injection, separation-treatment, well pumping) and
3 sub-systems non specific to oil industry (energy/rotating machinery-
safety/reliability-automation sytem).
- The gas-lift-sub-system study has 3 parts : conception and construction of a
2-phase flowmeter - automatic start-up of a well in high pressure gas-lift -
real time optimization of gas-lift injection.
- The sea water injection sub-system study aims at selecting sea water
treatment processes before injection and at specifing the best means of lifting
and pumping.

- The separation-treatment sub-system study aims at specifing a separation-treatment chain compact, optimal and adapted to the project.
- A study related to the various means of well pumping will enable us to select the (or the two) best means and to specify the automation works to be done.
- An on-condition plan will be worked out ; the main lines of the future on-condition system (including probably an expert system) will be defined.
- A comprehensive statistical investigation of rotating machines will be carried out. Besides, the development of new technologies in this field will be followed and directed in connection with the manufacturers ; the main planned actions are related to peripheral/centrifugal compressors, variable speed engines, magnetic bearings, and gas packing.
- The automation system includes sensors, making their measurements reliable, communication system and expert systems.
   The reliability of an automation system is mainly dependent on the reliability of sensors. The actual sensors will be tested then adapted, if needed for offshore oil conditions; their measurements will be made reliable by means of coherence treating numerical system. Prototype systems of optic fiber transmission will be tested; the development of optic sensors by manufacturers will be followed and guided.
- The safety/reliability module includes regulation studies, reliability studies related to equipment or systems safety equipment and platform surveillance (with possibly a future robot development).

STATE OF ADVANCEMENT :
****
Ongoing.
The project is still roughly at the study stage and the preparation of prototype tests. The gas-lift module is more advanced : a prototype has been tested and and a software achieved.
Comprehensive investigations have been made internally, into other oil companies and various industries.

RESULTS :
****
Two modules are being studied : gas-lift and sea water injection. Besides, for the other modules, investigations have been made and future work orientated.
- GAS-LIFT : A prototype 2-phase flowmeter has been built and tested; some modifications were needed and have been completed, but the improvement is not as great as expected. Further modifications will have to be done.
The automatic start-up has been defined.
In connection with the real time gas-lift optimization, a well simulation software has been completed; the laboratory test results are satisfactory.
- SEA WATER INJECTION : The principles of sea water treatment processes (existing or protoypes) have been assessed; this evaluation allowed us to select all the processes to be tested (prototypes or small-scale industrial units); these processes are related to filtration, deoxygenation, chlorination and bacterial treament. The tests will start in January 87. On the other hand, reliable and accurate enough control analysers have been found (filtration quality, bacteria content, O2, Cl2 content...); some will be tested, at least one type will need to be developed.
- ROTATING MACHINES : A comprehensive investigation has been made into our subsidiaries, oil companies and other industries. It is pointed out that a lot of existing equipment or facilities have failure rates inconsistent with PLATINE. This investigation will be the starting point of 1987 works.

Besides, the development of new technologies by manufacturers has been followed and guided (peripheral/centrifugal compressors, magnetic bearings, gas packings and variable speed engines).
- MAINTENANCE : A list of curative or preventive works and trouble shooting on existing platforms has been drawn up. The works to be performed in connection with static or small non-static equipment will be listed early 87. Besides, the optimal electric scheme is being studied (comparing for example offshore generation to onshore generation with cable transportation).
- AUTOMATION SYSTEM : The test program of existing sensors has been worked out. Prototype transmission systems of optic fiber multiplexing have been purchased and will be tested in 87-88.
- MISCELLANEOUS : A preliminary study of the interest of a surveillance robot has been made.
The main American oil companies operating in the Gulf of Mexico have been visited and questioned about their automation policy.

```
********************************************************************************
* TITLE : WATER PRODUCTION IN OFFSHORE WELLS       *       PROJECT NO        *
*                                                  *                          *
*                                                  *   TH./03174/85/FR/..    *
*                                                  *                          *
********************************************************************************
* CONTRACTOR :                                     * PROGRAM :               *
*   GERTH                                          *   HYDROCARBONS          *
*   AVENUE DE BOIS PREAU 4      TEL 1 47.52.61.39  *                          *
*   FR - 92502 RUEIL-MALMAISON TLX 203 050         *                          *
*                                                  * SECTOR :                *
*                                                  *   PRODUCTION SYSTEMS    *
* PERSON TO CONTACT FOR FURTHER INFORMATION :      *                          *
*   MR. KOHLER                                     *                          *
*                                                  *                          *
*                                                  *                          *
********************************************************************************
                                                        VERSION : 01/01/87
```

AIM OF THE PROJECT :
****
The aim of the project is to develop in the laboratory and apply on the field,
processs based on water soluble polymers in order to reduce in a selective
manner the water production in production wells without affecting the oil
production.

PROJECT DESCRIPTION :
****
The injection of water soluble polymers in production wells is one of the most
performing processes for reducing water production. The process consists in
adsorbing in the vicinity of the production wells a given amount of polymer
capable of considerably reducing the water production without affecting the oil
or gas production.
As water production from reservoirs present different characteristics both, of
reservoir rock composition and permeability and also of produced brine salinity
and temperature, it is proposed to search for the appropriate solutions for
various environmental conditions and to apply them on production wells. Indeed,
existing processes, based mainly on crosslinked polymer systems are considered
to reduce in a great extent the oil or gas production and are applied
indifferently on all cases of excess water production independently of
reservoir characteristics.

STATE OF ADVANCEMENT :
****
Completed laboratory work on phase 1.1. "Water production from high temperature
wells" by selecting a process based on a non-ionic polysaccharide.
Ongoing laboratory work on phase 1.2 "Medium salinity and temperature water
production".
Delayed field application corresponding to phase 2 due to a postponed drilling
program.

RESULTS :
****
PHASE 1.1 - Two original processes, one based on a hydrolyzed polyacrylamide;
the other on a non-ionic polysaccharide were compared in order to solve a real
field case. Polysaccharide type polymers were found to resist better than
polyacrylamides to elevated temperatures and to mechanical degradation.
Numerical simulation showed that it is preferable to inject the polymer near
the oil water contact zone rather than through existing perforations at the top
of the oil zone.
PHASE 1.2 - Field handling and injectivity of a non-ionic polysaccharide were
tested both on field cores in the laboratory and on an abandoned well in order
to choose the best conditions to perform a real field case. Results are
encouraging for this process.

```
********************************************************************************
* TITLE : DIPHASIC COMPRESSION                       *      PROJECT NO       *
*                                                    *                       *
*                                                    *    TH./03175/85/FR/..  *
*                                                    *                       *
********************************************************************************
* CONTRACTOR :                                       * PROGRAM :             *
*   GERTH                                            *   HYDROCARBONS        *
*   4, AVENUE DE BOIS PREAU      TEL 1 47 52 61 39   *                       *
*   FR - 92500 RUEIL-MALMAISON TLX 203 050           *                       *
*                                                    * SECTOR :              *
*                                                    *   PRODUCTION SYSTEMS  *
* PERSON TO CONTACT FOR FURTHER INFORMATION :        *                       *
*   MR. P. DOUINEAU             TEL 1 42 91 40 00     *                       *
*                              TLX 615 700           *                       *
*                                                    *                       *
********************************************************************************
                                                         VERSION : 15/10/86
```

AIM OF THE PROJECT :
****
The aim of this project is to develop a new production system able to boost the
stream of gas wells which are producing some liauids in addition to the gas.

PROJECT DESCRIPTION :
****
The production system is based on the use of a new equipment : a diphasic
compressor capable to boost the gas as well as the liauidiquid. The selected
type of compressor is a screw compressor.
The project applies to a programme which purpose is mainly to design, execute
and test at field conditions a screw compressor.
The project applies to a programme which purpose is mainly to design, execute
and test at field conditions a screw compressor which accepts gas containing a
liquid fraction.
Slugging will be also a matter of concern during the project as slug could
seriously damage the machine: a slug breaker will also be tested.
Attention is also paid to hydrates problems, which formation condition has to
be predicted as accurately as possible.
As regards liquid content, the project considers gas containing 2 percent
(volume) at suction conditions.

STATE OF ADVANCEMENT :
****
A gas field has been selected to be the one on which the studies will be
concentrated and where the new equipment could be tested.
Operating conditions have been estimated and preliminary design of the machine
has been performed.
A bibliography study has been made on hydrates formation condition in diphasic
streams.

```
*****************************************************************************
* TITLE : NEW RANGE OF COMPLETION EQUIPMENT        *       PROJECT NO       *
*                                                  *                        *
*                                                  *    TH./03176/85/FR/..  *
*                                                  *                        *
*****************************************************************************
* CONTRACTOR :                                     * PROGRAM :              *
*   GERTH                                          *   HYDROCARBONS         *
*   4 AVENUE DE BOIS PREAU        TEL 1 47.52.61.39 *                       *
*   FR - 92502 RUEIL-MALMAISON TLX 203 050         *                        *
*                                                  * SECTOR :               *
*                                                  *   PRODUCTION SYSTEMS   *
* PERSON TO CONTACT FOR FURTHER INFORMATION :      *                        *
*   MESSRS BASSE/JACOB            TEL 59.33.10.33   *                       *
*                                 TLX 541 891       *                       *
*                                                  *                        *
*****************************************************************************
```
                                                             VERSION : 01/03/87

AIM OF THE PROJECT :
****
This project consists of studying a new line of completion equipment to meet
four main objectives :
- safety implementation when bringing wells into production
- increment of production capacity
- long exploitation life of equipment, despite hostile environments (corrosion
and abrasion by effluent)
- reduction of exploitation costs by using different production modes.
presently the line of products on the market is all US brand and of old design,
while in this project, a new technology is used, the originality of which is
bound to the polyfunctional features of its different components, in particular
latching and sealing systems. The latter will be similar for each equipment
within this new line of completion equipment.

PROJECT DESCRIPTION :
****
The main completion equipment aimed at in this project are the following :
- Safety valve associated with a concentric gas lift system
- Subsurface controlled subsurface safety valve (SCSSV) wireline retrievable
- Safety valve for rod pumping wells
- Annular safety packers
- Circulating valve
- Tubing retrievable safety valve.
The project will consist of integrating polyfunctional systems in each
equipment, to meet safety, reliability and simplicity of use in the field of
requirements. The project development will consist of 3 stages :
- Engineering study
- Prototype manufacturing
- Cell and field testing
Prototypes of this equipment are made for the purpose of destructive tests and
to obtain API certificate (14A specifications).

STATE OF ADVANCEMENT :
****
Status of equipment as follows:
- safety valve associated with a concentric gas-lift system: first design completed: long duration field testing. Prototype manufactured.
- SCSSV: wireline retrievable designed, prototypes manufactured. Long duration cell tested and on its way to API qualification (14A specifications).
- safety valve for rod pumping wells: designed, prototype manufactured and cell tested.

RESULTS :
****
SAFETY VALVE ASSOCIATED WITH A CONCENTRIC GAS-LIFT SYSTEM
A certain number of prototypes are presently field tested in the Total Indonesia field, in Handil. The results obtained have revealed a few problems when latching and unlatching the system, and also the impossibility to have a good sealing with the upper pack off. Some sealing modifications were made but the locking device (made in high resistance steel) is no longer protected by hydraulic oil from the control line. This is why we are presently completely redesigning the safety valve in order to make it more reliable when running and latching and to overcome the sealing problems. One prototype of the new system is presently under design to be field tested next year.
SURFACE CONTROLLED SUBSURFACE SAFETY VALVE (SCSSV) WIRELINE RETRIEVABLE
Several prototypes have been made and long and severe duration cell tests have been run both in our facilities and in Elf Aquitaine Test Centre in Le Fourc (near Boussens), especially with sand slurry circulation (14%). Good results were obtained. we are presently preparing the equipment to perform the API test (14 specifications) in San Antonio. It is noticeable that the safety valve associated with concentric gas lift system and wireline retrievable safety valve have a lot of common parts and will be latched and unlatched by the same hydraulically actuated device.
SAFETY VALVE FOR ROD PUMPING WELLS
Prototype has been successfully cell tested in Elf Aquitaine Test Centre in Le Fourc.
We are waiting a well to field test the system.
ANNULAR SAFETY VALVES
Study not started.
CIRCULATING VALVE
First ideas are put on paper and advice is asked to users.
TUBING RETRIEVABLE SAFETY VALVE
A first design has been made but has to be reviewed, following remarks of future users.

REFERENCES :
****
SIX PATENTS HAVE BEEN APPLIED FOR BY DIAMANT BOART IN FRANCE ON JULY 29, 1986, UNDER THE FOLLOWING NUMBERS : 8611417 THROUGH 8611422, FOR THE WIRELINE RETRIEVABLE SURFACE CONTROLLED SUBSURFACE SAFETY VALVE. USA AND CANADA EXTENSIONS ARE PRESENTLY UNDER PROCESS.

```
******************************************************************************
* TITLE : IN-WELL LONG DRIVE OFFSHORE PUMPING     *        PROJECT NO        *
*         UNIT                                     *                          *
*                                                  *     TH./03178/85/FR/..   *
*                                                  *                          *
******************************************************************************
* CONTRACTOR :                                     * PROGRAM :                *
*   MAPE                                           *   HYDROCARBONS           *
*   BOULEVARD DU MARECHAL JUIN TEL 040 433411      *                          *
*   FR - 44100 NANTES         TLX 711436           *                          *
*                                                  * SECTOR :                 *
*                                                  *   PRODUCTION SYSTEMS     *
* PERSON TO CONTACT FOR FURTHER INFORMATION :      *                          *
*   MR. CHARDONNEAU                                *                          *
*                                                  *                          *
*                                                  *                          *
******************************************************************************
```

VERSION : 01/01/87

AIM OF THE PROJECT :
****
The object of this project is to develop a thorough pumping equipment for drill
strings from the bottomhole pump to the surface unit, specially adapted to
offshore pumping, but also suitable for onshore pumping.
Therefore, the surface will have an integrated jack at the wellhead in such a
manner that its dimensions are minimized compared to other types of existing
beam units. This technology will allow :
- grouping of wells with very near cross axes (offshore platforms or onshore
  cluster);
- easy intervention on a well whithout having to interrupt the nearby well(s)
  in the case of clustered wells;
- considerable masking of pumping unit compared to other types of units in
  the case of protected offshore or onshore sites;
- improvement, compared to conventional units, of the pumping drive (long
  drive) without handicaping the driving part (beam unit reducer), and thus,
  a more economical and reliable adaptation to the production of viscous
  and/or gas containing crudes.

PROJECT DESCRIPTION :
****
1. STUDY AND REALISATION OF A "MINIMUM" SURFACE UNIT PROTOTYPE aimed at
controlling the validity of the basic principle of this unit. This prototype
consits in :
    - a hydraulic jack temporarily fitted, for practical reasons, directly
      on the standard wellhead, thus higher than the wellhead.
    - an electrically driven hydraulic unit mounted on skid laved on the floor
      independly of the jack
    - a unit/jack connection formed by flexible and hydraulic piping.
    Its main characteristics are the following :
    - Maximum pumping feel = 25 600 Lbs
    - Three possible drive lengths = 100", 125", 144"
    - Six rythms also distributed up to 9 cp/mn maxi, for the 144" drive length
    - Power = 100 HP
    Hence, its name : V 256.144
    This prototype has no means of balancing the dead weight of the pumping
    drill string.
2. ON-SITE MEASUREMENTS OF PROTOTYPE PUMPING PARAMETERS AND DEVELOPMENT OF

COMPUTATION METHOD
- of the pumping efforts at the jack rod
- of the required engine power to pump a given flowrate at a given depth, without having to balance the deadweight of the pumping drill string.
3. STUDY AND CONSTRUCTION OF JACK HOUSED IN THE WELLHEAD TUBING, height above which it may only exceed with the hydraulic supply unit.
4. STUDY AND CONSTRUCTION OF BOTTOMHOLE PUMPS TO API REQUIREMENTS, BUT WITH DISMOUNTABLE BARREL OF IMPORTANT LENGTH (NOT WITHIN API STANDARDS)
   On-site tests of different prototypes.
5. STUDY AND CONSTRUCTION OF BOTTOMHOLE PUMP, long drive with piston beyond API standards.
   Tests on site.
6. STUDY AND CONSTRUCTION OF SUCKER-RODS 1 1/4 at API standards, in special steel with dispersoids
   - Previous realisation of prototypes of existing dimensions (3/4" - 7/8" - 1")
   - Fatigue tests on test bench, comparison with rods commercially known
   - Realisation of 1 1/4" prototypes
   - Fatigue tests on test bench under loading conditions similar to previous fatigue tests. Commercial rods cannot be used as this dimension does not exist
7. STUDY AND CONSTRUCTION OF SUCKER RODS IN COMPOSITE MATERIALS
   - Construction of prototypes
   - Dynamic tests on fatigue bench, comparison with best quality American rods (FIBERFLEX)
   - Fiability test of complete drill string on significant site.
8. STUDY AND CONSTRUCTION OF RAPID DISCONNECTION DEVICE FOR PUMPING DRILL STRING
   Safety envisaged in the case of offshore pumping from a floating unit.
9. INTEGRATION OF ABOVE SUBASSEMBLIES AND PREPARATION OF TESTS FOR A COMPLETE SYSTEM ON THE SITE
STATE OF ADVANCEMENT :
****
1 & 2 : completed
3 : pending
4 : partially completed
5 : study completed - ramp under construction awaiting for test site decision
6.1 & 6.2 : completed
6.3 & 6.4 : will not be made (cannot be financially justified)
7.1 & 7.2 : underway
8 : temporarily delayed
9 : pending

RESULTS :
****

1. The overall prototype unit operates satisfactorily with the long drive
(144"), at the maximum rythm of 9 cp/mn, with a power of 80/90 HP. However, its
reliability still needs to be confirmed. Criticism is made on the relatively
high engine power which is required due to the lack of dead weight balancing
contrary to the beam unit. This provides an important energy consumption. This
problem is being studied.

2. The computation method of the pumping with this type of jack unit has been
developed thanks to the measurements taken on the site. It derives from the API
method used for the beams, but corrected by the consideration :
- for pumping feed charges, of dynamic effects specific to its displacement
  function
- for the power, of the weight of the non-balanced dril string.

4. The nr 1 problem is the construction of barrels in Europe. The technical
alternative according to the minimum range of operations of the "American" type
is presently being solved and will be followed by a financial report.
Parallel to this, other components of the different types of pumps (rod pumps
and tubing pumps) and of different manufacture (SPRAY METAL lining of pistons,
valves in stellite,..) are being studied and progressively manufactured. They
are integrated in pumps with barrel of American origin for the moment.
Different pumps are being tested on site in France (bassin Parisien) and in
Italy recently (offshore site of ROSPO MARE).

5. Study completed. Ramp will be manufactured after Elf Congo's approval on the
technical specification for the tests under examination at the moment.

6.1 & 6.2 Prototype rods 3/4", 7/8", 1" have been made conform to API
specifications and carefully controlled at all stages. They were tested on the
fatigue bench of SNEA(P) at Boussens and compared with relatively good rods
existing on the market. They proved to be satisfactory.
Technically, our manufacturing range using a special steel with dispersoids is
now appropriate. However, the programme has been interrupted at this stage for
financial reasons.

7.1 & 7.2 prototype rods of 3/4" and 1" have been manufactured and tested on
the fatigue bench of SNEA(P) and compared with FIBERFLEX rods. The result is
still not satisfactory as regards the connection of the threaded end with the
main element of the rod in produced composite.

```
*******************************************************************************
* TITLE : CONCEPT FOR SUBSEA SYSTEMS IN SHALLOW    *      PROJECT NO         *
*         WATER                                    *                         *
*                                                  *    TH./03179/85/DK/..   *
*                                                  *                         *
*******************************************************************************
* CONTRACTOR :                                     * PROGRAM :               *
*   AALBORG VAERFT A/S                             *   HYDROCARBONS          *
*   P.O. BOX 661              TEL 45 8 163333       *                         *
*   DK - 9100 AALBORG         TLX 69705            *                         *
*                                                  * SECTOR :                *
*                                                  *   PRODUCTION SYSTEMS    *
* PERSON TO CONTACT FOR FURTHER INFORMATION :      *                         *
*   MESSRS JOERGENSEN/OLSEN                         *                         *
*                                                  *                         *
*                                                  *                         *
*******************************************************************************
                                                       VERSION : 19/03/87
```

AIM OF THE PROJECT :
****
The aim of the project is to develop a number of subsea systems which can be
installed in the North Sea, the Baltic Sea or in surrounding waters for the
production of hydrocarbons from economical marginal fields. The systems shall
be regarded as alternatives to light weight platforms which may not be
attractive of economical and environmental reasons in the deeper areas and in
areas with heavy ship traffic and large ice forces.

PROJECT DESCRIPTION :
****
A large number of known fields have not been developed due to their marginal
economics. The main objective of this project is to investigate the technical
and economical feasibility of marginal field development concepts based on
subsea systems.
Marginal fields can in general be defined as fields, whose recoverable reserves
are minimal and cannot be developed economically utilizing conventional
structures and processing facilities.
The definition applies in principle to two different categories of fields :
a. A peripheral marginal field allowing the produced oil/gas to be linked to an
existing facility for processing or tied into existing pipelines for
transportation to said facility.
b. A geographical marginal field remotely located from existing facilities or
pipelines and believed to hold insufficient reserves to make conventional
production methods economical.
In order to improve the economics of hydrocarbon production from marginal
fields in the more shallow parts of the North Sea, the Baltic Sea and
surrounding waters and in order to decrease the field size limit for
economically viable field developments a review of possible development options
will be carried out. This results in the identification of a number of concepts
for the subsea systems which can compete favourably with more traditional
solutions.

Further it is to be expected that removal of platforms will be made compulsory in the future when an oil/gas field is abandoned. This will add to the cost of these solutions, where as subsea systems will not have this disadvantage. Therefore, it is proposed to perform design verification for a number of alternatives. Emphasis will be placed on features requiring development or redesign beyond existing technology in order to meet the specific application.

STATE OF ADVANCEMENT :
The project has been divided into 30 activities, each covering a well defined number of objectives to be studied.
At the end of December 1986, 21 of these activities were more than 90% completed, bringing the overall progress to 67% completion.

RESULTS :
The working groups made up from the contractors have reported on the following results.
- Environmental DTA, conditioning, criterias, and loads.
- Well data for oil and gas wells.
- Assessment and comparison of risks for platform and subsea system, ice, ship impact, fishing gerr, etc.
- Design cases and standards.
- Economic and technical review.
- Template, wellheads, and manifold.
- Protection, geometry and structure.
- Two phase flow, oil and gas condensate pipelines.
- Pipelines and control lines, overall review.
- Installation, drilling, and completion incl. tie-back.
- Tie-in, diver assisted and remote controlled.
- Proces facilities, offshore and shore based.
- Control systems, valves and actuators.
- Power supply and signal transmission.
- Operation of systems.
- Maintenace, methods, equipment, and personel.
- Risk analysis, failure mode and effect analysis.
- Preliminary economic analysis.

REFERENCES :
TECHNICAL MEETINGS :
- SEMINAR ON SUBSEA SYSTEMS.
  DONG HOERSHOLM. DECEMBER 1986.

```
**********************************************************************
* TITLE : DEVELOPMENT OF VARIABLE DRAUGHT SEMI-    *     PROJECT NO      *
*         SUBMERSIBLE CONCEPTS                     *                     *
*                                                  *                     *
*                                                  *   TH./03180/85/UK/.. *
*                                                  *                     *
**********************************************************************
* CONTRACTOR :                                     * PROGRAM :           *
*   WIMPEY OFFSHORE ENGINEERS AND CONSTRUCTORS LTD *   HYDROCARBONS      *
*   FLYOVER HOUSE, GREAT WEST   TEL 01 560 3100    *                     *
*   BRENTFORD                   TLX 933861         *                     *
*   UK - MIDDLESEX TW8 9AR                         * SECTOR :            *
*                                                  *   PRODUCTION SYSTEMS *
* PERSON TO CONTACT FOR FURTHER INFORMATION :      *                     *
*   DR J.R. WILLIAMS                               *                     *
*                                                  *                     *
*                                                  *                     *
**********************************************************************
```

VERSION : 07/01/87

AIM OF THE PROJECT :
****
To develop a standard design for a new type of floating platform suitable for
providing a number of functions (such as drilling, workover and production) in
a range of water depths including deep water. The platform is a variable
draught semi-submersible STAbilised PLAtform (STAPLA) and the results from the
marriage of two existing technologies, namely semi-submersibles and jack-up
platforms. By incorporating a variable draught facility, it is intended to
combine the restricted motion response characteristics normally associated with
purpose built deep draught vessels and the inshore maintenance/modification
capabilities of conventional semi-submersible designs. The benefits of STAPLA,
compared to similar alternative floating systems, are thus being investigated
and evaluated in terms of: improved motion behaviour; wider operating
capabilities and reduced downtime; and lower capital and operating costs.

PROJECT DESCRIPTION :
****
PHASE I - FEASIBILITY STUDIES AND DEFINITION OF DESIGN PREMISE
Phase I consits essentially of four parallel but inter-related tasks :
- Defining the design premise. This includes establishing design information on
field specifications, environmental criteria, topsides facilities, risers,
storage, hull equipment, structural configuration (deck, hull, legs and jacking
mechanism), design codes and marine operations. The design premise is to be
updated and refined during Phase I in the light of information from the model
testing and analysis described below as well as from in-depth studies on :
playload, layout and storage scenarios; mooring systems and alternatives; riser
systems; and statutory requirements.
- Carrying out hydrodynamic model testing to demonstrate the anticipated
improved motion response characteristics at an early stage in the project and
investigate the sensitivity to changes in pontoon configuration, payload and
draught.
- Setting-up and using stability and vessel response analysis software to carry
out a limited parametric study.
- Analysing critical stuctural components including the pontoon raising-
lowering mechanism, member stability of columns in their extended state,
bracing configurations and additional fatigue problems originating from the
introduction of the variable draught facility.

PHASE II - PARAMETRIC STUDIES AND COMPONENT REFINEMENT
- Phase IIA : this consits of analytical parametric studies to evaluate vessel
response under the envelope of environmental conditions and operating
conditions identified in Phase I and thus optimise the STAPLA configuration.
Phase IIA also includes stability cheks to ensure that intact/damage statuatory
requirements are met and an evaluation of vessel behaviour whilst pontoons are
raised/lowered.
- Phase IIB : this will be established from the conclusions of Phase I and may,
if necessary, consist of structural tests on components which are unique to the
STAPLA concept due to its variable draught.
PHASE III - CONCEPTUAL DESIGN AND MARKETING
Phase III entails development and finalisation of the design to the level
required for conceptual engineering purposes.
It includes the formulation of construction sequences and the production of
drawings of sufficient detail to obtain cost/timescale estimates for
fabrication.
Although much marketing effort will have already been carried out by this stage,
 the detailed planning of a marketing strategy is part of Phase III with back-
up provided from comparative technical and economical appraisals between STAPLA
and alternative floating platforms.

STATE OF ADVANCEMENT :
****
Ongoing. Main project activities commenced on 1 September 1986 and work is
being carried out on all of the four preliminary design/testing tasks in Pase I
as described above.

RESULTS :
****
From a review of existing floating production systems, initial topsides design
criteria have been established to enable model testing to proceed and to
provide a basis for discussions with operators, certifying authorities, etc.
The typical weight and space profile produced is for a facility capable of
producing 100,000 bpd of oil in a hostile environment. To be self sufficient
within supply boat cycles, the platform topsides includes drilling (based on 4,
900 metre deep wells), wellhead - moonpool, separation of crude oil, water
injection, gas compression for gas lift, generation, utilities, living quarters
for 120 people with helicopter access, flexible riser platform, diving back-up
and seawater/fire water lift pumps. The deck area is approximately 90 metres x
90 metres and the total topsides operating weight of the initial vessel STAPLA
vessel is approximately 20,000 tonnes (including deck steelwork).
Various column and pontoon configurations are now being investigated during the
hydrodynamic model testing. From the tests it is also intented to establish a
suitable draught for STAPLA under the harsh environmental conditions likely to
be experienced in some of the sitesd of potential application in European
waters. A reference water depth of approximately 500 metres has been adopted
with survival storm conditions of Hs (significant wave height) = 18 metres, Tz
(zero crossing period) = 15 secs and operating conditions of Hs = 7.5 metres,
Tz = 9 secs. Although the full range of tests have not yet been completed,
initial results appear to be promising and indicate heave motion responses
similar to those of a tension leg platform.

```
*******************************************************************************
* TITLE : SUBSEA WELLHEAD SEPARATION SYSTEM        *        PROJECT NO        *
*                                                  *                          *
*                                                  *    TH./03182/85/UK/..     *
*                                                  *                          *
*******************************************************************************
* CONTRACTOR :                                     * PROGRAM :                *
*   BRITISH OFFSHORE ENGINEERING TECHNOLOGY LTD     *   HYDROCARBONS           *
*   CHESTERGATE HOUSE            TEL 01 630 0792     *                          *
*   253 VAUXHALL BRIDGE ROAD     TLX 261 821        *                          *
*   UK - LONDON SW1V 1HD                            * SECTOR :                 *
*                                                  *   PRODUCTION SYSTEMS     *
* PERSON TO CONTACT FOR FURTHER INFORMATION :       *                          *
*   MR. W. G. EDWARDS                               *                          *
*                                                  *                          *
*                                                  *                          *
*******************************************************************************
```

VERSION : 27/11/86

AIM OF THE PROJECT :
****
To investigate the potential market for subsea separation of wellhead crude oil
and to design and engineer a subsea separator for offshore oil fields.
Innovative aspects of the 3-stage separator are:
- the seabed location
- the concentric layout whereby the 3rd stage vessel is inside the 2nd stage
  vessel
- the use of the 3rd stage as a storage vessel as an additional function
- the possible use of steel/concrete composite structure for the vessel walls.
The subsea separator would permit fields to be developed that would otherwise
be uneconomic.
The project will aim to provide a pilot scale unit, tested on land and suitable
for offshore trials.

PROJECT DESCRIPTION :
****
The project will comprise a feasibility study, conceptual design and the
design/build/test of a pilot scale trials unit.
It will be undertaken by British Offshore Engineering Technology Ltd at 253
Vauxhall Bridge Road, London SW1V 1HD, from October 1986, for a 31-month
duration at a total cost of UKL 2,118,000.
PHASE 1   STAGE 1   FEASIBILITY STUDY   3 MONTHS
Activity: to identify and evaluate applications which include the following
field cases:
a. small isolated field; gas flared; tanker export
b. larger isolated field; gas conserved; tanker export
c. step-out field; pipeline to existing platform
d. platform with limited topsides weight-space
e. early production system
The sensitivity of product, location and design parameters on the field
development cost will be examined.
PHASE 1   STAGE 2   CONCEPT DESIGN   6 MONTHS

Activity: conceptual design of an optimised reusable unit for North Sea
marginal and currently uneconomic fields. Main design areas to address expected
to be: process design (gas, hydrates, temp.); pay size/layout; control;
inspection and maintenance; composite structure; logistics and
installation/recovery; pumping and power system specifications.
PHASE 2   STAGE 1   PILOT UNIT DESIGN   6 MONTHS
Design of pilot scale trials unit, selection of fabricator.
PHASE 2   STAGE 2   FABRICATION   6 MONTHS
Fabrication by subcontractors for pressure vessel, structure and controls.
PHASE 2   STAGE 2   LAND TEST   6 MONTHS
Trial of the pilot unit on an onshore site where live crude is available,
seabed environment is simulated, continual data collection is possible and
oilfield operator is interested and cooperative.

STATE OF ADVANCEMENT :
****
Feasibility study phase is about to commence, with an initial cost comparison
made with turret moored tanker (FPF) alternatives, for 35 million barrel
recoverable reserve cases in North Sea conditions.

RESULTS :
****
Details not available at this time but initial cost comparisons for 35 million
barrel field in 150 water depth suggest the following initial capital costs:
Case a.   Tanker FPF with 6 wells and tanker export   UKL 140 MM
          Subsea separator instead of tanker   UKL 78 MM
Case b.   Tanker FPF as Case a.   UKL 140 MM
          Subsea separator and moored semisub with gas
          compression/reinject facility instead of tanker   UKL 98 MM.

```
****************************************************************************
* TITLE : FLOATING PRODUCTION SYSTEM FOR THE     *      PROJECT NO        *
*         EXPLOITATION OF DEEP WATER             *                        *
*         MEDITERRANEAN OIL FIELDS               *   TH./03183/85/IT/..   *
*                                                *                        *
****************************************************************************
* CONTRACTOR :                                   * PROGRAM :              *
*   AGIP SPA                                      *   HYDROCARBONS         *
*   AGIP-TEIN               TEL 02 5205969        *                        *
*   P.O. BOX 12069          TLX 310 246           *                        *
*   IT - 20120 MILANO                             * SECTOR :               *
*                                                 *   PRODUCTION SYSTEMS   *
* PERSON TO CONTACT FOR FURTHER INFORMATION :     *                        *
*   MR. P. TASSINI                                *                        *
*                                                 *                        *
*                                                 *                        *
****************************************************************************
```
VERSION : 27/11/86

AIM OF THE PROJECT :
****
The scope of the research is the development of a floating production system
suitable for the exploitation of hydrocarbon fields in very deep waters in the
Mediterranean Sea. The system to be developed in the research consists of a
Tension Leg type platform and the relevant production riser.
The following are believed to be the main innovative aspects of the project:
- firstly, this project must be considered as the basis of a possible
industrial application of the Tension Leg solution in very deep waters (830 m,
Aquila field).
- secondly, the installation procedures foreseen are innovative for what it
concerns assembling and deployment of tethers (it is this the only one detailed
study of a TLP with welded anchoring lines) and for all the technical and
operative aspects influenced by the high water depth.

PROJECT DESCRIPTION :
****
The present contract follows a previous one (TH.03121/82) related to stage 1 of
this research project. The present phase (Stage 2) aims at testing critical
components and procedures, as well as at re-examing and perfecting the system
design in the light of results obtained during Stage 1 and at preparing design
procedures and computer programs required for the engineering phase.
The following components and systems will be tried and tested:
- Automatic welding machine (including set up of welding and control process);
for development and qualification of an automatic welding unit suitable for
anchor leg assembling.
Tether deployment equipment (pipe handling, positioning, centring, welding,
control); for verification and setting up of the critical installation
procedures.
- Lower anchor connector (on scale model); to verify its correct working and in-
service behaviour.
- Stress joint; to test the material and the welding process.
- Corrosion and cathodic protection deep water experimental station; to study
the effects of deep marine environment on corrosion phenomena.
- Foundation piles; to verify the behaviour tension piles under cyclic loads.
- Completion system; to verify the capability and behaviour of component(s).

The critical analysis of the design will be directed mainly towards: hull configuration, tethers, installation procedures, risers and completion system. Results of basin tests, of particular studies and analysis presently in progress and precertification information will form the basis of the design review.
The design of certain components and aspects of particular importance will be greatly improved. Among these:
- most important structural modes, with special emphasis on fatigue analysis
- risers
- behaviour of tethers
- temporary mooring system
- installation operations
Special computer procedures will be prepared for Tension Leg type structures, mainly with regard to the overall calculation of the platform, the detailed structural analysis of hull, tethers and risers.
Design procedures indicating basic design criteria and methods, including procedures for weight control, will be elaborated as well as standards for certain structural components.

STATE OF ADVANCEMENT :
****
Ongoing project. Activities in progress are : critical revision and design optimization of the main platform components and procedures (Hull configuration, tethers, anchor connectors, tapered joints, construction and installation procedures); preparation of tests and engineering of prototypes to be tested (automatic welding system for anchor leg assembling, anchor connector)

RESULTS :
****
So far the following relevant results have been achieved:
- OPTIMIZATION OF HULL CONFIGURATION
The dynamic behaviour of a tension leg platform has been studied, in order to evaluate the influence of several hull configuration parameters (geometrical and not) on the dynamic load of the tethers. On this basis a design procedure to correctly establish the optimum geometrical parameters of theeters of the configuration has been developed. The governing criterion is minimizing the dynamic tension induced in the tethers by the waves and the algorythm is based on the long term distribution of the maximum tension in the tethers.
- DESIGN OF THE TETHERS LOWER TERMINATION
A re-design was carried-out according to latest available loads and was accomplished with a particular attention to critical points arisen from the fabrication study. For this work a computer programme, suitable to draw the "optimum" stress joint shape (minimum weight criterion) relevant to a defined static loading condition and according to some fixedstructural consints, was used.ed.
- EVALUATION OF THE ENERGY PRESENT IN THE HIGH FREQUENCY RANGE
The analysis has been performed looking at surge and heave motions and at the mooring forces; the results, obtained by means of the formula describing the maxima distribution, have been compared with the results of hundred simulations.

- DESIGN OF THE TETHER WELDING SYSTEM

A specification describing the required performances and the basic design of an automatic welding system, able to perform butt joints on tubular sections in vertical position by means of arc welding, has been produced.

In this specification are also stated the requirements for the engineering development and fabrication of a prototype of the system.

- PROCEDURES FOR TETHERS LAYING

A specification describing the tests to be performed during the simulated tether laying operations has been produced.

TETHERS LOWER CONNECTOR

A specification describing the tests to be performed on the prototype has been issued.

Due to the early stage of the project, at the moment, no other relevant results have been achieved.

```
******************************************************************************
* TITLE : SEA TESTS ON RISERS FOR FLOATING      *       PROJECT NO          *
*         PLATFORMS                              *                           *
*                                                *    TH./03184/85/IT/..     *
*                                                *                           *
******************************************************************************
* CONTRACTOR :                                   * PROGRAM :                 *
*   AGIP SPA                                     *   HYDROCARBONS            *
*   AGIP-TEIN            TEL 02 520 59 69         *                           *
*   P.O. BOX 12069       TLX 310 246             *                           *
*   IT - 20120 MILANO                            * SECTOR :                  *
*                                                *   PRODUCTION SYSTEMS      *
* PERSON TO CONTACT FOR FURTHER INFORMATION :    *                           *
*   MR. P. TASSINI                               *                           *
*                                                *                           *
*                                                *                           *
******************************************************************************
                                                    VERSION : 27/11/86
```

AIM OF THE PROJECT :
****
The scope of this research project is to organise and carry out sea tests on
instrumented risers, in order to acquire experimental data on the dynamic
behaviour of operating risers suitablefor verifying the capability of the
available theoretical calculation procedures of compliant stuctures.
oreover, aim of the project is to carry out a direct experimental verification
of the conditions which cause hydroelastic vibrations induced by vortex
shedding, both in the presence and absence of a second riser and the
consequences which such phenomena have on the stresses and the displacements of
the riser.

PROJECT DESCRIPTION :
****
This project constitutes phase one of a two phase project. The phase one is
dedicated to design of the experimental set up; the phase two, which will
follow this one, will be dedicated to the building of the test rig, to the sea
tests and to data analysis.
The basic objective is to install two instrumented risers on an already
existing structure (a fixed platform) in water depth of about 75 metres. Such a
structure will act as a support to the equipment required to carry out the
tests.
The test rig will consist of :
- support structure
- motion simulation trolley supporting a rotating table
- tensioner
The instrumentation and data acquisition system will consist of three parts :
- instruments to measure risers' dynamic behaviour
- instruments to measure environmental conditions
- equipment for data acquisition and storage
The measurement of the risers' dynamic behaviour will be achieved by means of a
sufficient number of instrumented pipe sections placed on each riser.
The signals from all the sensors will be collected by a data acquisition system
based on a computer.
The data acquired during the tests will be suitable processed and compared with
theoretical simulations.

On the basis of this comparison, verification of the theoretical calculation procedures will follow, in order to establish wether or not the available procedures are adequate and to produce modifications of the existing calculation procedures or to develop new ones, where appropriate.

STATE OF ADVANCEMENT :
****
Ongoing project. The research is in the preliminary design phase. Basic design of equipment, instrumentation, riser and data acquisition system is now nearly completed. Drawings relevant to the modified deck platform have been issued.

RESULTS :
****
The project will develop and validate the calculation procedures needed to design production risers for floating systems suitable for exploitation of deep water hydrocarbon fields located in the European contiental shelf.
Development will refer mainly to :
- Instrumentations systems
- Data acquisition and analysis procedures
- Calculation procedures
Due to the initial stage of the project, at the moment, no relevant results have been achieved.

REFERENCES :
****
AGIP MOVES ON DEEP WATER FACILITY DESIGNS.
OCEAN INDUSTRY, APRIL 1986 PP 156-158.

```
*********************************************************************************
* TITLE : TENSION LEG PLATFORM INVESTIGATION FOR    *      PROJECT NO         *
*         WATER DEPTH RANGING BETWEEN 200 AND        *                         *
*         1200 M.                                    *    TH./03185/85/HE/..   *
*                                                    *                         *
*********************************************************************************
* CONTRACTOR :                                       * PROGRAM :               *
*   ALFAPI                                           *   HYDROCARBONS          *
*   B. GEORGIOU B'41              TEL 01 7225872     *                         *
*   HE - 116 34 ATHENS           TLX 221809          *                         *
*                                                    * SECTOR :                *
*                                                    *   PRODUCTION SYSTEMS    *
* PERSON TO CONTACT FOR FURTHER INFORMATION :        *                         *
*   MESSRS ANGELOPOULOS/PAPANI                       *                         *
*                                                    *                         *
*                                                    *                         *
*********************************************************************************
```

VERSION : 22/05/87

AIM OF THE PROJECT :
****
To investigate in detail the application of a Tension Leg Platform (TLP) for
the Mediterranean Sea and, in particular, for the Greek area. Specifically, a
type of TLP pay load in the order of 20,000 t and for depths of 200 to 1,200 m
is being considered. The performance of this study is innovative and the
results will be very important for the application of such constructions in the
Greek area.

PROJECT DESCRIPTION :
****
The application of a TLP type platform, first at all, is associated with risks
due to its dynamic sinking and the environmental conditions in general. The
performance of this study includes the prediction of TLP behaviour in the
frequency and time domain with interaction between hydrodynamical and
structural design, including non-linearities in a wide range of the TLP
application. The estimation of the techno-economical feasibility of the
production, erection and multi-usage of a TLP in a peripheral European area is
also being considered. The project is divided in the following main stages:
1. Environmental and design specifications
2. Calculation of the behaviour of the TLP
3. Foundation concepts
4. Application in South Kavalla region
5. Economic analysis according to the Greek industry capabilities
6. Project management.

STATE OF ADVANCEMENT :
****
Ongoing. Work on phases I, II, III and VI has already begun and is still in
progress. Work phases IV and V will start according to the planned program.

RESULTS :
****
Final result of the project is the general techno-economical feasibility of the
production, erection and multi-usage of a high technology system as TLP for the
Mediterranean Sea and in particular for the Greek area. The project supports
the exploitation of large areas such as the Mediterranean thus contributing to
the independence of the EEC countries regarding the energy resources problem.

```
**********************************************************************************
* TITLE : DEVELOPMENT OF A SEMISUBMERGED,        *        PROJECT NO           *
*         TENSIONED CONCRETE CASING FOR THE      *                             *
*         OFFSHORE PRODUCTION OF LIQUID AND      *    TH./03186/85/DE/..        *
*         LIQUEFIED HYDROCARBONS                 *                             *
**********************************************************************************
* CONTRACTOR :                                   * PROGRAM :                   *
*   SALZGITTER AG                                *   HYDROCARBONS              *
*   ABTEILUNG FORSCHUNG UND EN TEL 030 88 42 97 15 *                           *
*   POSTFACH 15 06 27          TLX 185 655       *                             *
*   DE - 1000 BERLIN 15                          * SECTOR :                    *
*                                                *   PRODUCTION SYSTEMS        *
* PERSON TO CONTACT FOR FURTHER INFORMATION :    *                             *
*   DR-ING G. PIETSCH                            *                             *
*                                                *                             *
*                                                *                             *
**********************************************************************************
```

AIM OF THE PROJECT :
****
From 1978 to 1984, the project group made up of HDW, LGA, Zueblin and
Salzgitter AG developed two production systems for use in deep-water areas
which are particularly suited to produce natural gas from offshore sites which
have to be termed marginal even today. The support structures for the necessary
process facilities comprise either a tension leg platform (TLP) or a concrete
casting situated on the sea-bed.
The aim of the new project is to take the fundamental components of these two
research and development projects as a basis on which to draw up a new concept.
The idea behind this new project is to reduce investment costs to such a degree
that, even in the face of decreasing prices for energy., it will be possible to
produce oil and gas from marginal offshore fields economically.
One particular advantage of the project described here is that the system can
also be used in areas where ice can be expected on the surface of the water.

PROJECT DESCRIPTION :
****
The purpose of the technical development under study is to design a production
system for small deposits in deep-sea areas. In this connection, priority is
given to keeping the invstment and system costs as low as possible, so as to
make the "marginal" fields in the northern part of North Sea accessible to the
European consumer, also at the present level of energy prices.
The main items of the project:
- Development of a semisubmerged tensioned concrete casing, which serves
simultaneously as a buoyancy body, a production casing and a storage casing as
well as the development of the foundation components (gravity foundation, piled
foundation) suited to the particular buoyancy conditions and the various
conditions on and in the sea-bed.
- Construction of a tensioning system to anchor the concrete casing to the
foundation body with regard to particular buoyancy conditions.
- Adaptation of the process plant with regard to the determined casing to the
foundation body with regard to particular buoyancy conditions.
- Design of an integrated intermediate store for different products for
intermittent tanker transport.
- Development of a system for transferring the separated hydrocarbons from the
submerged casing to the tanker.
- Adaptation of the developed riser technique to the submerged production
casing.

-   Modification of the overall system for application in arctic offshore areas (behaviour in driftnice, pack ice, etc.).
Determination of the capital investment, estimation of the operating costs and final considerations of economic efficiency.
-   Performance of model tests.
STATE OF ADVANCEMENT :
****
Ongoing the project is in the design and beginning construction phase.

RESULTS :
****
The area chosen for installation is the sea off the Norwegian coast to the north of the 62nd degree of latitude. The facility will be designed merely to accommodate the equipment required to process the oil and gas. Drilling facilities will not be installed in the casing. It is assumed that the underwater field has already been developed and that the processing facility can be connected up to an existing piping system. The quality characteristics of the auxiliary and ancillary systems needed for the self-sufficient running of the plant have been specified.
Due to the depressed market situation processing facilities for methane liquefaction are not considered. It is, however, planned to re-inject this constituent with associated gas to stimulate oil production. A liquefaction plant could be considered in a separate concrete casing if demand should require it.
In order to achieve the largest degree of flexibility for installation of the processing equipment, it is considered to accommodate the facility in one integrated room.
To use the production system in deep water, an intermediate store on the sea bed would result in large cross-sectional dimensions in order to withstand the high hydrostatic pressure which in turn would call for high investment costs.
When placing the intermediate store in the tensioned concrete casing, special significance is placed on the ballast system, for it would have to compensate variations in buoyancy arising when filling and emptying the intermediate store.
 In such a case, the hydrodynamic characteristics of the overall structure should alter as little as possible.
In contrast to the static requirements which requires resistance to high pressure, the production procedure calls for easy-to-assemble elements, i.e. with flat surfaces. A compromise must be found between both requirements.
The following most important characteristics values were determined:
- effective cubic capacity
- buoyancy volume
- intermediate storage volume
- ballast space volume
- immersion depth in a floating mode.
Parallel to the casing design, fundamental studies have been conducted for installation of tensioning and coupling systems of the tension legs, in the concrete casing.
Flex joints (30.000 KN) connect the tension legs with the tensioning arrangement anchored in the concrete casing have been studied.
Investigations about the intermediate store to be integrated into the overall system consider the following parameters:
- production rate
- strategy for tanker operation taking into consideration adequate stand-by capacity
- buoyancy differences between tanks filled with water and those filled with products.

```
*****************************************************************************
* TITLE : DEVELOPMENT OF REUSABLE CONCRETE       *       PROJECT NO         *
*         PLATFORMS FOR MARGINAL FIELDS IN THE   *                          *
*         NORTH SEA                              *     TH./03187/85/DK/..    *
*                                                *                          *
*****************************************************************************
* CONTRACTOR :                                   * PROGRAM :                *
*   CHRISTIANI & NIELSEN A/S                     *   HYDROCARBONS           *
*   VESTER FARIMAGSGADE 41      TEL 45 114 12 33  *                          *
*   DK - 1501 COPENHAGEN        TLX 22336         *                          *
*                                                * SECTOR :                 *
*                                                *   PRODUCTION SYSTEMS     *
* PERSON TO CONTACT FOR FURTHER INFORMATION :    *                          *
*   MR. T. MORDHORST                             *                          *
*                                                *                          *
*                                                *                          *
*****************************************************************************
```

VERSION : 27/11/86

AIM OF THE PROJECT :
****
On the basis of the tender project for a direct founded concrete platform on
the marginal Rolf field in the Danish sector of the North Sea it is proposed to
develop methods and to carry out project details in such a way that the
platform may be reused on another marginal field after the first one has been
depleted.
The need for a reusable off-shore platform has been furthered by the
development in the North Sea for exploitation of earlier found hydrocarbon
reservoirs which at the time of exploration were not considered to be
commercially exploitable.

PROJECT DESCRIPTION :
****
The proposed concrete platform may be refloated and moved to another location
provided that the following activities can be carried out:
- Disconnecting pipe systems etc.
- Refloating of platform.
- Inspection/cleaning of skirt and skirt departments.
- Inspection/cleaning of pipe systems etc. for ballasting/injection.
- Transport to another location.
- Sinking of platform.
- Injection of skirt departments.
All the activities mentioned are actually based upon proven techniques but may
impede on another. However, it is assumed that working procedures may be
developed and that suitable materials may be selected, which allow for the
above order of operations. As an example it may be mentioned that an injection
system which has been used for cement mortar is not reusable right away.
Likewise, comprehensive studies are required to ensure that the material to be
used for filling the void between the platform bottom and the sea bottom can be
removed in an economical way. Cement mortar is presumably not suitable and
therefore it may be required to find another material.
In connection with the refloating of the platform, it is necessary to analyse
again and possibly change the proposed sinking procedure to make it fully
reversible with due regard to an assumed 2-10 years' interval between the two
operations.

Likewise, the original platform design will have to be analysed for possible
new loading cases.
The above studies and investigations shall result in the description of
procedures, technical calculations and drawings.

STATE OF ADVANCEMENT :
****
Ongoing

RESULTS :
****
The study started out with a brief review of the literature describing the
geology and basic soil parameters for the North Sea, which will form the basis
for the geotechnical examination of reusability of the concrete platform. As a
basis two gravity platforms were suggested, one for deep water and one for less
deep water, dimensions, weights and loading were calculated and the platform
force response was checked against the basic soil parameters. Apart from
certain restrictions due to poor soil conditions there seems to be no
substantial technical obstacle to the study.

```
*****************************************************************************
* TITLE : DEVELOPMENT AND APPLICATION OF        *         PROJECT NO       *
*         COMPOSITE STRUCTURAL SYSTEM FOR        *                         *
*         OFFSHORE PLATFORMS.                    *     TH./03189/85/UK/..   *
*                                                *                         *
*****************************************************************************
* CONTRACTOR :                                   * PROGRAM :               *
*   TAYLOR WOODROW CONSTRUCTION LTD              *    HYDROCARBONS          *
*   TAYWOOD HOUSE, 345 RUISLIP TEL 01 578 2366   *                         *
*   SOUTHALL                 TLX 24428           *                         *
*   UK - MIDDELSEX UB1 2QX                       * SECTOR :                *
*                                                *    PRODUCTION SYSTEMS    *
* PERSON TO CONTACT FOR FURTHER INFORMATION :    *                         *
*   MR. J.R. SMITH                               *                         *
*                                                *                         *
*                                                *                         *
*****************************************************************************
                                                    VERSION : 28/04/87
```

AIM OF THE PROJECT :
****
The principal objective of the investigation is to develop the application of
composite steel/concrete sandwich construction for offshore engineering in
order to exploit the potential benefits of faster construction, more efficient
use of material and greater resistance to local loads.

PROJECT DESCRIPTION :
****
The programme will develop two reference designs of platforom structure, one
for an Artic environment and the other for a more benign environment. A design
method for composite construction will be developed and the validity checked by
a programme of confirmatory structural testing.Certification of the design
method will be sought.

STATE OF ADVANCEMENT :
****
Ongoing

```
*******************************************************************************
* TITLE : EXAMINATION OF FLEXIBLE OFFSHORE PIPE    *       PROJECT NO        *
*         SYSTEMS                                  *                         *
*                                                  *    TH./03190/85/DE/..   *
*                                                  *                         *
*******************************************************************************
* CONTRACTOR :                                     * PROGRAM :               *
*   PAG-O-FLEX                                     *    HYDROCARBONS         *
*   IN DEN DIKEN 16           TEL 0211 6505 378    *                         *
*   DE - 4000 DUESSELDORF 30  TLX 8588857          *                         *
*                                                  * SECTOR :                *
*                                                  *    PRODUCTION SYSTEMS   *
* PERSON TO CONTACT FOR FURTHER INFORMATION :      *                         *
*   DR. M. PEUKER                                  *                         *
*                                                  *                         *
*                                                  *                         *
*******************************************************************************
                                                        VERSION : 13/11/86
```

AIM OF THE PROJECT :
****

Within this project the applicability of flexible pipes for riser systems in
the offshore industry will be proven. Therefore the behaviour of the pipes is
investigated under various static and dynamic load conditions. Furthermore S/N-
curves for flexible pipes which are not yet available will be determined and by
further developed computer programs it will be possible to calculate lifetime
of risers.

PROJECT DESCRIPTION :
****

This project is a joint industry programme with the LMT/Aachen and six oil
companies. For a standard flexible pipe for subsea application (6in/6000 psi)
the static data (burst pressure, stiffness data, external pressure resistance)
as well as the dynamic data (tensible and bending fatigue data with different
load amplitudes) is calculated in order to evaluate fatigue failure criteria.
The fatigue test data is used to develop S/N curves for estimatiom of lifetime
for flexible riser systems by computer analysis. For the evaluation of fatigue
failure criteria the main components of a flexible pipe, i.e. the rubber
material and the reinforcement steel cord wire, are investigated in dynamic
tests. Furthermore the adaptability of the rubber material as liner material is
checked in chemical tests with various fluid media.
Upon completion of the dynamic tests of the standard flexible pipe, alternative
designs (varying inner diameter, pressure rate and liner type) are also
investigated in static and dynamic tests.

STATE OF ADVANCEMENT :
****

Ongoing project is still within time schedule and will probably end in March
1987. The static test of the flexible, as well as a part of the dynamic tests,
and the chemical tests are completed. Compound tests and development of
computer sofware are still running, while tests on pipes with alternative
designs have just begun.

RESULTS :
****

The theoretical static qualities of the standard flexible pipe have been
confirmed by test results. The expected number of cycles in the dynamic tests
has been exceeded and therefore some tests are still running with no failure of
the pipes having been reached thus far. The compound tests of the elastomer
material is completed with satisfying results, while the compound tests of the
steel cord is still running. Also the expected behaviour of the different
elastomer materials in chemical tests has been achieved.

```
*********************************************************************************
* TITLE : ROTATING TURRET ASSEMBLY FOR OIL       *       PROJECT NO           *
*         RECEIVING, HANDLING AND SHIP MOORING    *                           *
*                                                 *                           *
*                                                 *   TH./03191/85/UK/..       *
*                                                 *                           *
*********************************************************************************
* CONTRACTOR :                                    * PROGRAM :                 *
*   GEC MECHANICAL HANDLING LTD                    *   HYDROCARBONS            *
*   BIRCH WALK                    TEL 03224 36933  *                           *
*   ERITH                         TLX 263237       *                           *
*   UK - KENT DA8 1QH                              * SECTOR :                  *
*                                                 *   PRODUCTION SYSTEMS      *
* PERSON TO CONTACT FOR FURTHER INFORMATION :      *                           *
*   MR. N.B. CRABTREE                              *                           *
*                                                 *                           *
*                                                 *                           *
*********************************************************************************
```

VERSION : 13/11/86

AIM OF THE PROJECT :
****
To provide an engineered design of a turret system which can be incorporated
into floating production system for the exploitation of marginal oil fields.
The turret will allow production to continue despite changing wave and wind
direction by allowing the ship to rotate around its centre line, the turret is
moored to the seabed.
The turret will contain much of the necessary production equipment and will
incorporate a multi-path system for the production, injection, inspection and
test lines.

PROJECT DESCRIPTION :
****
A rotating turret assembly for oil receiving, handling and ship mooring and
designed for a floating production vessel used for the exploitation of offshore
oil reserves incorporates a number of innovative features.
The turret concept is covered under British Patent Specification No. 1447413 in
the name of GEC Mechanical Handling Ltd. and the design has already been
progressed through the preliminary stages by this company in collaboration with
two other major U.K. organisations, Foster Wheeler Petroleum Development Ltd.
and YARD Ltd.
Based on this conceptual work a turret design will now be established to meet
the current specification of requirements and design criteria.
The areas of technological innovation to be developed during the project
include the following:
a)   Turret structure.
b)   Support bearings and drive system.
c)   Mooring line handling equipment.
d)   Riser handling equipment.
e)   Multipath fluid transfer system.
In each of the above areas of the overall system the work will be carried
forward to the preparation of detailed assembly drawings.
Interface systems will also be worked on as required to develop a complete
turret assembly.

STATE OF ADVANCEMENT :
****
Ongoing: Design phase.

RESULTS :
****
Preliminary arrangement drawings have been prepared covering several of the
above listed areas. Successful negotiations have been carried out with a number
of specialised suppliers - e.g. high pressure hoses for the multipath fluid
transfer system.

```
*****************************************************************************
* TITLE : DEVELOPMENT PROJECT CONCERNING        *        PROJECT NO        *
*         EXPLORATION AND PRODUCTION OF MARGINAL *                          *
*         HYDROCARBON DEPOSITS                   *    TH./03192/85/DK/..    *
*                                                *                          *
*****************************************************************************
* CONTRACTOR :                                   * PROGRAM :                *
*   DANSK OLIE & GASPRODUKTION A/S               *   HYDROCARBONS           *
*   SLOTSMARKEN 16           TEL 02 57 20 44     *                          *
*   DK - 2970 HOERSHOLM      TLX 21259           *                          *
*                                                * SECTOR :                 *
*                                                *   PRODUCTION SYSTEMS     *
* PERSON TO CONTACT FOR FURTHER INFORMATION :    *                          *
*   MR. LARS H. GAD                              *                          *
*                                                *                          *
*                                                *                          *
*****************************************************************************
```

VERSION : 27/11/86

AIM OF THE PROJECT :
****
To develop a new mobile low cost process for disposal of gases with a high
inert content and hydrogen sulphide/ mercaptan impurities during exploration
and production of marginal hydrocarbon deposits. This new technology should
encourage exploration of certain geological layers where above gases are
expected.

PROJECT DESCRIPTION :
****
PHASE I, FEASIBILITY STUDY (Duration : 12 Months)
Comparison of possible processes for disposal of low grade sulphur contamined
gases. Selection of the most economical and technical feasible solution. Basic
design of the selected process. Economical evaluation of commercial feasibility.
 Budget and detailed planning of project for Phase II and III.
PHASE II, DESIGN AND CONSTRUCTION OF A MOBILE DEMONSTRATION UNIT (Duration : 12
months)
OBJECTIVE
Design and construction of a mobile demonstration unit, transportation to the
destination and site construction/installation.
MAIN ACTIVITIES
II.a Engineering of testing unit
II.b Construction of testing unit
II.c Site construction and installation.
PHASE III, TESTING AND DEMONSTRATION
Duration : 16 months.
Costs : 6 mill.DKR (current cost)
OBJECTIVE
The unit will be placed in connection with an existing well. The well will be
opened and made ready for production. The necessary production equipment will
be installed in the well and, if necessary, supplementary perforation and acid
treatment will be performed in order to increase the capacity of the well.
Drilling and well preparation work is excluded from the budget estimate, which
includes production tubing, production wellhead equipment, and connections to
the unit.

MAIN ACIVITIES
III.a Well opening, preparation and tie-in unit
III.b Testing of unit
III.c Preparation of final report.

STATE OF ADVANCEMENT :
****
The ongoing Phase I, Feasibility Study is completed with a feasible process selection.
Decision regarding project continuation with Phase II & III will be taken ultimo 1986.

RESULTS :
****
The most feasible process for disposal of high inert contamined gases from hydrocarbon production seems to be a combined catalytical/thermal incineration process.

```
*****************************************************************************
* TITLE : FLOATING PRODUCTION SYSTEM DEVELOPMENT,    *      PROJECT NO      *
*         DYNAMICALLY POSITIONED TANKER              *                      *
*                                                    *   TH./03194/86/UK/.. *
*                                                    *                      *
*****************************************************************************
* CONTRACTOR :                                       * PROGRAM :            *
*   BRITISH PETROLEUM SHIPPING LTD                   *   HYDROCARBONS       *
*   BRITANNIC HOUSE              TEL 01 920 8000      *                      *
*   MOOR LANE                    TLX 888811           *                      *
*   UK - LONDON EC2Y 9BR                             * SECTOR :             *
*                                                    *   PRODUCTION SYSTEMS *
* PERSON TO CONTACT FOR FURTHER INFORMATION :        *                      *
*   MR. I.M. BARRETT                                 *                      *
*                                                    *                      *
*                                                    *                      *
*****************************************************************************
```

VERSION : 16/03/87

AIM OF THE PROJECT :
****
A new system based on a dynamically positioned (DP) tanker vessel, progressing
from previous work on the SWOPS concept is to be developed to be used for
extended well testing and for production and delivery of oil from marginal
fields.
The primary aims are to increase the efficiency and to reduce the capital and
operating costs of oil extraction in the North Sea, and to extend the concept
for worldwide application.

PROJECT DESCRIPTION :
****
The system involves a converted tanker connected via flexible risers to one or
more subsea wellheads, or to a riser base and flowline manifold system,
handling systems for deployment of risers, a riser swivel unit, process
equipment and a novel DPP thruster system concept for maintening the vessel on
station.
The programme of work for the project includes development work on the new DP
thruster system with model simulation, tank testing and control system design;
development of the technology of multiple, flexible riser systems amd methods
of riser suspension; and design studies of devices and systems to accommodate
rotation of the riser and swivel-less capabilities particular to this concept.
It will be necessary to verify the operational behaviour of the process units
such as separators, water separation, injection and lift systems, as subjected
to the motions of the vessel. Development work will also be required to ensure
cost reduction through reconsideration of space and weight of process and
related plant and economic utilisation of other types of production system. A
thorough examination of the design will be made to reduce the costs of all the
major system components.
The feasibility of the DP Tanker Scheme has been studied by BP using a purpose
designed computer simulation. Using this tool, a novel arrangement of thruster
units was conceived which promised lower power requirements than the
conventional designs. Model ship tank testing and further simulation work is
necessary to confirm and extend the concept.

Design work during 1985 developed the conventional DP tanker concept for extended well testing (EWT). More recently, it became apparent that early production could be achieved and hence extended production, using the DP Tanker.
  The advantages over single point mooring systems are essentially due to the mobility of the vessel from one oilfield to another, the system's independence in environmental survival conditions and its particular suitability for deeep water oilfields.
The project has been divided into two phases; basically 1986 work including model testing and DP simulations, process and riser handling design studies, followed by 1987 work on riser analyses, offloading subsea and DP position reference systems, preliminary engineering work on process related equipemnt, safety, costs and marketing.

STATE OF ADVANCEMENT :
****
Ongoing. Phase 1 development work has been completed according to the schedule.

RESULTS :
****
The novel dynamically positioned tanker concept has been developed under Phase 1 by successful model testing following computer simulations and calibration work by BP Shipping Ltd.
Preliminary engineering design of elements of the process and well injection systems have also been conducted by subcontractors. A unit for flexible riser handling at the ship's bow has been developed and designed together with a patent for a novel device as an alternative to fluid/gas swivels. A design of fiscal meter has also been studied under this Phase 1 work.
The major conclusion resulting from the work to date was that the freely weathervaning DP system allows the vessel to obtain its most favourable heading and is feasible from the point of view of dynamic positioning in severe weather conditions.

```
********************************************************************************
* TITLE : ON-FIELD MEASURES OF RESIDUAL OIL         *      PROJECT NO        *
*         SATURATIONS                               *                        *
*                                                   *   TH./03195/86/FR/..   *
*                                                   *                        *
********************************************************************************
* CONTRACTOR :                                      * PROGRAM :              *
*   AGELFI C/O GERTH                                *   HYDROCARBONS         *
*   4, AVENUE DE BOIS PREAU      TEL 1 47.52.61.39  *                        *
*   FR - 92502 RUEIL-MALMAISON TLX 203 050          *                        *
*                                                   * SECTOR :               *
*                                                   *   PRODUCTION SYSTEMS   *
* PERSON TO CONTACT FOR FURTHER INFORMATION :       *                        *
*   MR. ROCHON                   TEL 59.83.40.00    *                        *
*                                TLX 560 804        *                        *
*                                                   *                        *
********************************************************************************
                                                         VERSION : 01/01/87
```

AIM OF THE PROJECT :
****
The project consists of developing methods to determine residual oil saturation
in water-flooded reservoirs. Several methods will be tested and compared in
different application cases: logs or tracer experiments.

PROJECT DESCRIPTION :
****
The project is to be developed in two main phases:
PHASE 1 - preliminary studies, laboratory activities and development of
computer programs.
The research will start with an extensive survey of technical literature on the
different methods employed for evaluating the oil saturation in situ. At the
same time an analysis of the reservoirs operated by SNEA(P) and AGIP will be
carried out to find out the wells allowing experiments under the best operating
conditions. The aim of the lab activity will be the development of an
experimental methodology for selecting and characterizing the most suitable
tracer for each field where the SOR measurements are to be carried out. The
modelling activity is intented to develop computer program and to adapt
existing software packages for the interpretation of logs or tracer experiments.
PHASE 2 - preparation, execution and interpretation of the field test.
In the selected well sites, facilities will be installed.
Logs or log-inject-logs will be the first technique to be experimented.
By using the surface facilities installed at the well location, the well
tracing tests will be completed.
The software packages developed in the first phase will be used for the
quantitative log analysis, the modelling and the interpretation of the tracing
tests.

STATE OF ADVANCEMENT :
****
The first phase of the project is underway. The second phase will be developed
during 1987.

```
********************************************************************************
* TITLE : CLUSTOIL (PHASE 1)                            *      PROJECT NO      *
*                                                       *                      *
*                                                       *    TH./03200/86/FR/..*
*                                                       *                      *
********************************************************************************
* CONTRACTOR :                                          * PROGRAM :            *
*   GERTH                                               *   HYDROCARBONS       *
*   4, AVENUE DE BOIS PREAU     TEL 1 47.52.61.39       *                      *
*   FR - 92502 RUEIL-MALMAISON TLX 203 050              *                      *
*                                                       * SECTOR :             *
*                                                       *   PRODUCTION SYSTEMS *
* PERSON TO CONTACT FOR FURTHER INFORMATION :           *                      *
*   MR. J. PHILIPPOT            TEL 59.83.40.00         *                      *
*                              TLX 560 804              *                      *
*                                                       *                      *
********************************************************************************
```

VERSION : 01/02/87

AIM OF THE PROJECT :
****
The project is aimed at studying subsea stations for oil production. The study
is based on the following fundamental assumptions : modular design, diverless,
installation and maintenance, handling operations from light surface supports.
In addition, the concepts will use the results of previous studies and should
lead to a significant cost reduction, both in investments and in development
and maintenance costs.
PROJECT DESCRIPTION :
****
The whole study includes three phases :
- A feasibility phase
- A test phase for components and subassemblies
- An overall test phase on the offshore pilot.
The present project only applies to the first phase and concerns two diffrent
subsea stations.
- The first one, called CLUSTOIL, will be designed for a North Sea application
with the following fundamental assumptions : cluster of twelve, total
production : 30000 m3/d, water depth : 350 meters, distance from treatment site
: about 10 km, production life time : 15-20 years.
- The second one, called SAPHIR, will be designed for a Gulf-of-Guinea
application with the main following requirements : cluster of six boosted wells,
 total production : 2000 m3/d, water depth : 160 meters, distance to onshore
treatment site : 8 km, soft foundation, production life time : 12 years.
For both stations, the feasibility phase will include two stages :
1 Definition of subsea station design
The design will be specified according to exploitation conditions, production
techniques, reliability of components and maintenance means.
2 Detailed study of subsea station
Once the layout is specified, the different modules and their links and
connections will be the subject of detailed studies.
Modules and related connections studies will be carried out while keeping in
mind the following objectives :
- acquisition of maximum reliability for the overall system
- optimization of dimensions and weight of different modules to facilitate
handling operations.
The main subassemblies are : template, production wellhead, manifold,
telecontrol system, production and electro-hydraulic connections.

STATE OF ADVANCEMENT :
****

Ongoing. During the last quarters of 1986, the studies focused on the Subsea station designed for a Gulf-of-Guinea application : SAPHIR. The general station design has been specified and the detail engineering of some modules has been carried out.
The design of the station intended for a North Sea application (CLUSTOIL) is under way.

RESULTS :
****

SAPHIR

The station is a six-well cluster. X-trees are installed around an octagonal guide base supported by a central conductor pipe. Each X-tree is divided into two parts : the safety block tree and the production block tree. The sensitive equipment such as pressure transducers, remote-controlled choke and the most prompted valves are housed in the production block tree. In case of failure, this block can be entirely removed and replaced by a new one using a dedicated vessel. During this operation, the safety block tree insures well safety.
A common inert manifold gathers the fluids from the six wells and allows automatic connections between the X-trees and flowlines.
The subsea station is remotely controlled from the surface control room. Coded messages are sent to the central control module through an underwater cable. These messages activate the relevant pod installed on the X-tree and activate the hydraulic pressure system driven from the surface through a hydraulic bundle.
Three lines are tied to the cluster, the 10" production line, the 6" test line and the 4" service line.
The installation of the whole station is carried out from the drilling rig through its moon pool.
The detailed studies of modules are in progress. The file concerning the manifold will be ready early in 1987.
The production wellhead, which is the main module of the station will be studied in 1987.

CLUSTOIL :

The station is a twelve-well cluster. the solutions used for the SAPHIR design are fitted and re-used for the CLUSTOIL design.

```
************************************************************************
* TITLE : DEEPWATER SUBSEA PRODUCTION SYSTEM AND    *     PROJECT NO      *
*         MAINTENANCE DEVICE - PHASE 2 - STAGE A    *                     *
*                                                   *    TH./03201/86/IT/..  *
*                                                   *                     *
************************************************************************
* CONTRACTOR :                                      * PROGRAM :           *
*   AGIP SPA                                        *   HYDROCARBONS      *
*   C.P. 12069              TEL 02 5201             *                     *
*   IT - 20120 MILANO       TLX 310 246             *                     *
*                                                   * SECTOR :            *
*                                                   *   PRODUCTION SYSTEMS *
* PERSON TO CONTACT FOR FURTHER INFORMATION :       *                     *
*   MR. P. TASSINI - TEIN                           *                     *
*                                                   *                     *
*                                                   *                     *
************************************************************************
                                                    VERSION : 09/03/87
```

AIM OF THE PROJECT :
****
Scope of this project is the development of a new generation of subsea systems
specifically conceived for hydrocarbon production in deep (200 - 600 m w.d.)
and very deep beyond 600 m) waters including a dedicated maintenance device.

PROJECT DESCRIPTION :
****
During Phase 1, the engineering of the production system and the maintenance
device has been performed.
During Phase 2 the prototype of the production system and the maintenance
vehicle will be manufactured.
Phase 2 is split in two stages. During Stage A qualified manufacrurers will be
selected and the construction engineering will be performed. During Stage B the
subsystems will be manufactured. This contract is relevant to Phase 2 - Stage A.
 Integration of the systems, dry tests, installation on a live well and long
term tests are the scope of Phase 3.

STATE OF ADVANCEMENT :
****
Phase 2 - Stage A will start in May 1987.

RESULTS :
****
A prototype will be installed and tested on a live well. This constitutes Phase
3 of the project.

```
********************************************************************************
* TITLE : DEVELOPMENT AND MATHEMATICAL ANALYSIS    *       PROJECT NO         *
*         OF HIGH PERFORMANCE RESERVOIR            *                          *
*         SIMULATORS BASED ON ADVANCED MODELS      *    TH./03202/86/ES/..    *
*                                                  *                          *
********************************************************************************
* CONTRACTOR :                                     * PROGRAM :                *
*   HISPANICA DE PETROLEOS SA                      *   HYDROCARBONS           *
*   PEZ VOLADOR, 2            TEL 1 274.72.00       *                          *
*   ES - 28003 MADRID         TLX 49544            *                          *
*                                                  * SECTOR :                 *
*                                                  *   PRODUCTION SYSTEMS     *
* PERSON TO CONTACT FOR FURTHER INFORMATION :      *                          *
*   MR. L. PEREZ MANZANERA                         *                          *
*                                                  *                          *
*                                                  *                          *
********************************************************************************
```

VERSION : 18/03/87

AIM OF THE PROJECT :
****
The main objective of this project is to improve the present technology in oil
reservoirs numerical simulation, measured in terms of computer efficiency,
investigating the behaviour of various numerical models for multicomponent flow
equations in porous media.

PROJECT DESCRIPTION :
****
The project will develop advanced numerical simulation prototypes based on a
compositional model of the state equation (Peng-Robinson).
These prototypes will include the numerical models developed as follows :
- Investigation of time approximation methods, improving the adaptative
implicit type.
- Investigation of space approximation methods. Three methods will be
investigated :
  . Standard finite element method.
  . Mixed finite element method with method of characteristics.
  . Particle method.
Three main phases can be differentiated during the project development :
Phase 1. Numerical models development.
1.1 Analysis and development of numerical models which will be further
implemented.
1.2 Models behaviour analysis using mathematical tools.
Phase 2. Software development.
2.1 Algorithm design.
2.2 Software design.
2.3 Software implementation.
2.4 Software tests and validation.
Phase 3. Models tests.
3.1 Time approximation.
3.2 Space approximation.

STATE OF ADVANCEMENT :
****
The project started on March 1st, 1987.

```
*****************************************************************************
* TITLE : INVESTIGATION OF THE DYNAMIC        *        PROJECT NO          *
*         PERFORMANCE OF FLEXIBLE RISERS      *                            *
*                                             *     TH./03213/86/UK/..     *
*                                             *                            *
*****************************************************************************
* CONTRACTOR :                                * PROGRAM :                  *
*   BHRA THE FLUID ENGINEERING CENTRE         *   HYDROCARBONS             *
*   CRANFIELD                TEL 0234 705422  *                            *
*   UK - BEDFORD MK43 OAJ    TLX 825 059      *                            *
*                                             * SECTOR :                   *
*                                             *   PRODUCTION SYSTEMS       *
* PERSON TO CONTACT FOR FURTHER INFORMATION : *                            *
*   DR. R. KING/MR. A. HUMPHRI                *                            *
*                                             *                            *
*                                             *                            *
*****************************************************************************
                                                   VERSION : 06/04/87
```

AIM OF THE PROJECT :
****
The aim of the project is to undertake a closely controlled parametric
experimental study of the dynamic response of flexible pipes due to current and
slug flow. The information obtained will help form the basis of an improved
understanding off the behaviour of this type of pipe technology.

PROJECT DESCRIPTION :
****
The project is in two stages each taking one year to complete.
Stage 1 will investigate the vortex induced response of flexible pipes
including influence of bending stiffness and mass. Both modelled and
commercially available pipe will be tested for a range of configurations and
conditions. The influence of imposed top motions will also be investigated.
Stage 2 will investigate the effect of internal slug flow up both the modelled
and commercially available pipe. Changes in slug flow length and frequency will
be made to examine if flexible risers will/be susceptible to transient or
resonant motions.

STATE OF ADVANCEMENT :
****
Ongoing. Preliminary work has commenced on the project. The results will be
available in forms of technical reports detailing the design, test equipment,
test programme and findings. The data will be reduced to give frequency,
amplitude and nature of the response, as well as loads on the pipes.

```
*********************************************************************************
* TITLE : GA-SP PROJECT                           *        PROJECT NO          *
*                                                 *                            *
*                                                 *     TH./03217/86/UK/..     *
*                                                 *                            *
*********************************************************************************
* CONTRACTOR :                                    * PROGRAM :                  *
*   GOODFELLOW ASSOCIATES LTD                     *   HYDROCARBONS             *
*   71 ECCLESTON SQUARE         TEL 01 821 1377   *                            *
*   UK - LONDON SW1V 1PJ        TLX 739439        *                            *
*                                                 * SECTOR :                   *
*                                                 *   PRODUCTION SYSTEMS       *
* PERSON TO CONTACT FOR FURTHER INFORMATION :     *                            *
*   MR. J.L. CHASSEROT                            *                            *
*                                                 *                            *
*                                                 *                            *
*********************************************************************************
```

VERSION : 01/01/87

AIM OF THE PROJECT :
****
The project will develop equipment forming integrated subsea systems that can
be applied to a range of production scenarios on existing and future
developments of small marginal fields and isolated accumulations.
Low capital cost, early production, low maintenance/operating costs,
reliability and safety will be principal objectives of the systems.
Innovation will be in subsea control and monitoring systems, processing
equipment, remote maintenance systems and module interface connections.

PROJECT DESCRIPTION :
****
In order to validate the performance of the crucial components of a subsea
production system, the project will entail the design, construction and testing
of a prototype system. The design and testing will be complemented by computer
analyses, particularly in the area of reliability engineering. The principal
modules of the system are considered to be :
- production and riser manifolds
- vertical and horizontal separators
- remote pigging module
- pumping/filtering module
- maintenance module
Engineering the interfaces between modules will comprise a major part of the
programme.
The GA-SP project is divided into three phases as follows, of which only the
Phase 1 is supported by the Community within the project TH.03.217/86.
Phase I : FRONT END ENGINEERING & PRELIMINARY DESIGN
This phase will define the sytem and identify the components to be integrated
within the prototype riser base/test rig structure. Unreliable equipment will
be identified using failure records and Failure Mode and Effect Analysis (FMEA).

Preliminary concepts, Design Schemes and confirmation of system components
prior to Scheme Enhancement will also be developed. Finally, preliminary detail
design and model making will be undertaken. Phase I will be concluded with
reporting, confirmation meetings and discussions, prior to moving into Phase II.
Phase II : DETAIL DESIGN, ENGINEERING & PROTOTYPE MANUFACTURE
Phase II will commence with Detailed Engineering including detailed
manufacturing drawings and general arrangements including model making and
testing. Manufacture, assembly and preliminary shop floor tests will complete

Phase II work.

Phase III : PROTOTYPE & SYSTEM TESTING

The system components will be initially tested prior to assembly onto the test rig. The complete system will then be assembled to ensure correct interfacing of components in a dry dock facility. The system will then be connected to a well simulation and processing system on the dock side. Testing will then commence first in the dry and then in the flooded dock. During testing, procedures for installation and retrieval of modules, along with maintenance operations, will be prime considerations in the validation of the system.

STATE OF ADVANCEMENT :

****

Ongoing. The project is in the early stages of front end engineering, with activity on initial concept design and preliminary reliability studies.

```
****************************************************************************
* TITLE : IMPROVEMENT AND TEST OF A COMPACT        *       PROJECT NO      *
*          SEPARATOR ON AN OFFSHORE PLATFORM       *                       *
*                                                  *    TH./03219/86/FR/.. *
*                                                  *                       *
****************************************************************************
* CONTRACTOR :                                     * PROGRAM :             *
*    BERTIN & CIE                                  *    HYDROCARBONS       *
*    B.P. 3                    TEL 34.81.85.00      *                       *
*    FR - 78373 PLAISIR CEDEX   TLX 696 231         *                       *
*                                                  * SECTOR :              *
*                                                  *    PRODUCTION SYSTEMS *
* PERSON TO CONTACT FOR FURTHER INFORMATION :      *                       *
*    MESSRS. DEYSSON/REYBILLET                     *                       *
*                                                  *                       *
*                                                  *                       *
****************************************************************************
```
VERSION : 02/04/87

AIM OF THE PROJECT :
****
Improvement and test of a new compact vertical three phased separator for
offshore oil production reducing weight and saving floor area on top-sides.

PROJECT DESCRIPTION :
****
- Inspection of the state of the 15/25 000 bbl/d compact separator on the OBAGI
site (ELF-Nigeria)
- Definition and implementation of several modifications in order to reduce
liquid carry over in the gas and improve oil/water separation.
- Test of the modified separator on the OBAGI site
- Transport and installation of the separator on an offshore production
platform
- Long duration testing of the separator.

STATE OF ADVANCEMENT :
****
Starting date of the program : April 1st 1987.

```
*************************************************************************
* TITLE : FLEXTECH : PROGRAM FOR ANALYSIS AND     *     PROJECT NO      *
*         DESIGN OF FLEXIBLE PIPES AND RISERS.     *                    *
*                                                  *   TH./03224/86/IR/.. *
*                                                  *                    *
*************************************************************************
* CONTRACTOR :                                     * PROGRAM :          *
*   MCS INTERNATIONAL                              *   HYDROCARBONS      *
*   SCIENCE PARK            TEL (091) 24373        *                    *
*   UPPER NEWCASTLE         TLX 50094              *                    *
*   IR - GALWAY                                    * SECTOR :           *
*                                                  *   PRODUCTION SYSTEMS *
* PERSON TO CONTACT FOR FURTHER INFORMATION :      *                    *
*   MR. J.F. MCNAMARA                              *                    *
*                                                  *                    *
*                                                  *                    *
*************************************************************************
                                                    VERSION : 14/05/87
```

AIM OF THE PROJECT :
****
To develop a sofware package which will model all aspects of the complex
behaviour of flexible pipes and risers as applied to the offshore oil industry.

PROJECT DESCRIPTION :
****
Features of the FLEXTECH system will include finite element modelling,
sophisticated numerical solution schemes, capability to examine different pipe
construction types, monitoring of failure modes, and availability of efficient
pre and postprocessors to facilitate data handling.

STATE OF ADVANCEMENT :
****
Following a comprehensive literature survey and a detailed examination of
present state of the art techniques, work on the finite element formulation
commenced. All terms essential to accurate modelling of flexible pipe elements
are included. A highly efficient automatic time step system and an enhanced
solution scheme are under development. A three dimensional frequency domain
theory is also under development, with the objective of providing a complete
time/frequency domain facility.

RESULTS :
****
A prototype three-dimensional time domain program is now operational, and
validation tests demonstrate a high level of accuracy and efficiency.

SECONDARY AND ENHANCED RECOVERY

```
****************************************************************************
* TITLE : METHOD OF OIL RECOVERY BY CO2 INJECTION   *      PROJECT NO      *
*         IN THE COULOMMES-VAUCOURTOIS FIELD         *                      *
*                                                    *                      *
*                                                    *    TH./05023/81/FR/..*
*                                                    *                      *
****************************************************************************
* CONTRACTOR :                                       * PROGRAM :            *
*   PETROREP                                         *   HYDROCARBONS       *
*   42, AV.POINCARE         TEL 1 45 05 14 00        *                      *
*   F - 75116 PARIS         TLX 611036               *                      *
*                                                    * SECTOR :             *
*                                                    *   SECONDARY AND      *
* PERSON TO CONTACT FOR FURTHER INFORMATION :        *   ENHANCED RECOVERY  *
*   MR.E. COUVE DE MURVILLE                          *                      *
*                                                    *                      *
*                                                    *                      *
****************************************************************************
```

                                                    VERSION : 31/12/86

AIM OF THE PROJECT :
****
(1) check that the upper reservoir, of very low permeability, can be produced
by carbon dioxide injection and (2) find out if the production of the nearly
depleted lower reservoir can be extended economically by injecting carbon
dioxide and nitrogen.

PROJECT DESCRIPTION :
****
The project is divided into three phases:
Phase i:

Theoretical studies including a geological updating of the structure with
emphasis on fracture system, a laboratory study on miscibility and fluids
behaviour, a reservoir analysis based on production data and simulation.
Phase ii:

Injectivity and production test in the upper reservoir of the same well.
Phase iii :

Test the displacement efficiency and profitability with a four-spot pattern in
the lower reservoir.

STATE OF ADVANCEMENT :
****
The project was terminated in December 1984. Nevertheless various measurements
and analysis were performed during 1985 and are included in the final appraisal
of the project.

RESULTS :
****
Phase I : Showed that, at bottom hole conditions, the miscibility oil/carbon dioxide is effective and that additional oil recovery by displacement could be satisfactory.
Phase II : The injection-production test in the same well shut down a few years before, has proved disappointing, probably due to the very low permeability of the upper reservoir.
Phase III : 2600 tons of carbon dioxide and 1800 tons of nitrogen were injected. Gas breakthrough occured from 10 weeks to 4 months after starting injection depending on the distance and location of producing wells.
A total of 1000 tons of incremental oil was produced over one year period.
This yields a ratio of 0.23 tons of additional oil per ton of gas injected.
This low ratio condemns any industrial development.
The pilot has also confirmed that extreme caution has to be used when planning to push a slug of CO2 with nitrogen.
The CO2 gets trapped in reservoir liquids due to its high solubility and one may end up with a nitrogen displacement front.
A high permeability streak was evidenced between the injector and one well outside the four producer pattern.
A new project called "testing a new method to improve sweeping by injection of foaming agent and nitrogen (Phase I)" (TH./05073/86) is under way.
Its aim is to attempt to block high permeability paths and thus increase matrix productivity.

REFERENCES :
****
E. COUVE DE MURVILLE ET AL "INJECTION DE CO2 DANS UN RESERVOIR AVEC ACQUIFERE FORTEMENT ACTIF - ESSAI PILOTE DE COULOMMES-VAUCOURTOIS" 3RD EUROPEAN MEETING ON IMPROVED OIL RECOVERY, ROMA, APRIL 16-18, 1985. PROC. 2, pp 109, PUB BY AGIP S.P.A.,
E. COUVE DE MURVILLE ET AL "INTERPRETATION OF THE CO2/N2 INJECTION FIELD TEST IN A MODERATELY FRACTURED CARBONATE RESERVOIR "SPE 14 942, 5TH SPE/DOE SYMPOSIUM ON EOR, TULSA, APRIL 20-23, 1986.

```
********************************************************************************
* TITLE : RESTORING PROCESS FOR SCHANDELAH        *      PROJECT NO           *
*         OILSHALE.                               *                           *
*                                                 *    TH./05026/81/DE/..      *
*                                                 *                           *
********************************************************************************
* CONTRACTOR :                                    * PROGRAM :                 *
*   VEBA OEL AG                                   *   HYDROCARBONS            *
*   ALEXANDER VON HUMBOLDT STR TEL 0209 366 7968  *                           *
*   DE - 4650 GELSENKIRCHEN-HA TLX 824 881 - 95 VOE *                         *
*                                                 * SECTOR :                  *
*                                                 *   SECONDARY AND           *
* PERSON TO CONTACT FOR FURTHER INFORMATION :     *   ENHANCED RECOVERY       *
*   DR. R. HOLIGHAUS                              *                           *
*                                                 *                           *
*                                                 *                           *
********************************************************************************
                                                     VERSION : 01/01/87
```

AIM OF THE PROJECT :
****
The experimental demonstration of an oil shale retorting process to produce
marketable oil and gas products. The test results will be used for the
engineering and cost estimate of a demonstration plant.

PROJECT DESCRIPTION :
****
VEBA OEL AG has developed a pyrolysis process for oil shales with low Kerogen
(oil) contents called "Cyclone Pyrolysis". The project activities are the
engineering, construction and operation of a pilot plant with an hourly
throughput of 300 kg oil shale. The plant is located at the RUHR OEL Refinery,
Gelsenkirchen-Scholven.
In the process developed by VEBA OEL AG, the oil shale is heated up to a
pyrolysis temperature of approx. 500 deg.C by direct heat exchange with
recycling of hot pyrolysis gas. The pyrolysis is carried out primarily in a
cyclone and subsequently in a rotating drum. This drum is directly heated by
burning natural and/or proces off-gases.
The consideration of the pyrolysis vapors is carried out in two steps. In the
first step the pyrolysis vapors are quenched with oils generated in the
pyrolysis slage from 500 deg.C to about 300 deg.C. Simultaneously, fine dust
and soot are washed out from the pyrolysis vapors. In a second condensation
step the dust-free pyrolysis vapors are cooled from 300 deg.C to about 30 deg.C.
 The condensate passes to a separator to remove water from the light oil.
The necessary heat for the process is produced by burning the carbon containing
pyrolysis residue in a fluidized bed oven.

STATE OF ADVANCEMENT :
****
Ongoing. The pilot plant for pyrolysis of 300 kg/h oil shale was designed and
constructed. The construction of the pilot plant is completed except the
installation of the oil shale storage, the oil shale dosage into the hot
recycle gas and the pyrolysis cyclon.
To test the installed part of the plant, i.e. the rotating drum in recycle gas
mode, condensation of the heavy oil vapors; flue gas and residue cooling, first
pyrolysis test runs with liquid distillation residues were done.

RESULTS :
****
The tests showed difficulties, particularly concerning the condensation of the
heavy oil vapors, the separation of solids from the pyrolysis gas and the
recycle gas heating system. By modifications of the process scheme,
constructive changes of individual components, these difficulties were solved.
The difficulties mentioned above led to a considerable delay in the time
schedule and a overrun in the calculated budget. The current market situation
for primary energy resources excludes a commercial use of the oil shale
reserves in central Europe in medium term.
Therefore it is considered to abandon the project.

```
*****************************************************************************
* TITLE : PILOT PLANT FOR INCREASING THE RECOVERY   *      PROJECT NO      *
*          OF HEAVY OIL IN THE VALLECUPA FIELD.      *                      *
*                                                    *   TH./05027/81/IT/.. *
*                                                    *                      *
*****************************************************************************
* CONTRACTOR :                                       * PROGRAM :            *
*   AGIP SPA                                         *   HYDROCARBONS       *
*   PO BOX 12069 S. DONATO MIL TEL 02 520 5884       *                      *
*   I - 20120 MILANO             TLX 310246          *                      *
*                                                    * SECTOR :             *
*                                                    *   SECONDARY AND      *
* PERSON TO CONTACT FOR FURTHER INFORMATION :        *   ENHANCED RECOVERY  *
*   MR. G. SCLOCCHI                                  *                      *
*                                                    *                      *
*                                                    *                      *
*****************************************************************************
```

VERSION : 12/02/87

AIM OF THE PROJECT :
****
To develop a pilot plant of Steam Injection to enhance the recovery of heavy
oil from the Vallecupa field.

PROJECT DESCRIPTION :
****
The main phases of the project were:
1. Collection of reservoir and well basic data
2. Study of the reservoir for the selection of the best injection/production
   scheme
3. Drafting and construction of the pilot plant (based on a five-spot
   geometry and with central system injection)
4. Test of the process on the pilot plant.

STATE OF ADVANCEMENT :
****
Completed

RESULTS :
****
The Vallecupa field was an abandoned fractured carbonate reservoir at a
moderate depth (500 to 650 m) with a medium-heavy type oil (16 deg.C to 26 deg.
C API gravity). In a block isolated by faults from the rest of the field, five
wells were drilled (in a five spot pattern). The steam injection started in
November 1984. The total amount of steam injected was 10534 t. The cumulative
production of the wells at the abandonment of the pilot was 161 m3 of oil and
2200 m3 of water. The very poor efficiency of the process was due to the very
high heterogeneity of the reservoir rock (fractured carbonate) which exhanced
the water movement towards the producing wells and prevented any further
recovery of oil.

REFERENCES :
****
G.L. CHIERICI, A. DELLE CANNE, O. PROPERZTI, :"STEAM DRIVE IN A FRACTURED
CARBONATE : THE VALLECUPA, ITALY, PILOT PLANT", PAPER PRESENTED AT THE 3RD
EUROPEAN MEETING ON IMPROVED OIL RECOVERY, ROM, APRIL 16TH-18TH 1985.

```
********************************************************************************
* TITLE : TAR SHALES: THE TRANQUEVILLE IN SITU    *      PROJECT NO         *
*         COMBUSTION PILOT PROJECT                *                         *
*                                                 *    TH./05031/81/FR/..   *
*                                                 *                         *
********************************************************************************
* CONTRACTOR :                                    * PROGRAM :               *
*    GERTH                                        *    HYDROCARBONS         *
*    AV DE BOIS PREAU 4         TEL (1) 47.52.61.39 *                       *
*    FR - 92502 REUIL MALMAISON TLX (1) 47.52.69.27 *                       *
*                                                 * SECTOR :                *
*                                                 *    SECONDARY AND        *
* PERSON TO CONTACT FOR FURTHER INFORMATION :     *    ENHANCED RECOVERY    *
*    MR. J.E. VIDAL             TEL (1)42.91.40.00 *                        *
*                               TLX TCFP615700    *                         *
*                                                 *                         *
********************************************************************************
                                                     VERSION : 01/01/87
```

AIM OF THE PROJECT :
****
The aim of the Tranqueville in-situ combustion pilot project was full-scale
verification of the feasibility of this method of exploiting the potential
energy of tar shales and, if necessary, overcoming the new operational problems
arising from this type of exploitation.

PROJECT DESCRIPTION :
****
Two wells were drilled in the layer of shales (depth 200 m - thickness 25 m -
distance 60 m). They were in communication by a horizontal crack brought about
by hydraulic fracturation from the injection well.
Air was then injected into this crack. Combustion was ignited by electric
heating at the bottom of the injection well. The second well enabled the
products of combustion/pyrolysis to be collected.

STATE OF ADVANCEMENT :
****
The work started on 1st of July 1980. The project was completed by the 30th of
September 1984.

RESULTS :
****
Following two attempts to set-up a communication between an injection well and a production well, a third fracturation test was run in September 1982. This proved to be a partial success. The fracture is practically closed at the production well level during the production but opens when the well is under pressure and continues to propagate any further. Thus, it seems impossible to set-up a better communication without witholding the fracture.

Nevertheless, this communication has been judged sufficient to carry out an in-situ combustion test by the end of 1983.

From 1 December 1983 to 9 January 1984, 260 000 m3 of reconstituted air were injected in well Tr 1 at different flowrates between 100 and 1000 Nm3/h, following an ignition period of 36 hours. 5% of the injected volume were produced in well Tr 2.

The gas effluents were analyzed and showed that the combustion had been maintained during the whole air injection operation.

It was impossible to ensure a total oxygen consumption (about 30 000 Nm3 of oxygen were burnt by the combustion, that is to say 50% of injected O2) and thus to avoid the total consumption of the pyrolisis products. The shale volume required by combustion can be estimated at 50 m3.

Study of these cores sampled in the combustion area allowed to confirm the conclusions drawn after the analysis of the gaseous effluents.

Although it was impossible to avoid burning all the pyrolysis products, the first goal has been achieved:

- it was possible to ignite a combustion front
- to propagate it during about 40 days.

According to the balance of oxygen consumption, about 50 m3 of shale was burnt. Calculations show that the combustion front was propagated over about 10 meters. This result was also confirmed by a simplified analytic model of the combustion adjusted on curves representing the temperature raise at the bottom of well Tr 1 during the back-flows.

As far as the shale oil production is concerned, the pilot has outlined two problems:

- the poor performance of the production well, due to the high impedance of the fracture production connection;
- the difficulty of controlling the oxygen consumption rate and the front progress velocity, two fundamental parameters depending in a complex manner on the specific characteristics of the environment (particularly the scattering properties).

Although the first aspect does not seem inhibitory (it is possible to envisage sustaining the fractures near the production wells and to choose a pattern with a high production/injection ratio), it is more difficult to conclude on the second aspect.

The production of poor gas that had been envisaged as an alternative is fraught with the same difficulties. Nevertheless, the test has allowed to show that it was possible to start-up and maintain a combustion, and consequently, that this method allowed to consider a heat production.

```
*************************************************************************
* TITLE : NITROGEN INJECTION IN NORTH SEA         *      PROJECT NO       *
*          RESERVOIRS                             *                       *
*                                                 *   TH./05034/82/UK/..   *
*                                                 *                       *
*************************************************************************
* CONTRACTOR :                                    * PROGRAM :             *
*   BRITOIL PLC                                   *   HYDROCARBONS        *
*   301 ST VINCENT STREET      TEL 041 204 2525   *                       *
*   UK - GLASGOW G2 5DD        TLX 777633         *                       *
*                                                 * SECTOR :              *
*                                                 *   SECONDARY AND       *
* PERSON TO CONTACT FOR FURTHER INFORMATION :     *   ENHANCED RECOVERY   *
*   MR. K.J. WELLS                                *                       *
*                                                 *                       *
*                                                 *                       *
*************************************************************************
                                               VERSION : 01/01/87
```

AIM OF THE PROJECT :
****
To evaluate whether it would be technically feasible and economically
attractive to utilise immiscible gas displacement (hydrocarbon gas/Nitrogen) as
an enhanced oil recovery method for a field in the North Viking Graben, Brent
Oil province of the North Sea.
Using reservoir modelling and history matching techniques the data from a
hydrocarbon gas injection scheme that had been in progress in an oilfield for
some years was analysed. If the performance of this reservoir confirmed the
merit of an immi__ible gas displacement process then gas injection in a pilot
area of an oilfield would be studied using reservoir simulation. If shown to be
attractive then a full field application study would be undertaken.

PROJECT DESCRIPTION :
****
TOPSIDES ENGINEERING
Four methods of making nitrogen available to the North Viking Graben were
reviewed :
- Installation of a cryogenic plant on an existing platform
- Onshore generation with pipeline distribution to a platform(s)
- From turbine exhaust gases using inert generation
- Transportation of cryogenic liquid nitrogen by tanker
Using preliminary vendor data equipment, capital and operating cost estimates
were prepared, these were then subject to economic evaluation.
RESERVOIR ENGINEERING
Gas injection had been in progress for over 6 years in the Tarbert sand of a
northern North Sea oilfield. The area of interest is bounded by faults and
include 2 producers and injector (alternating water/gas injector).
RESERVOIR ENGINEERING CONT'D
The Tarbert Sand in this area can be regarded as two dispositional cycles, good,
 clean hgih permeability sands overlain by low permeability units. Furthermore,
a strong vertical flow barrier is present at the top of unit 3, making two
separate reservoir systems.
In the period of interest a total of 4.3 bslf of gas and 7 mmbbl water had been
injected.

To analyse the performance of this reservoir under gas and water injection, a 3D, 3 phase simulation model of the reservoir was constructed and fully history matched. In order to quantify the incremental oil recovery, due to gas injection, this history matched model was re-run using the volumetric equivalent of water to replace the injected gas into the reservoir.

STATE OF ADVANCEMENT :
****
Completed project.
The Tarbert sand in the main field is still in early development stages and cannot be considered for gas injection now, however, it may be a suitable candidate for gas injection in the future. This is being monitored.

RESULTS :
****
Results of preliminary studies indicated that nitrogen injection was technically feasible. it was also concluded that, for a single installation, the onshore plant option was less attractive due to higher capital and operating costs, and that the preferred method for supplying nitrogen offshore is by cryogenic distillation of air. Should a number of gas injection projects share a central source of nitrogen the economics may favour an onshore plant. The preliminary studies also indicated the need for a more detailed reservoir model, hence it was decided to analyse the data from an existing immiscible gas injection scheme which was underway into the Tarbert formation of a Northern North Sea field using reservoir modelling studies.
This study is now complete and indicates that gas injection has resulted in an incremental oil recovery of around a half a million barrels. However, this reservoir sand in the rest of the field is not yet developed and gas injection under these conditions is not recommended. For this reason the pilot study was not pursued. This horizon may be considered for gas injection at a later date. The waterflooding sweep efficiency on other horizons is sufficiency high to make immiscible gas injection unattractice.

```
*****************************************************************************
* TITLE : RECOVERY OF HYDROCARBONS FROM        *       PROJECT NO          *
*         BITUMINOUS ROCKS                      *                           *
*                                               *     TH./05037/82/IT/..    *
*                                               *                           *
*****************************************************************************
* CONTRACTOR :                                  * PROGRAM :                 *
*   AMMONIA CASALE                              *   HYDROCARBONS            *
*   VIA LAMPEDUSA 13          TEL (02)84401     *                           *
*   20141 MILANO              TLX 321203        *                           *
*   ITALIA                                      * SECTOR :                  *
*                                               *   SECONDARY AND           *
* PERSON TO CONTACT FOR FURTHER INFORMATION :   *   ENHANCED RECOVERY       *
*                                               *                           *
*                                               *                           *
*                                               *                           *
*****************************************************************************
                                                 VERSION : 12*03*85
```

AIM OF THE PROJECT :
****
Development of a retorting process for the extraction of hydrocarbons from
bituminous rocks and/or sands.

PROJECT DESCRIPTION :
****
The new process to be developed is characterized by the way in which retorthing
heat is supplied in a special reactor where no intermediates are used (whether
gas or inert solid).
The main phases of the project are:
1. basic study (identification of the necessary chemical physical parameters,
optimization of the pilot plant capacity, planning of laboratory tests,
preliminary financial calculation)
2. construction of a pilot plant
3. laboratory and pilot plant tests.
If the results of phase 3 are positive, a semi-industrial plant (500 to 1 000
tpd)could be envisaged in a later stage.

STATE OF ADVANCEMENT :
****
Abandoned. The following activities were performed:
- analysis of existing retorting processes
- definition of the proposed Ammonia Casale process flowsheet and main units
- study of a static, bench scale, pilot plant and its construction
- analysis of laboratory experiment data
- continuous test on a rotary drum.

RESULTS :
****
A patent application concerning the new retorting process was drawn up and
filed. The tests on a rotary drum have led to a definition of the criteria for
the design of a pilot plant with a capacity of 48 MTPD.
The basic study of the industrial plant (50 000 MPTD) has been completed. It
concerned material and heat balances. A rotary drum and two main furnaces were
also designed. Laboratory tests showed that pyrolysis of bituminous rock could
be carried out with 13 kg of HC per 200 kg of impregnated rock.

```
*******************************************************************************
* TITLE : EXAMINATION OF THE INTERACTION BETWEEN   *      PROJECT NO        *
*         STEAM AND THE DISTILLATES OF HEAVY       *                        *
*         CRUDE IN ENHANCED OIL RECOVERY           *   TH./05038/82/NL/..   *
*                                                  *                        *
*******************************************************************************
* CONTRACTOR :                                     * PROGRAM :              *
*  TECHNICAL UNIVERSITY OF DELFT                   *  HYDROCARBONS          *
*  COLLEGE VAN BESTUUR          TEL 784378         *                        *
*  JULIANALAAN 134              TLX 38151          *                        *
*  NL - 2628 BL DELFT                              * SECTOR :               *
*                                                  *  SECONDARY AND         *
* PERSON TO CONTACT FOR FURTHER INFORMATION :      *  ENHANCED RECOVERY     *
*  DR. J. BRUINING                                 *                        *
*                                                  *                        *
*                                                  *                        *
*******************************************************************************
                                                          VERSION : 18/03/87
```

AIM OF THE PROJECT :
****
The main objective of the project was to evaluate the possibility to enhance
the recovery from oil reservoirs by addition of distillable oil to the injected
steam during steamdrive recovery. This evaluation was effected in terms of a
preliminary design of a field test for distillation enhanced steam drive
recovery.

PROJECT DESCRIPTION :
****
Small amounts of distillable oil added to the steam injected in a tube
containing an oil sand are capable of enhancing the recovery efficiency, which
may approach 100% in the steamed out region. Without this addition the
comparable efficiency is 80%. We have studied this effect in detail to allow
extrapolation to field conditions. The complexity of the process made it
necessary to confine ourselves to the microscopic displacement process i.e.
disregard sweep efficiency effects. This aspect will be addressed in a follow
up contract "The beneficial effects of distillation during oil recovery by
steamflooding."
In this project we have elaborated at length on the competition between film
flow effects and the distillable oil bank to lower the "residual oil
saturation" in the steam zone. Film flow effects, which were discovered in this
project, tend to lower oil saturations in the steamed out zone in the absence
of distillation effects.
Distillation effects have been attributed to the distillable oil bank, which is
formed near the steam condensation front. This also leads to a lower (heavy)
oil saturation in the steam swept zone.
In this project we have disregarded the sweep enhancement owing to a more
favourable mobility ratio when a distillable oil bank is present.
The project consisted of the following phase :
i)   Design and building of high pressure equipment.
ii)   Continuation of experiments at atmospheric pressure.
iii)   Experiments with transparent models.
iv)   Experiments at high pressure.
v)   Preliminary design of a field test.

194

STATE OF ADVANCEMENT :
****
Completed. High pressure equipment has been designed and built. Numerous experiments have been performed at atmospheric pressure. Visual observations in transparent models leaded to the discovery of oil film flow in the steam zone. This film flow results in a different relative permeability behaviour of oil in the steam zone. The equipment was tested and a number of experiments have been performed at high pressure. A preliminary design for a field test was made.

RESULTS :
****
The idea to enhance distillation effects to improve oil recovery was first studied in tube experiments. It was verified experimentally that small amounts of distillable hydrocarbons added to the steam were indeed capable to enhance oil recoveries from 80% to 95%. We have experimentally verified that a distillable oil bank is indeed formed near the steam condensation front when liquid hydrocarbons are coinjected with the steam. Numerous experiments carried out on behalf of the present project allowed us to correlate the recovery with mixing effects which reduced the efficiency of the distillable oil bank. Dispersion is a possible mixing mechanism. Low oil saturations in the steamed out zone are in general attributed to distillation effects. Another effect, which reduces the residual oil saturations to extremely low values, is the film flow effect. Film flow in the steam zone was discovered in this project. Film flow also occurs in the absence of distillation. Film flow means that films of oil are formed between the steam and the water, which envelops the pore matrix. The result of film flow is that oil remains mobile at extremely low oil saturations.
We have quantified the film flow effect by measuring relative permeabilities in gravity drainage experiments.
Hardware requirements to implement distillation enhanced steam drive recovery in the field are simple. This is related to the fact that distillable oil can be injected as a liquid owing to its favourable phase behaviour. A schematic design of the surface equipment necessary for the addition of distillable hydrocarbons to the steam is given in the end report of Project TH/05038/82 "Distillation enhanced steamdrive recovery" by J. Bruining, D.W. van Batenburg and H. Ronde.
Extrapolation of the experimental data to a prototype field showed that only up to three barrels of oil can be recovered for each barrel of distillable oil coinjected. The extra recovery, however, will be less when film flow is effective.
On the other hand we have not, as yet, addressed the recovery improvement owing to the favourable mobility ratio, which results from the presence of the distillable oil bank. This aspect will be elaborated in our next project (TH 05064/85) "The beneficial effects of distillations during oil recovery by steam flooding "as part of a broader objective i.e. the economic competion of steamflooding and waterflooding of medium viscosity oil.

REFERENCES :
****
NUMEROUS REFERENCES AVAILABLE FROM PROJECT LEADER

```
****************************************************************************
* TITLE : CO2 INJECTION AND RECOVERY OF CRUDE IN    *      PROJECT NO       *
*         THE PISTICCI RESERVOIR.                   *                       *
*                                                   *                       *
*                                                   *    TH./05040/82/IT/.. *
*                                                   *                       *
****************************************************************************
* CONTRACTOR :                                      * PROGRAM :             *
*   AGIP                                            *   HYDROCARBONS        *
*   SAN DONATO MILANESE         TEL 02 520 5884     *                       *
*   CP 12069                    TLX 310246-ENI      *                       *
*   IT - 20120 MILANO                               * SECTOR :              *
*                                                   *   SECONDARY AND       *
* PERSON TO CONTACT FOR FURTHER INFORMATION :       *   ENHANCED RECOVERY   *
*   MR. G. SCLOCCHI                                 *                       *
*                                                   *                       *
*                                                   *                       *
****************************************************************************
```

VERSION : 12/02/87

AIM OF THE PROJECT :
****
Scope of the project was to ascertain through laboratory research and a field
pilot test, whether a huff-n-puff process using supercritical carbon dioxide as
"solvent" is able to massively improve well productivity in an existing heavy
oil field, which has been exploited for some twenty years at a very low oil
production rate.

PROJECT DESCRIPTION :
****
Three main phases.
- Reinterpretation of all existing data.Study of reservoir oil and of its
  mixtures with carbon dioxide.
- Work over of pilot well.
- Huff-n-puff with CO_2 for some cycles (200 tons). All relevant parameter
  will be measured and recorded and results evaluated.

STATE OF ADVANCEMENT :
****
Completed

RESULTS :
****
Pisticci 13 well, producing since 1964, was selected to test the "huff-and-
puff" technique with CO_2, in the Pisticci reservoir. PVT tests and laboratory
displacement tests with CO_2 in slim tubes at reservoir conditions excluded any
possibility of miscibility in the CO_2/reservoir oil system.The first cycle
started on Sept.1985. Two complete cycles (injection, soaking and production)
were run, by injecting 190 tons per cycle of CO_2 at supercritical conditions.
During the production period after each injection,gas coning,connected with the
production increase, was observed. The coning was due to the presence of free
CO_2 in the reservoir, because of the incomplete solution of CO_2 in the
reservoir oil.
The benefits of lowering the reservoir viscosity was thus balanced by worsened
flow conditions due to the presence of free CO_2.

```
******************************************************************************
* TITLE : STEAM INJECTION PILOT ON OFFSHORE       *      PROJECT NO        *
*         EMERAUDE: TESTS                         *                        *
*                                                 *   TH./05042/82/FR/..   *
*                                                 *                        *
******************************************************************************
* CONTRACTOR :                                    * PROGRAM :              *
*   GERTH                                         *   HYDROCARBONS         *
*   4, AVENUE DE BOIS PREAU     TEL 1 47.52.61.39 *                        *
*   FR - 92502 RUEIL-MALMAISON TLX 203 050        *                        *
*                                                 * SECTOR :               *
*                                                 *   SECONDARY AND        *
* PERSON TO CONTACT FOR FURTHER INFORMATION :     *   ENHANCED RECOVERY    *
*   MR. R. COTTIN               TEL 59.83.40.00   *                        *
*                               TLX 560 804       *                        *
*                                                 *                        *
******************************************************************************
                                                     VERSION : 01/04/87
```

AIM OF THE PROJECT :
****
The present project is a continuation of the project TH 05033/81 which aimed at
establishing the technologic feasibility of the pilot and led to the
installation of two platforms (mid 82 - mid 83) and the drilling of wells from
July 1983 on.
The object of the present project is to prepare, follow up and interprete all
the test which precede the steam injection :
- water and steam injectivity test
- test of surface installations
- tests of interference between wells
- primary pumping tests
- elaboration of a geological model of the pilot area by interpreting cores,
logs and previous tests.

PROJECT DESCRIPTION :
****
The project comprises 4 phases :
PHASE 1 :
1.1 Productivity and injectivity tests
   - water and steam injectivity tests in the first three wells
   - production test on these wells after having drilled the eight other wells
   - interference tests on R1 and R2 (between central injection well and
peripheral production wells) by long water injection
   - production tests on production wells
   - depending on the results of these tests, modification of the completion
programme in view of eventually making additional perforations and carry out
treatment specific to this type of reservoir (selective acidification or
cementations..)
1.2 Testing the installations
   - Testing of heat exchangers, separators and crude processing units
   - Testing of pumps
   - Testing of desalting units and of the two steam generators
   - Testing the utilities
   - Start-up of gas feed unit for the steam generator
   - and more generally, all that is necessary to start-up the installations.

PHASE 2 : Primary production through pumping
Evaluation of the potential of the wells and of each layer by long drive
pumping over a long period with regular measurement of the production
parameters on all wells successively.
PHASE 3 : Preparation
- Optimization of steam injection parameters (flowrate, pressure) and of
production parameters in view of modifying the production and injection
programmes
- Steam tracing (titrated water in R2)
- Adjustment of units depending on injection and production patterns
PHASE 4 :
Utilisation of the analysis of cores, logs, interference tests and of
production and injection tests to define the image of the reservoirs. Creation
of a geological model.
STATE OF ADVANCEMENT :
****
The first three wells EMV 01, EMV 02 and EMV 03 were drilled and tested between
July and December 1983. The different tests were run in 1984 and early 1985.
The steam injection began in March 1985 (contract TH./05050/83 entitled
"Emeraude - Evaluation of results").

RESULTS :
****
a) Modification of drilling and completion programmes following the
interpretation of results of the first three wells:
- abandonment of the cyclic injection on the GIK level which is dynamically
connected to the R1 and will be used for the injection in the five-spot R1;
- abandonment of the two observation wells, problems of plugging losses and
cementations encountered during the first drilling operations proving that it
was difficult to guarantee simultaneously their efficiency and the absence of
cross flows between the two five-spots.
b) Injectivity/productivity
The flows obtained during injection and production which were lower than those
of the initial project, are imputable to the low height perforations. The
difficulty of recovering and the risk of creating new short-circuits by
continuous drains have led to postpone all new perforation.
The steam will be injected at a rate of 80 t/d in R1, 150 t/d in R2 and 50 t/d
in R3 with a maximum head pressure of 30/35 bars so to avoid any risks of
fracturation in cement zones or in sub-jacent beds.
The average oil flow before steam injection is of 10 m3/d/well.
c) Interference tests
For R1, the water injections carried out in the EMV 01 injection well showed
rapid connections with the EMV 05 production well and with the wells of
neighbouring platforms CC, DD and E. For R2, the water injections carried out
in the EMV 03 injection well provided only one neat response in the EMV 09 well.
d) Reservoir image
The results of the punctual pressure measurements were excellent and made it
possible to show in a very significant manner that the intermediate levels GIF
were dynamically connected to R1. These measures also showed that the present
gradient of pressures in R2 is parallel to the initial hydrostatic gradient,
which proves a good vertical permeability of the reservoir, whereas in R1, the
gradient is inversed, which shows a very low permeability value.

The SHDT logs proved to be extremely helpful to detect the drains in the lime and their continuity.

e) Injection patterns

In R1, the steam injection was carried out at the base of the reservoir whereas the production wells were perforated at the top so to favour the vertical crossing of the steam in the silts.

In R2, where the raise of the steam by gravity was carried out without problem due to the high vertical permeability, the injection well and the production wells werre perforated at the base of the reservoir.

f) Testing the installations

Tests relative to the installations were run in parallel with the drilling operations and the well tests. Modifications were made on the air feed systems of the boilers, on the assembling of the compressors and on one of the desalting units to ensure their functioning under conditions analog to nominal conditions.

REFERENCES :
****

"STEAM INJECTION PILOT PROJECT INTO EMERAUDE OFFSHORE RESERVOIR" B. SAHUGUET, D. MONFRIN
PROCEEDINGS OF THE 2ND E.C. SYMPOSIUM HELD IN LUXEMBOURG 5-7 DECEMBER 1984, VOL 2, P 709 GRAHAM & TOTMAN EDITORS

```
*******************************************************************************
* TITLE : TESTING OF NEW PRELIMINARY TECHNOLOGIES    *      PROJECT NO       *
*         ON HEAVY OIL FIELDS                         *                       *
*                                                     *   TH./05043/82/FR/..  *
*                                                     *                       *
*******************************************************************************
* CONTRACTOR :                                        * PROGRAM :             *
*   GERTH                                             *   HYDROCARBONS        *
*   4, AVENUE DE BOIS PREAU    TEL 1 47 52 61 39      *                       *
*   FR - 92502 RUEIL MALMAISON TLX 203050 F           *                       *
*                                                     * SECTOR :              *
*                                                     *   SECONDARY AND       *
* PERSON TO CONTACT FOR FURTHER INFORMATION :         *   ENHANCED RECOVERY   *
*   MR. J.F. LE PAGE              TEL 1 47 49 02 14   *                       *
*                                                     *                       *
*                                                     *                       *
*******************************************************************************
```

VERSION : 01/01/87

AIM OF THE PROJECT :
****

The overall project covered the construction and testing of demonstration units
combined together onto Solaize platform, in order to study the various methods
of upgrading heavy oils on the production field. The first part of the project
was a feasibility study, whilst the second covered the construction proper of
the platform. The third part of the overall project is the subject of the
present contract.

This project is a continuation of the contracts TH 05.030/81 and TH 05.043/82
concerning the ASVAHL platform set-up in Solaize near Lyons.

PROJECT DESCRIPTION :
****

The programme comprises two distinct phases of work :
- startup of the platform units, implying a number of prior technological test;
the main units involved are
  desalination, atmospheric distillation, vacuum distillation, deasphalting,
hydroprocessing, visbreaking
  and hydrovisbreaking.
- hydrotreating on fixed or moving beds; design, engineering and the
construction follow-up of a special
  device to realize a circulation of catalyst through one hydrotreating reactor.

STATE OF ADVANCEMENT :
****

The project started on 01.January.82 and was achieved on 30 April 1985.

RESULTS :
****
Phase 1 - This phase involved the oiling of the various units and then the test
and possibly the modifications of many original devices or equipment set-up on
several units. Mention must be made of the new water-oil mixing devices on the
desalter, a new settler for asphalt separation in deasphalting, a new falling
film evaporator, a new extruding machine for hard asphalt. All the units have
been oiled within a period of six months. Most of the original devices have
proved as interesting as foreseen except for the falling film exchanger less
efficient and less easily opewreated than expected.
Phase 2 - Design, engineering and construction of the catalyst loop at a scale
of 0,5 (relatively to an industrial unit) has been set-up without any major
problems and is today successfully working. For this purpose, it has been
necessary to design and develop new valves, sensors and a computer program to
have the catalyst injected into and removed from the reactor automatically at
450 deg.C and 200 bars.

REFERENCES :
****
NEW PROCESSING ROUTES FOR HEAVY OILS AND RESIDUES
J.F. LE PAGE AND ALL. OPAEC SEMINAR - JUNE 1986
AOSTRA SEMINAR 4.12.1986 - TEXTS AVAILABLE AT THE IFP 2 & 4 AVENUE DU BOIS
PREAU 92502 RUEIL MALMAISON
UPGRADING ARABIAN HEAVY RESIDUE
G. CHOUX - TOTAL CRD - SAUDI OIL SHOW DAHRAN 1.4.1984

```
********************************************************************************
* TITLE : INJECTION OF CO2 INTO PECORADE    *        PROJECT NO          *
*          RESERVOIR                        *                            *
*                                           *   TH./05044/82/FR/..       *
*                                           *                            *
********************************************************************************
* CONTRACTOR :                              * PROGRAM :                  *
*   GERTH                                   *   HYDROCARBONS             *
*   AV DE BOIS PREAU 4          TEL (1) 47.52.61.39 *                    *
*   F - 92502 RUEIL MALMAISON  TLX (1) 47.52.69.27 *                     *
*                                           * SECTOR :                   *
*                                           *   SECONDARY AND            *
* PERSON TO CONTACT FOR FURTHER INFORMATION : *  ENHANCED RECOVERY       *
*   MR. J.L. MINEBOIS           TEL 59.05.24.50 *                        *
*                               TLX 560053 PETRAKI *                     *
*                                           *                            *
********************************************************************************
```

VERSION : 01/01/87

AIM OF THE PROJECT :
****
This project concerns the implementation at PECORADE reservoir of pilot
injection of CO2 to maintain the reservoir pressure above the critical pressure
whilst at the same time enhancing oil recovery in a part of the reservoir
difficult to exploit using conventional methods. It follows a feasibility study
performed under contract TH 05.29/81.

PROJECT DESCRIPTION :
****
The CO2 is extracted from lacq gas, liquefied and stored at - 20 deg.C and at
pressure of 20 bars. It contains less than 1,000 ppm of H2S and is then carried
by tank trucks to the injection site 65 kilometres away, where it is again
stored into 175 m3 containers. Injection takes place in liquid form at an
average daily rate of 60 tons of CO2 for 420 bars at the injection wellhead
(pce 13).
The conditions in the three production wells (PCE 04, PCE 22 and PCE 23) about
200 metres away from the injection wells, are followed by gauges and weekly
samples of effluents, so as to be able to follow the action of the CO2 in the
reservoir.

STATE OF ADVANCEMENT :
****
The operations of starting up the surface installations and pilot project wells
took place from 1st July 1982 to 15th March 1983.
Injection of CO2 was started in mid-March 1983 and ended on 7th March 1984,
injecting a total of 10,700 tons of CO2.

RESULTS :
****
The technical phase of the project (production, liquefaction, transport and injection of CO2) took place satisfactorily, except for a failure of the CO2 injection pumps, which resulted in a lengthy interruption of injection from 18th october 1983 to 6th february 1984.
The very poor petrophysical characteristics of the reservoir have required a high injection pressure in the range of 420 bars at the welhead (PCE 13), to maintain an injection of an acceptable level (60 tons per days). These pressure conditions have resulted in fracturation inside the reservoir at the injection well, whereas initially injection was to take place at pore communication level.
 The result of this fracturation has been very quick breakthrough of the CO2 into the production wells, so that the surrounding matrix was only affected little by the injection.

```
********************************************************************************
* TITLE : SURFACTANTS FOR MICELLAR/POLYMER            *      PROJECT NO        *
*          FLOODING IN HIGH SALINITY ENVIRONMENTS.    *                        *
*                                                     *    TH./05045/83/IT/..  *
*                                                     *                        *
********************************************************************************
* CONTRACTOR :                                        * PROGRAM :              *
*   AGIP SPA                                          *   HYDROCARBONS         *
*   C.P. 12069                TEL 02 5201             *                        *
*   IT - 20120 MILANO         TLX 310246              *                        *
*                                                     * SECTOR :               *
*                                                     *   SECONDARY AND        *
* PERSON TO CONTACT FOR FURTHER INFORMATION :         *   ENHANCED RECOVERY    *
*   PROF. G. CHIERICI                                 *                        *
*                                                     *                        *
*                                                     *                        *
********************************************************************************
                                                         VERSION : 20/02/87
```

AIM OF THE PROJECT :
****
To evaluate various families of commercial surfactants and to synthesize
completely new molecules able to form stable micellar solutions in high-
salinity reservoir environment where enhanced oil recovery by chemical flooding
should be carried out.

PROJECT DESCRIPTION :
****
Commercial surfactants and polymers (for mobility control of the micellar
solution) and new surfactant and polymer molecules synthesized specially for
the purpose of this research will be screened. They will be tested through
standard lab procedures and by sand-pack oil displacement tests at reservoir
conditions. Simulation of the process by numerical model will also be performed.

STATE OF ADVANCEMENT :
****
Completed

RESULTS :
****
Several families of commercial surfactants were tested. new families of
sulfonated aliphatic alcohols, alkan-sulfonates and hydroxyalkansulfonates were
sintesized. The phase diagrams of all these chemicals in mixture with oils of
various EACN, brines, and cosurfactants were determined. Microemulsion
viscosity maps were also determined. Hydroxypropyl; carboxymethyl;
hydroxypropyl-carboxymethyl derivatives of guar gum were sinthesized and tested
as water thickeners. flooding tests at reservoir conditions were run on
sandpacks; oil recovery, surfactant and polymer absorption were measured. A
compositional 2D micellar/ polymer streamline reservoir simulator was developed
and tested.
Problems arose in phase behaviour simulation when in presence of very high
salinity brines. Although none of the surfactants and polymers tested proved to
be economically viable for EOR processes, the experience gained will be helpful
in designing new projects.

```
*******************************************************************************
* TITLE : PRIMARY UPGRADING OF HEAVY OIL PHASE 3.   *        PROJECT NO       *
*                                                   *                         *
*                                                   *      TH./05046/83/DE/..  *
*                                                   *                         *
*******************************************************************************
* CONTRACTOR :                                      * PROGRAM :               *
*   VEBA AG/LURGI                                   *   HYDROCARBONS          *
*   POSTFACH 45                    TEL 0209/3667968 *                         *
*   ALEXANDER-VON-HUMBOLD-STRA TLX 824881 - OVOD    *                         *
*   D 4650 GELSENKIRCHEN 2                          * SECTOR :                *
*                                                   *                         *
* PERSON TO CONTACT FOR FURTHER INFORMATION :       *                         *
*   MR. HOLIGHAUS                                   *                         *
*                                                   *                         *
*                                                   *                         *
*******************************************************************************
                                                       VERSION : 01/12/85
```

AIM OF THE PROJECT :
****
Aim of the project "Primary upgrading of heavy oil" is the demonstration and
further development of VEBA OEL's residue hydrocracking technologies in a pilot
plant scale (24 t/d). VEBA OEL's resid cracking processes "VEBA-LQ-Cracking"
(VLC) and "VEBA-Combi-Cracking" (VCC) are derived from the former Bergius-Pier-
Technology for the liquefaction of coal which was used for residue upgrading
between 1951 and 1964 in a commercial scale. A further improvement of the
processes was evaluated in process development units since 1979 (phase 1 of the
project) and the resulting processes had to be tested in the larger scale. For
this purpose VEBA OEL together with LURGI GmbH erected the VLC/VCC Pilot Plant
at their Gelsenkirchen refinery (phase 2 of the project). The plant was ready
for start up in June 1983.

PROJECT DESCRIPTION :
****
Phase 3 of the project "Primary upgrading of heavy oil" covers the operation of
the pilot plant until end of 1984 with the target to confirm the evaluated
process improvement in the larger scale, to supply reliable layout data for a
commercial VCC plant, to train personal and to process residues from different
sources for the generation of guarantee data.

STATE OF ADVANCEMENT :
****
During phase 3 of the project the plant was on stream for appr. 8500 h under
hydroconversion conditions. In total 15 test runs were performed until end of
1984. All technical difficulties could be solved within the first periods of
operation In 1984 an average availability of >70% was achieved thus
demonstrating the reliability of the process. A further operation phase for the
years 1985 and 1986 is scheduled.

RESULTS :
****
The test runs of the plant prove that residue conversion levels up to 95 % can
be achieved for various feedstocks. In parallel a throughput increase and a
pressure reduction is feasable thus improving the economy of the process. The
use of a one way additive supports these process improvements. Product yields
and properties fit with the results gained in small scale units.
The results of a long term test lasting more than 1000 h were used as design
data for the engineering of a commercial size VCC upgrader (phase 4).

```
****************************************************************************
* TITLE : PECORADE EVALUATION OF RESULTS.            *      PROJECT NO      *
*                                                    *                      *
*                                                    *   TH./05048/83/FR/.. *
*                                                    *                      *
****************************************************************************
* CONTRACTOR :                                       * PROGRAM :            *
*   GERTH                                            *   HYDROCARBONS       *
*   AV. DE BOIS PREAU 4          TEL (1)47.52.61.39  *                      *
*   FR - 92502 RUEIL-MALMAISON TLX (1)47.52.69.27    *                      *
*                                                    * SECTOR :             *
*                                                    *   SECONDARY AND      *
* PERSON TO CONTACT FOR FURTHER INFORMATION :        *   ENHANCED RECOVERY  *
*   MR. J.L. MINEBOIS           TEL 59.05.24.50      *                      *
*                               TLX PETRAKI 560053   *                      *
****************************************************************************
                                                      VERSION : 01/01/87
```

AIM OF THE PROJECT :
****
This project concerns the evaluation of the results from a CO2 injection in
PECORADE reservoir which has been performed in 1983 (contract TH 5044/82) to
maintain the reservoir pressure above the critical pressure. It follows a
feasibility study performed under contract TH 5029/81 and a CO2 injection under
contract TH 5044/82.

PROJECT DESCRIPTION :
****
The present project comprises the water sweeping of the injected CO2 to the
production wells. Two phases were to be carried out:
Phase 1 - Water injection
Phase 2 - Evaluation of results

STATE OF ADVANCEMENT :
****
The project was stopped by the 30th of June 1985.

RESULTS :
****
PHASE 1 - WATER INJECTION
At the end of the takeover operation of well PCE 13 (former CO2 injection well) in the upper bed B2, an injectivity test has provided relatively unsignificant results compared to the forecast, with about 500 bars at the wellhead for non-significant injection flowrates (July 1984).
After perforation cleaning production started (10 to 15 m3/d). This caused water drive to be postponed, but to respect the previsional schedule presented at the EEC, water injection was cancelled.
PHASE 2 - EVALUATION OF RESULTS CO2 PRODUCTION
Although early breakthroughs of CO2 and very short transit times were observed, only 10% of the CO2 injected was produced. There are two possible theories to explain the location of the remaining 90% within the reservoir.
Communication of fractures network during injection with the highly fissurized Northern zone forming the top of the reservoir. In this case, the CO2 very quickly migrated towards the gas cap without displacing much oil.
CO2 percolation into the matrix within and beyond the pilot zone. In this case, there is no gas front as is the case with a homogeneous circular radial sweep. The CO2 dissolved gradually in a large volume of oil, displacing this oil essentially by swelling action. Production of CO2, from PCE 13 and PCE 23 a year after ending the injection, the ingress of part of the CO2, into the drain and the absence of very high GORs are all factors in favour of this theory.
OIL PRODUCTION
The oil produced appears to have been slightly affected by the injection of CO2 and can be related directly to the effects of pressure caused by injecting water into PCE 18.
Extremely poor characteristics of the lower porous zone of B2, which turned out to have productivity and injectivity indices much poorer than had been suggested by the petrophysical analyses have led to the creation of widely extending fractures during injection and the impossibility to drain through the production wells of the pilot project.
However, when well PCE 13, which was intended as a water injection well, was placed on production, water drive had to be postponed and did not prove possible as part of the CO2 project (in order to comply with the schedule). This resulted in total and final shutdown of all the Pecorade storage and injection facilities.
CONCLUSION
Implementation of CO2 injection at Pecorade has enabled mastering of the method, from elaboration of the fluid at the Lacq plant units until injection of carbon dioxide gas into the well. Whilst this objective can be considered by and large to have been reached, one specific point has not been solved so far, with repeated incidents of cracking of the hydraulic blocks of the CO2 injection pumps. The conclusion of the experts examining these blocks was that gradual fatigue cracking had occured, though without revealing any corrosive action.

REFERENCES :
****
PRESENTATION TO THE E.E.C. SYMPOSIUM (LUXEMBURG, 5-7 DEC. 1984)

```
*****************************************************************************
*  TITLE : PREPROCESSING OF HEAVY OILS:           *      PROJECT NO          *
*          EXPERIMENTATION                        *                          *
*                                                 *    TH./05049/83/FR/..    *
*                                                 *                          *
*****************************************************************************
*  CONTRACTOR :                                   *  PROGRAM :               *
*    GERTH - ASVAHL                               *    HYDROCARBONS          *
*    AV. DE BOIS PREAU 4         TEL 1 47.52.61.39 *                         *
*    FR - 92502 RUEIL-MALMAISON TLX 203 050 F     *                          *
*                                                 *  SECTOR :                *
*                                                 *    SECONDARY AND         *
*  PERSON TO CONTACT FOR FURTHER INFORMATION :    *    ENHANCED RECOVERY     *
*    MR. J.F. LE PAGE            TEL 1 47.49.02.14 *                         *
*                                                 *                          *
*                                                 *                          *
*****************************************************************************
                                                        VERSION : 01/01/87
```

AIM OF THE PROJECT :
****
The object of this project is to test on the Solaize platform, the technology,
equipment, methods and process diagrams advocated by ASVAHL to process heavy
oils in view of making them transportable and processable by conventional
refining means.

PROJECT DESCRIPTION :
****
The project is divided into two important phases :
Phase 1 - Experimentation of heavy oils through thermophysics(indirect route)
Phase 2 - Experimentation of heavy oils through thermocatalysis(direct route)
These test campaigns were to be run successively on various types of heavy oils
coming from different reservoirs (Boscan, Rospo Mare, Saragomare, Athabasca
Laguna Once etc...) or from heavy residues of crudes (Safaniya).
Two sets of units were scheduled to operate simultaneously:
- desalting, atmospheric distillation
- deasphalting
- visbreaking, hydrovisbreaking, catalytic hydrotreatment
in order to reach the following objectives:
- comparative study of process diagrams for the preprocessing of heavy oils
  in a direct or indirect manner
- behaviour of different heavy oil charges in the various processes
  and optimization of operating conditions
- tests at representative scale of the innovations and technical
  improvements brought by some of the methods
- study of materials,of inhibitors
- production of representative samples for subsequent tests in the pilot.

STATE OF ADVANCEMENT :
****
The work programme is to be achieved by end of 1986.

RESULTS :
****
1. The first runs have been carried out to check the technical feasibility of
various improvements concerning the different processes. These improvements
involved new concepts, new equipments, additives or catalysts.
- in heavy oil deasphalting the new technology involving original precipitating
drum and washing device are worth being mentioned as well as the recovery of
solvent in opticritical conditions. With this new process, capacities of
210.000.000 T/year for a single line can be reached and solid asphalts with a
softening point of 210 deg. can be produced.
- in heavy oil desalting and distillation, new special mixing values have been
selected, operating conditions have been optimized and sets of additives
selected.
- in thermal treatments, visbreaking and specially hydrovisbreaking, the
operating conditions and the geometry of the requested soakers have been
optimized in order to avoid, as far as possible pitch production.
- in hydrotreatments a new set of catalysts has been developed exhibiting a
metal retention as high as 100% of the catalyst weight. Moreover, the
demetallation catalyst has proved to produce very low amounts of coke. This set
of catalysts allows to aim the following objectives :
- ordinary refining at low severity
- production of acceptable cat-cracking charges at mean severity
- thorough conversion at high severity;
with this set of proprietary catalysts,heavy oils with a metal content as high
as 300 ppm (Ni + V) can be treated in a fixed bed; over this value the feeds
can be treated in a counter-current moving bed technology available on the unit.
2. Various heavy oils have been treated with the aim of testing the processes,
selecting the processing schemes and producing industrial samples of products
such as deasphalted oils, asphalts, thermal and catalytic residues: Boscan,
Rospomare, Safaniya, Morichal, Athabasca have been thoroughly studied.
3. As far as the processing schemes are concerned, many alternatives have been
checked and compared. For the indirect route, mention must be done of the
following processing schemes:
- desaphalting, hydrovisbreaking, DAO hydrotreatment
- hydrovisbreaking,deasphalting,DAO hydrotreatment
A comparison of the schemes has shown the advantage of the second one.
As an example of the direct route the advantage of coupling a hydrovisbreaking
step with a catalytic hydrotreatment is worth being mentioned.Many other
processing schemes have been studied and estimated on a technico-economical
basis. As a summary it must be noted that the deasphalting process, the
visbreaking and hydrovisbreaking processes, the hydrotreatment process are now
commercially proposed with the following trade marks, Solvahl,opticritical,
Thervahl T and Therval H, Hyvahl F and Hyvahl T.

```
****************************************************************************
* TITLE : EMERAUDE - EVALUATION OF RESULTS.        *      PROJECT NO       *
*                                                  *                       *
*                                                  *   TH./05050/83/FR/..  *
*                                                  *                       *
****************************************************************************
* CONTRACTOR :                                     * PROGRAM :             *
*   GERTH                                          *   HYDROCARBONS        *
*   AV. DE BOIS PREAU 4          TEL 1 47 52 61 39 *                       *
*   FR - 92502 RUEIL-MALMAISON TLX 203 050 F       *                       *
*                                                  * SECTOR :              *
*                                                  *   SECONDARY AND       *
* PERSON TO CONTACT FOR FURTHER INFORMATION :      *   ENHANCED RECOVERY   *
*   MR. B. COUDERC               TEL 59 83 40 00   *                       *
*                                TLX 560 804       *                       *
*                                                  *                       *
****************************************************************************
                                                      VERSION : 01/01/87
```

AIM OF THE PROJECT :
****
This project concerns a steam injection pilot for the EMERAUDE heavy oil field,
offshore Congo (ELF-CONGO/AGIP), in a water depth of 65 m.
The important accumulation (575 million tons) and the low recovery rate (3%)
obtained by primary production explain the efforts that were undertaken to make
a pilot and prove that an industrial development is possible by injecting steam
in spite of existing problems (shallow beds, reservoir, nature, oil quality...).

PROJECT DESCRIPTION :
****
Engineering works, started end 1980, within the scope of contract TH 5033/81,
have made it possible to define the platforms and equipment necessary for this
pilot. The equipment was set in-place between December 1981 and June 1983.
Drilling operations started in July 1983, without the help of the Community,
continued in 1984. The first water injection tests began in October 1983,
within the scope of contract TH 5042/82. Steam injection tests began in 1985.
The present contract is divided into three phases :
- preparation of a numerical model
- analysis of measures taken during tests
- final results of the pilot.

STATE OF ADVANCEMENT :
****
Ongoing

RESULTS :
****
The first phase was the subject of studies :
A new thermal model to suit the specific characteristics of the EMERAUDE field
was modified. Calibration of the state of reservoirs observed during the EMV
drilling operations during a 10 year primary depletion and of different tests
ran in 1983, has made it possible to test these modifications.
Phase 2 - Interpretation of the results after an 18 month continuous steam
injection in five spots R1 and R2 and the two cycles ran in the huff and puff
well R3, led to the following conclusions :
- in R1 : continue the injection until a thermal reaction is obtained allowing
to quantify the production of tertiary oil to be expected from a system
operating under an almost pure drive. However it may take some time before an
answer can be obtained. Completions will be adapted so as to improve the
probability of such a reaction during the lifespan of the pilot test.
- in R2, where production of tertiary oil represents today about 120 m3/d
(taking into account the reactions in peripheral wells), the interest of a
steam injection has already been proven and simulations have allowed to master
the phenomena. Continuation of the steam injection and of the numerical
simulations until significant degradation is obtained would allow to draw finer
conclusions.
- in R3 : abandon cyclic injection, a process which does not seem suitable for
this geological pattern.

```
******************************************************************************
* TITLE : NITROGEN GAS INJECTION                   *      PROJECT NO        *
*                                                  *                        *
*                                                  *   TH./05053/84/NL/..   *
*                                                  *                        *
******************************************************************************
* CONTRACTOR :                                     * PROGRAM :              *
*   TECHNISCHE HOGESCHOOL DELFT, AFD. MIJNBOUWKUNDE *   HYDROCARBONS         *
*   POSTBUS 5028                 TEL 015 781328     *                        *
*   NL - 2600 GA DELFT           TLX 38151          *                        *
*                                                  * SECTOR :               *
*                                                  *   SECONDARY AND         *
* PERSON TO CONTACT FOR FURTHER INFORMATION :      *   ENHANCED RECOVERY     *
*   DR. H. RONDE                                    *                        *
*                                                  *                        *
*                                                  *                        *
******************************************************************************
```

VERSION : 21/04/87

AIM OF THE PROJECT :
****
The aim of nitrogen gas injection into an oil reservoir is based upon the super
critical extraction properties of N2.
The advantages of N2 injection with respect to CO2 are :
1. Low-production costs
2. Inert properties with respect to injection wells
3. Availability.

PROJECT DESCRIPTION :
****
a. Theoretical study on gas injection in oil reservoirs
b. Theoretical study on N2-oil phase behaviour
c. Development of computer model
d. Design and construction of experimental rig
e. Investigation of phase behaviour of crude oil/nitrogen
f. Conducting displacement experiments at reservoir temperatures and
pressures
g. Formulation of screening criteria and a field test proposal
h. Economical evaluation of the process.
. A literature survey has been performed and an apparatus has been acquired to
perform in-sity sampling during phase behaviour measurements of N2-oil and N2-
crude oil samples.
. The development of a fully implicit computer simulation model, capable to
interpret our one-dimensional preliminary low-pressure experiments, has started.
. A software package, including a source list, has been bought. It allows to
describe the phase behaviour of oil-N2 mixtures particularly to determine the
minimum miscibility pressure, phase envelopes and other physical-chemical data
necessary for the development of our computer simulation study.
. Due to the long period to construct the experimental rig, experiments on
atmospheric pressure on liquid mixtures, having the same phase behaviour as N2-
oil mixtures with high pressure, will be started. It is envisaged to conduct
experiments in the existing high pressure rig (100 atm) with gases instead of
nitrogen, at a higher critical temperature.

STATE OF ADVANCEMENT :
****
Ongoing

RESULTS :
****
Fluid phase equilibria of nitrogen and aromatic compounds have been measured
and interpreted.
An introductary economic evaluation (in the case of a condensate reservoir) was
performed.
A two-phase high pressure sampling device was designed to be able to sample
before separation.

```
*******************************************************************************
* TITLE : TERTIARY RECOVERY BY CYCLIC STEAM       *       PROJECT NO          *
*         INJECTION IN THE DEPLETED HEAVY OIL      *                           *
*         FIELD OF TOCCO CASAURIA                  *     TH./05057/84/IT/..     *
*                                                  *                           *
*******************************************************************************
* CONTRACTOR :                                     * PROGRAM :                 *
*   AGIP SPA                                        *   HYDROCARBONS            *
*   CP 12069                      TEL 02 5201       *                           *
*   IT - 20120 MILANO             TLX 310 246       *                           *
*                                                  * SECTOR :                  *
*                                                  *   SECONDARY AND           *
* PERSON TO CONTACT FOR FURTHER INFORMATION :      *   ENHANCED RECOVERY       *
*   PROF. G.L. CHIERICI                            *                           *
*                                                  *                           *
*                                                  *                           *
*******************************************************************************
```

VERSION : 08/09/86

AIM OF THE PROJECT :
****
To evaluate an old, abandoned oil reservoir in Central Italy through
geophysical, geological and drilling investigations. To study rock and oil
samples obtained from drilling in order to evaluate the potential of alternate
(cyclic) steam injection and production as an EOR process in this fractured
carbonate reservoir rock. To carry some cycles of huff-n-puff in order to
validate the study.

PROJECT DESCRIPTION :
****
1. Shooting of shallow seismic and interpretation in order to evaluate the
shape and dipping of the structure and to locate the best position for the
wells to be drilled.
2. Drilling of two wells, coring and sampling of the fluids. Production testing
of the wells.
3. Study of reservoir rock and fluids and lab evaluation of the potential of
alternate (cyclic) steam injection.
4. Carrying out of some cycles of steam injection in at least one well and
interpretation of the results obtained.
5. Preparation of the final report.

STATE OF ADVANCEMENT :
****
Phase 1 has been completed. Two locations have been chosen. The wells are going
to be drilled in the next few months.

RESULTS :
****
The shape and dipping of the reservoir have been determined. Four "uphole"
velocity survey wells have been drilled and the cuttings examined. The surface
mapping of the faults system in the area has been completed. It results that
the reservoir dips, on the average, by 60 degrees, this being the limit of the
interpretation possibility of the seismic survey.

```
*******************************************************************************
* TITLE : PREPARATION OF GAS INJECTION IN GRAND    *        PROJECT NO        *
*         ALWYN DEPOSIT.                           *                          *
*                                                  *     TH./05058/84/FR/..   *
*                                                  *                          *
*******************************************************************************
* CONTRACTOR :                                     * PROGRAM :                *
*   GERTH                                          *   HYDROCARBONS           *
*   AVENUE DE BOIS PREAU 4      TEL 1 47 52 61 39  *                          *
*   FR - 92502 RUEIL-MALMAISON TLX 203050          *                          *
*                                                  * SECTOR :                 *
*                                                  *   SECONDARY AND          *
* PERSON TO CONTACT FOR FURTHER INFORMATION :      *   ENHANCED RECOVERY      *
*   MR. G. AUXIETTE             TEL 1 42.91.42.56  *                          *
*                               TLX 615700         *                          *
*                                                  *                          *
*******************************************************************************
```

VERSION : 20/02/87

AIM OF THE PROJECT :
****
The project consists in developing a set of thermodynamic models to provide a
really reliable restitution of the PVT laboratory experiments, and hence give a
better simulation of injections in a reservoir model, particularly as regards
miscible gas. This is essential to optimize the implementation of a pilot phase
in a project, which can be extremely expensive, especially in the North Sea
(e.g. ALWYN).

PROJECT DESCRIPTION :
****
This project consists of two phase :
PHASE 1 - MEASUREMENTS AND ANALYSES
The thermodynamic model at present available, requires data on the critical
properties of the components and the interaction coefficients between the
heaviest specific cuts and the lightest components. An apparatus has been
developed in order to determine the interaction coefficients between the heavy
components and the CO_2, by experimental determination of the phase envelopes
under given pressures and temperatures.
The heavy cut is represented by a pseudocomponent to which are attributed
critical properties, based on experimental data, such as density, molecular
weight, boiling point. The NMR determination of a typical average molecule,
makes it possible to know the molecular weight of the mixture with greater
accuracy, and the use of group contributions in the models, allowing a better
representation of this type of components.
PHASE 2 - MODELLING STUDY
The most commonly used models are derived from Van der Waals' equation (the
best known is that of Peng-Robinson). The adjustment of the parameters (Tc, Pc,
W, kij) of the pseudocomponent(s), which was almost manual, has been
systematized, determining from all the data contained in the PVT laboratory
reports, the properties of the heaviest pseudocomponent. The method uses the
principle of maximum probability, and integrates the experimental errors
through variance and covariance matrices.

Other representations of heavy components, such as that by group contributions, or continuous distributions, are being studied. With group contributions, the properties of the cuts will be determined, and in particular, the interaction coefficients can be calculated. In a first phase, the continuous distributions will be used to make the best choice of the number of pseudocomponents representing the heaviest cut, and to determine their characterisitics.

STATE OF ADVANCEMENT :
****
Phase 1 : underway, the study on CO2-Eicosane equilibria continues; the determination of molecular weights and average molecular models continues on Alwyn fluid with a view to application during phase 2.
Phase 2 : the model for adjusting the parameter has been achieved. Studies on group contributions and continuous distributions are underway, and already 80% complete.

RESULTS :
****
Phase 1 : the methodology apparatus for determining the measurement of pressure, volume, temperature, has been built and is widely used. It was proven on CO2-isobutane mixtures. The CO2 n-octane studies have been completed, and the CO2-eicosane studies are underway, plus research on the dew point, for strict methods, is underway (dew point detection by image analysis, luminous intensity).

Phase 2 : during the development of the parameter adjustement method, the algorithms for calculating the liquid-vapor equilibrium have been made thouroughly reliable; an algorithm for calculating three-phase equilibria (L1-L2-G) has been developed to allow the simulation of CO2 injection into shallow reservoirs. An industrial version will be available in October 1987.

```
******************************************************************************
* TITLE : ADDITIVES TO IMPROVE SWEEPING DURING    *      PROJECT NO         *
*         GAS INJECTION (PHASE I)                 *                         *
*                                                 *    TH./05059/84/FR/..   *
*                                                 *                         *
******************************************************************************
* CONTRACTOR :                                    * PROGRAM :               *
*   GERTH                                         *   HYDROCARBONS          *
*   AVENUE DE BOIS PREAU 4       TEL 1 47 52 61 39 *                        *
*   FR - 92502 RUEIL-MALMAISON TLX 203 050        *                         *
*                                                 * SECTOR :                *
*                                                 *   SECONDARY AND         *
* PERSON TO CONTACT FOR FURTHER INFORMATION :     *   ENHANCED RECOVERY     *
*   MR. ROBIN                    TEL 1 47.49.02.14 *                        *
*                                                 *                         *
*                                                 *                         *
******************************************************************************
                                                   VERSION : 01/01/87
```

AIM OF THE PROJECT :
****
When using an enhanced recovery method requiring gas injection, a low oil
residual saturation can be obtained in the swept areas. Unfortunately, due to
its density, this gas tends to flow in the upper part of the formation. Also,
in the case of heterogeneous or cracked reservoirs, the important mobility of
the gaseous phase favours an instability of the displacement and the passage of
the gas in higher permeability drains, thus provoking a premature breakthrough.
Once the injected fluid has reached the production well, the volume of the
reservoir subjected to sweeping only increases slightly.
 To improve the sweeping coefficient, it is envisaged to block the areas
preferentially flooded. This can be obtained by a combined injection of gas and
additives likely to form a foam within the porous medium. The gas would thus be
diverted towards non-flooded areas, which would allow to incrase the recovery.
ease the recovery.

PROJECT DESCRIPTION :
****
This project "Additives to improve sweeping during gas injection"comprises 3
phases:
Phase 1 - Methodology
The object of this phase is to evaluate the foaming capacity of commercial
products, or of original products provided by the chemical industry.
A physical property of surfactant solutions, characteristic of their aptitude
of forming foam is being sought. This property is being observed under
pressure and temperature up to values representing those existing in the
reservoirs.
 This phase is valid both in porous medium and out of porous medium.
Phase 2 - Combined injection of steam and foaming agents
A specific study has been carried out concerning the evolution of the foaming
properties with the temperature and the pressure. Operating conditions retained
are of 20 to 100 bars for the pressures and of 200 to 300 deg.C for the
temperatures (that is to say temperatures corresponding to the vaporization of
the water at the above mentioned pressures).

The experimental programme includes foam injection experiments in porous medium run in cylindrical laboratory cells,containing homogeneous or non-homogeneous porous media, formed by non-consolidated sand. The experiments will be run either under isothermal conditions or under adiabatic conditions.

Phase3 - Combined injection of gas and foaming agents

A specific study has been carried out concerning the compatibility of surfactant solutions with reservoir fluids. We are more particularly interested in the water salinity and in the influence of the presence of hydrocarbonate phase in the porous medium.

The experimental programme includes expriments in porous medium ran in cylindrical laboratory cells, containing consolidated porous media, whether homogeneous or not. These experiments are run under isothermal conditions.

STATE OF ADVANCEMENT :
****

The methodology for the evaluation of the surfactants is now finalized (Phase 1 completed).Studies relevant to phases 2 and 3 are underway.

For phase 2, original products provided by the chemical industry are presently being evaluated. For phase 3, a study is being carried out for commercial products.

RESULTS :
****

We have finalized a methodology allowing to evaluate the foaming properties of surfactant solutions, and to follow the evolution of this property under pressure and temperature.

The effect of the gravity or of the contrasts of the permeability on the gas sweep under ambient conditions has also been evidenced.Various types of surfactants have been used, some being of commercial origin,others still under experiment. The influence of the foam sensitivity in the presence of a hydrocarbonate phase has been more particularly evidenced. The action of foams deriving from these surfactants has been visualized in porous medium. We are equipped with devices and a technique allowing to evaluate, under pressure and temperature conditions, the efficiency of these different surfactants on oil recovery.

Experience shows that foam flows in porous medium are very long to stabilize (several tens of volumes of pores to inject before stabilization). Hence, it was not possible to exploit some of the tests owing to a lack of permanent flowing pattern.

Studies are being pursued on products likely to foam under pressure and temperature conditions,but the important fact is that the formed foam is not sensitive to the presence of a hydrocarbonate phase.

REFERENCES :
****

1 M.ROBIN
UTILISATION OF FOAMING AGENTS TO IMPROVE THE EFFICIENCY OF THE VAPOUR INJECTION.
3RD EUROPEAN COLLOQUE ON THE IMPROVEMENT OF ENHANCED OIL RECOVERY. ROME 16-18 APRIL 1985. TECHNIP PUBLICATIONS.
2. J.BURGER, P.SOURIEAU AND M.COMBARNOUS
ENHANCED OIL RECOVERY BY THERMAL METHODS.
TECHNIP PUBLICATIONS,PARIS 1985.
3. J.H.DUERKSEN
LABORATORY STUDY OF FOAMING SURFACTANTS AS STEAM DIVERTING ADDITIVES.
CALIF.REGION.MEETING OF SOC.PETROLEUM ENGRS.SPE PAPER 12785.APRIL 1984.
4.J.P.HELLER
RESERVOIR APPLICATION OF MOBILITY CONTROL FOAMS IN CO2 FLOODS.
SPE/DOE 4TH SYMP.ON ENHANCED OIL RECOVERY.SPE PAPER 12644;TULSA,APRIL 1984.

```
*****************************************************************************
* TITLE : ENHANCED OIL RECOVERY BY MISCIBLE GAS    *      PROJECT NO       *
*         INJECTION IN TERTIARY CONDITIONS         *                       *
*                                                  *    TH./05061/85/FR/..  *
*                                                  *                       *
*****************************************************************************
* CONTRACTOR :                                     * PROGRAM :             *
*   GERTH                                          *   HYDROCARBONS        *
*   AVENUE DE BOIS PREAU 4      TEL 1 47.52.61.39   *                       *
*   FR - 92502 RUEIL-MALMAISON TLX 203 050          *                       *
*                                                  * SECTOR :              *
*                                                  *   SECONDARY AND       *
* PERSON TO CONTACT FOR FURTHER INFORMATION :      *   ENHANCED RECOVERY   *
*   MISS. C. BARBOUX            TEL 1 30.52.92.92   *                       *
*                               TLX 696793          *                       *
*                                                  *                       *
*****************************************************************************
                                                      VERSION : 01/01/87
```

AIM OF THE PROJECT :
****
Many reservoirs have long been subjected to water drive. The aim of the project
is to specify, in particular through laboratory work, the conditions of
optimization of injecting dynamically miscible gas into a reservoir already
operated by water injection. This new tertiary enhanced recovery method should
enable 25 to 30% of the residual oil to be recovered, our objective being to
quantize more exactly this saving in recovery for different types of reservoirs.
It is of utmost interest to determine the economics of a tertiary miscible
recovery process, whose optimization will be reached by a detailed kmowledge of
the implementing mechanisms which is not actually available.

PROJECT DESCRIPTION :
****
The efficiency of miscible gas injection under tertiary conditions is related
to the formation of an oil bank by remobilization followed by coalescence of
the drops of oil trapped in dispersed fashion.
The following questions arise :
- how do the mechanisms of diffusion, dispersion, mass transfers implement the
creation and the growth of the oil bank?
- what is the influence of the petrophysical properties of the reservoir?
- what procedures must be complied with so that the laboratory results can be
considered as usable for the following reservoir simualtion?
The project is to take place as follows :
- PHASE 1 : Study of the mechanisms.
The purpose is to identify for each type of dynamic miscibility (frontal, rear,
partial) the critical parameters playing the major role to implement the
creation and the growth of the oil bank :
. mechanisms of dispersion, diffusion
. polyphasic flow mechanisms
The experimental part of the work consists mainly in displacement experiments
using synthetic fluids in order to facilitate the interpretation of the results.
 The numerical simulation of these experiments is to show if adaptations of the
available numerical simulators are necesary to match their history.

- PHASE 2 : Extension to actual cases.
This phase consists of verifying the statements obtained in phase 1 by studying
the sentivity of the tertiary recovery process on actual field conditions
(displacemnet experiments with fluids and rocks from different types of
reservoir).
- PHASE 3 : Feasibility of a pilot injection project.
The choice of the reservoir for this feasibility study. will be done according
to the results of the preceding steps. Much attention will be devoted in order
to select the best candidate among the different fields preferentially situated
in the European zone where the contractor, through participations or operations,
 has already got a good preliminary information. This phase will consist of a
complementary laboratory study, if needed, the reservoir engineering study, the
pre-development study and the economical aspects.

STATE OF ADVANCEMENT :
****
Completed phase 1 : displacement experiments consisting in waterflooding
followed by alternate injection of miscible gas and water, with a limitation of
the total quantity of gas to be injected.
Completed phase 2 : selection of th rock samples of a North Sea reservoir for
the constitution of a long core model. Selection of the miscible gas to be
injected in the first displacement experiment.
Ongoing phase 2 : the first displacement experiment using actual rock and
fluids of a North Sea reservoir

RESULTS :
****
Phase 1
Both frontal and near miscibility, as tertiary recovery process, were proven to
be very promising in terms of their efficiency to displace the oil trapped
during the water drive.
It is also proven that the critical parameters or mechanisms for the formation
of the oil bank differ following the type of miscibility :
The gas advance is completely stabilized by the mass transfers at the front,
and therefore is not very dependent on the mobility ratio in case of frontal
miscibility.
The gas advance is much more dependent on the mobility ratio in case of near
miscibility.
These conclusions arise from two sets of displacement experiments which were
realized with the same operating conditions (physical model, porous medium,
injection rate, position, dynamic miscibility..) the only difference being the
repartition of intermediate components between the oil and the gas.
This reason why alternate gas-water injection which ids thourgh to act on the
mobility ratio and to enable the optimization of the quantity of gas to be
injected, was tested first in case of near miscibility. If we define the
efficiency of the tertiary gas injection as the ratio oil recovery/volume of
gas injected, both expressed as a percentage of the pore volume, the first
results show that the efficiency is increased considerably by the alternation
gas-water.
We plan to study now the optimization of the size of the slugs.

REFERENCES :
****
DOCTORATE THESIS FROM UNIVERSITY OF BORDEAUX BY D. HADIATNO

```
******************************************************************************
* TITLE : ADDITIVES FOR THE IMPROVEMENT OF      *       PROJECT NO        *
*         SWEEPING DURING STEAM DRIVE - PHASE 2  *                         *
*                                                *    TH./05062/85/FR/..   *
*                                                *                         *
******************************************************************************
* CONTRACTOR :                                   * PROGRAM :               *
*   GERTH                                         *   HYDROCARBONS          *
*   4, AVENUE DE BOIS PREAU      TEL 1 47.52.61.39*                         *
*   FR - 92502 RUEIL-MALMAISON TLX 203 050        *                         *
*                                                 * SECTOR :               *
*                                                 *   SECONDARY AND         *
* PERSON TO CONTACT FOR FURTHER INFORMATION :     *   ENHANCED RECOVERY     *
*   MR. SAHUQUET                 TEL 59.83.40.00  *                         *
*                                TLX 560 804      *                         *
*                                                 *                         *
******************************************************************************
                                                  VERSION : 15/02/87
```

AIM OF THE PROJECT :
****
Enhanced oil recovery operations through continuous steam drive are often
penalized by the irregular displacement of the injected steam owing to the
reservoir heterogeneities. Hence, this phenomenon provokes sweeping anomalies
and often a final recovery rate that is lower than expected. The solution
consists in injecting, together with the steam, a foam which blocks the
preferential passages and thus improves sweeping.
The object of this project is to determine the optimal application conditions
of this technique while setting up a real size pilot on the field.

PROJECT DESCRIPTION :
****
This project includes 3 phases :
- Preparation of the pilot
- Execution of the pilot and tests
- Interpretation of tests.
The first phase consists of choosing a test site, where sweeping anomalies have
already been observed and located during steam drive, and thus a site where it
may be easier to assess the efficiency of the process. Once this site is chosen,
 laboratory tests will allow, under real reservoir conditions, to test the
foaming agents and their stability, and to choose the type of gas to inject so
to obtain good performances. These tests should also allow to specify the
optimal operating conditions.
The second phase will concern the realization of the pilot on the chosen site.
After a number of zone characterization tests (well tests, interference tests,
tracers...) it will be possible to begin the foam injection tests. According to
forecasts, the pilot will require a series of foam plug injections in order to
obtain a significant and lasting effect. Various formulations and operating
conditions are necessary to optimize the process.
The third phase will consist in interpreting these tests : evaluation of
sweeping and recovery rate improvement, technical and economical result of the
process.

STATE OF THE PROJECT
****
Ongoing.
The first phase may be considered as completed.

RESULTS :
****
Having chosen a site in California, on an operated field that showed evident sweeping anomalies, it was possible to run the preliminary tests which then allowed us to specify the initial operating conditions.
However, during 1986, owing to the oil price raise, steam injection on this field has had to be interrupted. Another site is presently being sought for to pursue the project and exploit the previous operations. This change of site should not effect the conclusions of the work carried out to date.

```
*****************************************************************************
* TITLE : THE BENEFICIAL EFFECTS OF DISTILLATION    *      PROJECT NO       *
*         DURING OIL RECOVERY BY STEAMFLOODING      *                       *
*                                                   *    TH./05064/85/NL/..  *
*                                                   *                       *
*****************************************************************************
* CONTRACTOR :                                      * PROGRAM :             *
*   BUREAU HOGESCHOOL                               *   HYDROCARBONS         *
*   COLLEGE VAN BESTUUR          TEL 015 784378     *                       *
*   JULIANALAAN 134              TLX 38151          *                       *
*   NL - 2628 BL DELFT                              * SECTOR :              *
*                                                   *   SECONDARY AND        *
* PERSON TO CONTACT FOR FURTHER INFORMATION :       *   ENHANCED RECOVERY    *
*   DR. J. BRUINING                                 *                       *
*                                                   *                       *
*                                                   *                       *
*****************************************************************************
                                                    VERSION : 13/11/86
```

AIM OF THE PROJECT :
****
The main objective of the project is to show that steamflooding, which is
applied routinely to the recovery of heavy oil, can be applied economically to
the recovery of medium viscosity oil i.e. can compete with waterflooding.
Another objective is to assess the beneficial effect of the distillable oil
bank. In our previous project we have elaborated at length on the competition
between film flow effects and the distillable oil bank to lower the "residual
oil saturation" in the steam zone. In our present project we shall focus our
attention on the sweep enhancement owing to a more favourable mobility ratio
when a distillable oil bank is present.

PROJECT DESCRIPTION :
****
Steam displacement applied to heavy oil reservoirs, with a low primary and
secondary potential, leads to an appreciable extra oil recovery. Indeed, steam
drive is routinely applied in heavy oil reservoirs.
Application of steamdrive to medium viscosity oil reservoirs (with a lower
primary and secondary recovery potential) leads to a higher sweep efficiency
than its application in heavy oil reservoirs. Also distillation effects will
give a larger contribution to the recovery efficiency or medium viscosity oils.
These distillation effects have been attributed to the distillable oil bank,
which is formed near the steam condensation front. This leads to a lower (heavy)
 oil saturation in thee steam swept zone. Furthermore, owing to an improved
mobility ratio the sweep efficiency will be enhanced by these distillation
effects.
The innovating aspect of the present project is to improve with the help of
(high pressure) experiments and visual studies existing mathematical/physical
models with the aim to quantify the sweep efficiency depending on oil viscosity
and oil composition.
In this way we want to investigate under which circumstances steam flooding can
be applied to the recovery of medium viscosity oils i.e. can compete with water
flooding. Where possible use will be made of field data for which we can call
upon the Nederlandse Aardolie Maatschappij.

The project consists of the following phases:
1. Design and building of a reactor with auxiliary equipment
2. Visual studies of the steamdrive process
3. Execution of the experiments
4. Theoretical description of the experimental results and extrapolation to field conditions
5. Economical evaluation between steam and water drive of medium viscosity oils.

STATE OF ADVANCEMENT :
****
The project is in its design phase. There is a high pressure rig available to perform experiments at elevated pressure. A rig of somewhat larger dimensions will be designed to allow sweep efficiency studies at medium pressure (10 bars). Visual studies and surface chemistry aspects will play a vital role in this present project.

RESULTS :
****
As the project is in its starting phase we can only report one result. A finite element programme has been developed which predicts the angle of inclination of the steam condensation front. The calculations agree well with an extended Dietz theory for the angle of inclination for water oil interfaces.
The final product will be an economic comparison between steam and water drive recovery of a schematic reservoir in which use can be made of an improved understanding of the displacement mechanisms.
The results of our previous project, related to the present project, has been described in 15 reports and articles, as described in its end-report entitled "Distillation enhanced steamdrive recovery".

REFERENCES :
****
J.BRUINING, D.N. DIETZ, W.H.P.M. HEIJNEN, G. METSELAAR, J.W. SCHOLTEN AND A. EMKE, "ENHANCEMENT OF DISTILLATION EFFECTS DURING STEAMFLOODING OF HEAVY OIL RESERVOIRS". NEW TECHNOLOGIES FOR THE EXPLORATION AND EXPLOITING OF OIL AND GAS RESOURCES, PROC. 2ND EC SYMPOSIUM LUXEMBOURG, 5-7 DECEMBER 907-913
J. BRUINING, D.N. DIETZ, G. METSELAAR, J.W. SCHOLTEN, A. EMKE AND F. HUBNER, "IMPROVED RECOVERY OF HEAVY OIL BY STEAM WITH ADDED DISTILLATES", 3RD EUROPEAN MEETING ON IMPROVED OIL RECOVERY PROC I (1985) 371-378.

```
******************************************************************************
* TITLE : CONVERSION OF HEAVY OIL, BITUMEN AND      *      PROJECT NO        *
*         REFINERY RESIDUE INTO LIGHT BOILING       *                        *
*         DISTILLATES                               *      TH./05067/85/DE/.. *
*                                                   *                        *
******************************************************************************
* CONTRACTOR :                                      * PROGRAM :              *
*   VEBA OEL AG                                     *   HYDROCARBONS         *
*   DE - 4650 GELSENKIRCHEN-HA TEL 0209 366-7968    *                        *
*                             TLX 824881-90         *                        *
*                                                   * SECTOR :               *
*                                                   *   SECONDARY AND        *
* PERSON TO CONTACT FOR FURTHER INFORMATION :       *   ENHANCED RECOVERY    *
*   DR. R. HOLIGHAUS                                *                        *
*                                                   *                        *
*                                                   *                        *
******************************************************************************
                                                       VERSION : 27/11/86
```

AIM OF THE PROJECT :
****
- Conversion of non-boiling crude oil fractions under hydrogen pressure into
light boiling distillates.
- Upgrading of natural bitumen and refinery residues with :
- high percentage of non-boiling components
    - high asphalt -, sulfur - and nitrogen contents
    - high metal contents
    - high viscosity and density.
The hydrogenation is performed in a cascade of liquid-phase reactors and gas-
phase reactors. A one-way additive in the liquid-phase reactors is normally
used.

PROJECT DESCRIPTION :
****
The VCC-Process was derived from the Bergius-Pier technology and was applied in
a precursor-process called "Scholven Combi-Chamber".
The VCC-Process developed by VEBA OEL in cooperation with LURGI is
characterized by its extremely high conversion efficiency up to 95 wt. % based
on non boiling residue.
The project will develop in the following phases :
Phase I   : Basic bench scale research
Phase II   : Design and construction of a pilotplant for a throughput of
                 1 t/h
Phase III   : Operating phase of the pilotplant
Phase IIIA   : Second operating phase of the pilotplant
Phase IV   : design and construction of an industrial-scale plant for
demonstration. The present state of development is the second operating phase
of the pilotplant (Phase IIIA).

The development covers the following fields of experiments and theoretical research
- Additive optimization
- Pressure reduction
- Establishment of design data
- Establishment of scale-up factors
- Extension of feedstock basis
- Changes in process configuration
- Screening of gase phase catalysts
- Longtern tests

The central objectives of these tests is to develop design data for commercial VCC-plants and to optimize the process for each application.

STATE OF ADVANCEMENT :
****

Ongoing the first operating phase (Phase III) in the present phase (phase IIIA) several plant modifications had been realized. The plant is in a good technical condition and the test program is performed as scheduled.

RESULTS :
****

By application of the VLC/VCC-technology several vacuum residues from conventional and heavy crudes (arab heavy, bachaquero, tia juana, morichal) and from a visbreaking plant were converted to an extreme high degree (greater than 90 perc.) into light distillates, some hydrocarbon gases and a small amount of hydrogenation residue which is suitable for the production of hydrogen via partial oxidation.

Feeding a vacuum bottom from a typical venezuelan crude with the following analytical data

Carbon		84,8	WTPERC.
Hydrogen		10,4	WTPERC.
Sulphur		3,3	WTPERC.
Nitrogen		0,6	WTPERC.
Vanadium	630		PPM
Nickel	75		PPM

to the process as an example 80 WTPERC. of a VCC syncrude is produced which contains 27 perc. of naphtha, 48 perc. of middle distillates and 25 perc. of vacuum gasoil. Due to the application of the so-called gas phase hydrogenation (a catalytic fixed bed reactor directly combined with the primary conversion step) this VVC syncrude having an relatively high hydrogen content is almost sulphur and nitrogen free (less than 200 ppm each). The VCC middle distillates can be sold directly, the VVC naphtha meets reformer feed spezification and the vacuum gasoil is an excellent feedstock for a FFC or a hydrocracker unit.

As these results can be generalized for all these residues processed up to now it can be concluded that the VLC/VCC process is a well advanced technology to convert less valuable bottoms into light distillates thus making maximum use of the "Bottom of the Barrel".

REFERENCES :
****

U. GRAESER, K. NIEMANN
OIL GAS JOURNAL 80, NR. 12, 121 (1982)
U. GRAESER, K. KRETSCHMAR, K. NIEMANN
ERDOEL UND KOHLE, ERDGAS, PETROCHIMIE 36, NR. 8, 362 (1983)
U. GRAESER, K. NIEMANN
PREPRINTS, DIV. PETR. CHEM. AM. CHEM. SOC. 28, NR. 3, 675 (1983)

```
********************************************************************************
* TITLE : PHYSICAL AND NUMERICAL MODELLING OF IN    *       PROJECT NO        *
*         SITU-COMBUSTION                           *                         *
*                                                   *   TH./05068/85/DE/..    *
*                                                   *                         *
********************************************************************************
* CONTRACTOR :                                      * PROGRAM :               *
*   INSTITUT FUER TIEFBOHRTECHNIK                   *   HYDROCARBONS          *
*   ABT. LAGERSTAETTENTECHNIK   TEL 05323/72 26 18  *                         *
*   AGRICOLSTR. 10              TLX 953813          *                         *
*   D-3392 CLAUSTHAL-ZELLERFEL                      * SECTOR :               *
*                                                   *   SECONDARY AND        *
* PERSON TO CONTACT FOR FURTHER INFORMATION :       *   ENHANCED RECOVERY    *
*   PROF. DR.MONT. GUENTER PUS                      *                         *
*                                                   *                         *
*                                                   *                         *
********************************************************************************
```
VERSION : 27/11/86

AIM OF THE PROJECT :
****
The objective of this project is to improve the physical modelling and
numerical simulation of oil recovery supported by in situ combustion. For this
purpose, the influence exerted by the oil composition (especially the resin and
asphaltene contents) on the generation of fuel during in situ combustion is to
be investigated. The reaction kinetic data are to be determined as functions of
the oil composition under isothermal conditions. With the aid of the data thus
ascertained, a physically more exact description of the in situ combution
process is envisaged on the basis of numerical models (at IFP, Paris). The
results concerning reaction kinetics are to be employed in the investigations
of the relationship between the oil composition and propagation velocity of the
combustion front. For this purpose, combustion tests are to be conducted in a
linear combustion cell under largely adiabatic conditions.

PROJECT DESCRIPTION :
****
The research project has been subdivided into two phases.

.
PHASE 1
This phase comprises three stages for investifation the combustion process.
During the first stage, the relationship between the oil composition and
generation of fuel is to be investigated. During the second stage, the effect
of the oil composition on the reaction kenetics during combustion is to be
examined. During the third stage, the influence of the oil composition on the
propagation velocity of the combustion front is to be investigated with the use
of the results from stages 1 and 2.

.
PHASE 2
During this phase, numerical models for in sity combustion are to be elaborated
in two stages.
Stage 1 comprises the development of a model without considering the fluid
transport.
In stage 2, a model is constructed with the fluid transport taken into account.

TECHNICAL DESCRIPTION

An appropriate crude oil is enriched with its own asphaltenes and resins for obtaining various types of model oil differing mutually in colloidal-chemical composition. The pyrolysis of the different model oil types is performed under inert gas, and the quantity of fuel thereby generated is determined. A differential reactor is being constructed for the purpose.An attempt will be made to derive a satisfactory correlation between the quantity of fuel thus formed and the oil composition. The combustion of the coke is to be investigated under isothermal conditions in the presence of air in a temperature range between 350 and 650 deg.C. The quantitative and qualitative analyses of the reaction gases, as well as the combustion temperature and pressure constitute the basis for the determination of the reaction rate constant for the oxidation. By means of linear combustion tests, the effect of the oil composition on the propagation velocity of the combustion front is to be investigated.

The construction of the numerical models begins with a simplified radial model without considering kinetics and transport processes. Heat exchange, reaction kinetics, and fluid transport are then taken into account, thus resulting in a two-dimensional model. The block size for the process description is thereby matched to the reservoir scale. The models are to be tested by means of field experiments conducted in the classical manner..

STATE OF ADVANCEMENT :
****

As scheduled in the working plan, the project is currently in the stage of preliminary experimentation, treatment of the crude oil, contruction of measuring cells and other apparatus for determining the kinetic data, as well as the experimental facility for the linear combustion tests. The intended project steps can be executed within the period proposed.

RESULTS :
****

As planned, results have not yet been obtained for the present status of the project.

```
*****************************************************************************
* TITLE : A FLOATING BARGE-MOUNTED PLANT,          *       PROJECT NO       *
*         PRODUCING HIGH-PRESSURE N2/CO2 GASES     *                        *
*         FOR EOR PROJECTS IN THE NORTH SEA        *    TH./05069/85/DE/..   *
*                                                  *                        *
*****************************************************************************
* CONTRACTOR :                                     * PROGRAM :              *
*   SALZGITTER AG                                  *   HYDROCARBONS         *
*   ABTEILUNG FORSCHUNG UND EN TEL 030 88 42 97 15 *                        *
*   POSTFACH 15 06 27          TLX 185 655         *                        *
*   DE - 1000 BERLIN 15                            * SECTOR :               *
*                                                  *   SECONDARY AND        *
* PERSON TO CONTACT FOR FURTHER INFORMATION :      *   ENHANCED RECOVERY    *
*   DR-ING G. PIETSCH                              *                        *
*                                                  *                        *
*                                                  *                        *
*****************************************************************************
                                                      VERSION : 13/11/86
```

AIM OF THE PROJECT :
****
As the world's reserves of crude oil diminish, the low yield of often less than
30% from a deposit becomes the central aspect of energy policies. The injecting
of CO2 or N2 represents one way of stimulating oil recovery. So far the use of
such tertiary production measures has been restricted to onshore fields. Such
measures, however, take on greater economic significance for offshore fields.
The investment necessary to develop an offshore site exceeds that required for
an onshore location by many times. This makes it even more important that the
amount of oil extracted from an offshore deposit should be as high as possible.
The proposed technical development aims at providing low-cost gas which will
then be used to recover the oil offshore by tertiary measures. In awareness of
the conditions specific to offshore sites, this means that the production
facilities will have to be installed on a mobile carrier, i.e. barges that can
be used at several locations irrespective of water depth.

PROJECT DESCRIPTION :
****
Assuming that in future large quantities of CO2/N2 and steam will also be
required for the tertiary production of oil offshore, a combined process is
proposed which features a particularly high degree of economy. The process is
made up of the following three main components:
a. production of an N2/CO2 mix by burning preferably gaseous hydrocarbons,
whereby high-pressure steam is also generated,
b. production of N2 in an air separation plant,
c. compression of the CO2/N2 mix by using the high-pressure steam produced
during combustion.
The combination of combustion and air separation plant results in a substantial
decrease in the cost of injection gas when drawing comparisons with each
separate process.
The main items of the project:
- CO2/N2 production using a boiler and steam turbine
- CO2/N2 production using an air separation plant with an increased delivery
  pressure for N2
- application of gas turbines to produce CO2/N2 with gas turbines driving
  the compressors directly
- basic engineering for the combined production of N2/CO2 and high-pressure
  steam when using steam and gas turbines

- determination of energy requirements and comparison of the capital investment and operating costs for each alternative
- development of a prestressed concrete barge as a floating carrier for a plant to produce N2/CO2 high-pressure gas
- design of a mooring system with central articulated joint and integrated gas conveyance for N2/CO2 high-pressure gas and high-pressure steam, as well as for supplying the plant with low-pressure fuel
- summary covering capital investment and costs, and final economic analysis.

STATE OF ADVANCEMENT :
****
The project is in the design phase.

RESULTS :
****
The Norwegian and British oil fields in the North Sea were examined with regard to their yearly production rate and the remaining production capacity. A possible use of the EOR barge can be envisaged for those fields whose regular capacity will be exhausted in the foreseeable future.
First the production capacity of the producing plant was determined. Since the concept of an EOR gas barge aims at a multiple use of the barge in various oil fields, the design value was selected higher than the average of the expected cases of application. The concept of a multiple use precludes from the outset the fixing of a determined pressure. The gas processing plant should therefore be conceived in such a way that an adaptation to various pressures can be made with as low an amount of alteration as possible. The plant should be conceived in the form of a basic outfit for 200 bar injection pressure, and should be laid out such that either and additional compressor set can be installed for compression to a higher pressure, or only one additional compressor which can be coupled onto the shaft of an existing set.
Associated gas is planned to be used as fuel which is obtained on the oil production platform.
In order to determine the most favourable shape of the barge under the aspect of its sea behaviour, two different pontoons were subjected to an investigation.
The size of the deck area is governed by the processing plant which is to be housed. The plant assemblies are partly located on the pontoon's bottom and partly on the deck.
It is intended to conduct the hydrodynamic evaluation in two stages. At first the barge will be tested while freely floating during various sea states without any anchoring (calculation for two types of barge designs). Thereafter an anchoring calculation will be conducted using the optimum barge design.
The purpose of the anchoring system is, in the first instance, to anchor the concrete barge together with the gas producing plant in a permanent way, and additionally it is used to anchor the gas supply lines leading from the seabed to the barge. The demands placed on the gas line could be fulfilled by using a tower for anchoring. Here the high-pressure line can be run within the tower so that repairs, maintenance and inspections can be made under atmospheric conditions.
At present the parameters for the hydrodynamic evaluation of the anchoring system are being determined. These include the geometry of the articulated tower and also the masses, i.e. the buoyancy volume of the various tower elements.

```
*********************************************************************************
* TITLE : ENHANCED OIL RECOVERY FROM LOW        *        PROJECT NO           *
*         PERMEABLE CHALK RESERVOIRS (FOLKA)     *                             *
*                                                *    TH./05070/85/DK/..        *
*                                                *                             *
*********************************************************************************
* CONTRACTOR :                                   * PROGRAM :                   *
*   COWICONSULT/CONSULTING ENGINEERS AS          *   HYDROCARBONS              *
*   TEKNIKERBYEN 45              TEL 2 85 73 11   *                             *
*   DK - 2830 VIRUM              TLX 37280        *                             *
*                                                * SECTOR :                    *
*                                                *   SECONDARY AND             *
* PERSON TO CONTACT FOR FURTHER INFORMATION :    *   ENHANCED RECOVERY         *
*   MR. A. BJERRUM                               *                             *
*                                                *                             *
*                                                *                             *
*********************************************************************************
```

VERSION : 27/11/86

AIM OF THE PROJECT :
****
The aim of the FOLKA project is to investigate some problems related to
waterflooding and gas injections in low permeable chalk reservoirs.
The project includes 4 phases:
- Planning and preliminary studies
- Gas injection miscibility conditions
- Chalk wettability conditions
- Reservoir modelling, field studies
The FOLKA project includes a laboratory program using samples from Danish and
Norwegian chalk reservoirs. The laboratory program is carried out by IKU in
Trondheim.

PROJECT DESCRIPTION :
****
PHASE 1 - PLANNING AND PRELIMINARY STUDIES
- Preparation of the work program
- Selection of an injection gas/reservoir fluid system (recombined samples)
from a Danish reservoir. Compilation of a PVT report based on available
laboratory data including viscosities and densities of the gas and liquid
phases. Determination of reservoir pressure and reservoir temperature.
PHASE 2 - GAS INJECTION MISCIBILITY STUDIES
- Perform phase behaviour studies on injection gas and reservoir fluid from a
Danish reservoir (multiple contact miscibility studies with saphire cell,
pressure-composition studies including swelling of the liquid phase).
Condensing gas drive shall be studied with gas bubling through the reservoir
fluid at reservoir conditions but pressure will be increased until miscibility
is obtained.
- Perform slim-tube displacement experiments upacked with chalk to determine
the minimum miscibility pressure for the given system.
In conjunction with the wettability studies (see phase 3) core displacements
will be performed using the same fluid system as in the miscibility studies.
The displacement efficency, will be determined.
Computer modelling of the displacement experiments. Two compositional models
shall be used: cell to cell model and finite element model to be developed.

PHASE 3 - CHALK WETTABILITY CONDITIONS
- Determine the wettability of chalk from a Danish reservoir both at low
pressure and reservoir conditions using Amott test, contact angle measurements
and cappillary pressure tests.
- Perform gas-oil displacement experiments at reservoir conditions using the
same fluid system as used in the miscibility studies.
- Computer modelling of the displacement experiments.
PHASE 4 - RESERVOIR MODELLING, FIELD STUDY FOR A SELECTED RESERVOIR
- Reservoir modelling using a 30 reservoir simulator based on a reservoir
description laboratory data obtained in phase 2 and 3 shall be used. Two cases
shall be considered for the present reservoir: gas injection, water injection.
- Development of a screening guide for improved recovery from chalk reservoirs
based on the rsults from 4.1 and on a technical/economic analyses of the
secondary recovery field development.

STATE OF ADVANCEMENT :
****
Ongoing. Laboratory experiments (phase 2) nearly completed.

REFERENCES :
****
- "ENHANCED OIL RECOVERY FROM LOW PERMEABLE CARBONATE ROCK RESERVOIRS PHASE 1 -
LITTERATURE REVIEW". (COWICONSULT,JULY 1983)
- "MISCIBLE GAS INJECTION PHASE 1 - DETERMINATION OF OIL AND GAS MISCIBILITY
PERFORMANCE". (COWICONSULT, JUNI 1985)

```
******************************************************************************
* TITLE : DEVELOPING AND TESTING A CHEMICAL        *      PROJECT NO         *
*         SYSTEM FOR POLYMER FLOODING IN OIL       *                         *
*         RESERVOIRS WITH HIGHLY SALINE FORMATION  *   TH./05072/85/DE/..    *
*         WATER                                    *                         *
******************************************************************************
* CONTRACTOR :                                     * PROGRAM :               *
*   BASF AG, WINTERSHALL AG                         *   HYDROCARBONS          *
*   JOINT WORKING GROUP          TEL 0621/60-47358  *                         *
*   DE - 6700 LUDWIGSHAFEN       TLX 46499-0        *                         *
*                                                   * SECTOR :                *
*                                                   *   SECONDARY AND         *
* PERSON TO CONTACT FOR FURTHER INFORMATION :       *   ENHANCED RECOVERY     *
*   DR. MARTISCHIUS                                 *                         *
*                                                   *                         *
*                                                   *                         *
******************************************************************************
                                                    VERSION : 09/03/87
```

AIM OF THE PROJECT :
****
The efficiency of polymer flooding in reservoirs with highly saline formation
water depends to a large extent on the polymer's stability towards salt.
Desalination or the use of a polymer that is efficient only at high
concentrations may adversely affect the economics or technical feasibility of
tertiary recovery. It is therefore necessary to develop a chemical system that
can be applied in highly saline formation waters.
Once a given field or part of a field has been selected for a pilot test, the
data relating to it must be compiled.

PROJECT DESCRIPTION :
****
A simulation study is intended to predict the optimum metering rate for the
polymer and the production pattern. For this purpose, polymer-specific data for
the field selected will be determined on samples of the reservoir rock.
POLYMERS FOR ENHANCED OIL RECOVERY. Polymers produced in the laboratory have to
be tested to determine whether they are suitable for use in reservoirs with
highly saline formation water. If they are found to be suitable, their
reproducibility must be checked on a pilot and a production scale.
PRODUCT SCREENING. The conditions for dissolving and shearing on a technical
scale are to be varied so that the optimum relationship between viscosity and
filterability is obtained under the conditions in a reservoir with a salinity
of about 150 g/l of total dissolved solids.
DETERMINATION OF THE PROPERTIES RELATING TO FLOODING ON WATER-WET MODEL CORES.
Flooding tests are performed on cores of different permeability in the range of
intended injection and production rates. Adsorption and retention effects of
the polymer are determined.
OPTIMIZATION OF THE POLYMERS' LONG-TERM STABILITY. The effects exerted by
various additives, additive concentrations and additive combinations on the
polymers' long-term stability under the conditions in the reservoir.
DETERMINATION OF THE PROPERTIES RELATING TO FLOODING ON OIL-WET MODEL CORES.
The stability of the solutions in the presence of an oil phase and the
chromatographic separation of the polymers are investigated.

SELECTION OF THE RESERVOIR. A field must be selected that might appear suitable for polymer flooding in the light of its data, e.g. mobility ratio, average permeability, permeability variation, formation thickness, mineralogy, stage of depletion, and the maximum degree of depletion attainable by waterflood. The next step is the selection of a suitable, well-delineated reservoir block, for which purpose geological and reservoir engineering data must be reviewed. The possibility of communication with wells in adjacent blocks must be investigated by pulse tests and/or pressure measurements.

PETROPHYSICAL INVESTIGATIONS. Coring in unconsolidated Valendis reservoir by special techniques and embedding the deep-frozen cores with inorganic solvent - resistant material allow determination of permeability, porosity, compressibility, capillary pressure, wettability, relative permeability, initial oil saturation, and residual oil saturation after waterflooding. In addition, mineralogical studies are performed.

SIMULATION STUDY. A history match and feasibility tests to optimize flooding geometry and polymer slug volumes have to be carried out.

DETERMINATION OF DATA RELATING TO THE POLYMER ON ORIGINAL CORES AT THE RESERVOIR TEMPERATURE. Mobility and permeability reduction are determined as well as retention and adsorption effects in large cores. Based on the injectivity of the polymer it has to be decided, whether a hydrofrac is necessary.

STATE OF ADVANCEMENT :
****
The project is still ongoing. The polymer development on a laboratory scale is finished, the scale up is going to be realized. An area in the eastern part of the field Dueste-Valendis is favoured, petrophysical investigations and original core tests as well as simulation studies and tracer tests still have to be performed.

RESULTS :
****
Anionic and nonionic synthetic polymers with different chemical characteristics show only a poor thickening behaviour in reservoir water of a high salinity. The effect exerted by polyacrylamides and acrylamide/acrylic copolymers in raising the viscosity is insufficient, even if their mol mass is extremely high.
 Using only nonionic monomers, e.g. vinyl pyrrolidone or acrylamide, results in a salt stable viscosity, but the absolute value of the viscosity is too low. However, polysaccharides are very effective thickeners in highly saline formation water. In contrast to xanthanes, glucanes have no anionic components and therefore show the largest viscosities. Classification of the glucane broth is possible by the combination of a centrifugation and a filtration process. Dueste-Valendis is selected as a field, where polymer flooding is suitable. Several geological and reservoir engineering data are obtained. Pulse tests have been performed, in order to see possible communication with wells in adjacent blocks.

Tracer tests are necessary to determine sweep geometry and length of dispersion.
 Basic research work resulted in the development of a tracer system, which is suitable in Dueste-Valendis.

```
****************************************************************************
*  TITLE : TESTING A NEW METHOD TO IMPROVE        *     PROJECT NO       *
*          SWEEPING BY INJECTION OF FOAMING AGENT *                      *
*          AND NITROGEN - PHASE 1                 *    TH./05073/86/FR/.. *
*                                                 *                      *
****************************************************************************
*  CONTRACTOR :                                   * PROGRAM :            *
*    GERTH                                        *   HYDROCARBONS       *
*    4, AVENUE DE BOIS PREAU     TEL 1 47.52.61.39 *                      *
*    FR - 92502 RUEIL-MALMAISON TLX 203 050       *                      *
*                                                 * SECTOR :             *
*                                                 *   SECONDARY AND      *
*  PERSON TO CONTACT FOR FURTHER INFORMATION :    *   ENHANCED RECOVERY  *
*    MR. DENOYELLE               TEL 1 47.49.02.14 *                      *
*                                TLX 203 050       *                      *
*                                                 *                      *
****************************************************************************
```

                                               VERSION : 01/01/87

AIM OF THE PROJECT :
****
The object of this project is to prepare the test, on-field and for the first
time, of a method which will considerably increase the sweeping performance of
a reservoir while using nitrogen, a non-miscible, readily available and non
expensive gas, combined with foaming agents in small quantity.
PROJECT DESCRIPTION :
****
The project is divided into four phases.
Phase 1 : injection facilities.
Two processes will be investigated :
- on-site separation of nitrogen from air using molecular sieve or membranes
and compression,
- transportation, storage and pumping of liquid nitrogen.
The study will take into account operating constraints and will include a
quickeconomic evaluation of each process.
Phase 2 : Tolerance study of crude oil oxygen.
This phase will evaluate the risks of letting some oxygen to be injected with
nitrogen, especially if on-site separation is chosen.
The oxydation/combustion parameters of the crude oil will be measured at two
oxygen partial pressures and two temperatures.
Phase 3 : Selection of a foaming agent.
Different manufacturers will supply samples of foaming agents. These products
will have to foam at 75 deg.C in a water with a saline content of 30g/l and in
the presence of oil. Adsorption on rock and aging in the presence of oxygen
will be also be studied. The products will be compared with one another.
Phase 4 : Definition of injection mode.
This consists in studying the injection mode of the foaming agent, wether inone
time or several plugs, and the influence of the injection mode on the
injectivity loss around the wellbore. Water supply will also be studied. Final
specifications and recommandations will be made.
STATE OF ADVANCEMENT :
****
Ongoing. The project has just started, with phases 1, 2 and 3 being looked at
simultaneously.
RESULTS :
****
After consulting foaming agent manufacturers and reviewing literature, 8
products were selected for the study. These products are either anionic
surfactants (mostly sulfonates or sulfates) or amphoteric surfactants used for
foam drilling or for manufacturing cleansing agents.

```
********************************************************************************
* TITLE : HYDROCONVERSION OF HEAVY OILS ON         *      PROJECT NO         *
*         PRODUCTION FIELDS                        *                          *
*                                                  *      TH./05074/86/FR/..  *
*                                                  *                          *
********************************************************************************
* CONTRACTOR :                                     * PROGRAM :                *
*   GERTH                                          *   HYDROCARBONS           *
*   4, AVENUE DE BOIS PREAU     TEL 1 47.52.61.39  *                          *
*   FR - 92502 RUEIL-MALMAISON TLX 203 050         *                          *
*                                                  * SECTOR :                 *
*                                                  *   SECONDARY AND          *
* PERSON TO CONTACT FOR FURTHER INFORMATION :      *   ENHANCED RECOVERY      *
*   MR. J.F. LE PAGE            TEL 1 47.52.66.69  *                          *
*                              TLX 203 050         *                          *
*                                                  *                          *
********************************************************************************
                                                    VERSION : 01/01/87
```

AIM OF THE PROJECT :
****
The object of this project is to render heavy oils transportable by reducing
their viscosity and their density through hydropyrolysis with the appropriate
catalyst.

PROJECT DESCRIPTION :
****
This project involves two separate phases :
- the first one aims at demonstrating the technical feasibility of a new
catalytic hydropyrolysis using an original and very active homogeneous catalyst
as hydrogenating agent.
- the second one covers an economic study of this new process when used to make
the heavy oils transportable from the production fields.
Concerning the first phase, tests will be carried out on the Solaize platform
from the beginning of March 1987 till the end of May. They include the
preparation of 10 tons of the catalyst precursor and the test in pilot plant of
this catalyst formulation in order to prepare in the best conditions the
further tests at a hourly flow, ranging between 2 and 4 T. Three extra heavy
oils have been ordered and two topped (Boscan, Athabasca, Pilon) for these
tests.

STATE OF ADVANCEMENT :
****
Ongoing

```
*******************************************************************************
* TITLE : PROMETHEE - PHASE 1                       *       PROJECT NO        *
*                                                   *                         *
*                                                   *   TH./05075/86/FR/..    *
*                                                   *                         *
*******************************************************************************
* CONTRACTOR :                                      * PROGRAM :               *
*   GERTH                                           *   HYDROCARBONS          *
*   4, AVENUE DE BOIS PREAU      TEL 1 47.52.61.39  *                         *
*   FR - 92502 RUEIL-MALMAISON TLX 203 050          *                         *
*                                                   * SECTOR :                *
*                                                   *   SECONDARY AND         *
* PERSON TO CONTACT FOR FURTHER INFORMATION :       *   ENHANCED RECOVERY     *
*   MR. AUXIETTE                 TEL 1 42.91.40.00  *                         *
*                                TLX 615 700        *                         *
*                                                   *                         *
*******************************************************************************
```

VERSION : 20/02/87

AIM OF THE PROJECT :
****
This project consists of the development of laboratory tools capable of
representing the exact reservoir conditions during combustion in-situ with air
or oxygen; e.g. the gas flows, temperature conditions, pressure conditions,
consolidated porous medium, etc...These tools will make it possible to obtain
results that may easily be extrapolated to a reservoir scale, or study of
combustion in unexplored domains such as light oil reservoirs.

PROJECT DESCRIPTION :
****
This project consists of three phases :
PHASE 1 - STUDY OF COMBUSTION WITH OXYGEN
The work consists of comparing the conditions of propagation of combustion with
oxygen and with air, and in determining the influence of the principle
parameters governing combustion with oxygen so as to identify the technical
domain in which the method could be applied.
PHASE 2 - STUDY ON COMBUSTION IN SITU ON AN ELABORATED MODEL
Simulation in the laboratory of combustion in reservoir conditions and a
consolidated environment implies the creation of equipment unique in the world.
All the developed technologies undergo routine tests before their inclusion in
the equipment. When the apparatus has been debugged, comparative tests are run
between the standard models and the latter, and provide indications as to its
limitations and the feasibility of combustion in unusual conditions.
PHASE 3 - ECONOMIC ESTIMATION
The experimental results obtained during the tests would make it possible to do
an economic assessment of combustion in non-standard applications.

STATE OF ADVANCEMENT :
****
Phases 1 and 2 are underway. Phase 3 has not yet started.

RESULTS :
****
So far there have not been any significant results; all the studies are
underway.

```
*******************************************************************************
* TITLE : GEOPHYSICS INTERPRETATION OF SUBSIDENCE   *      PROJECT NO         *
*         MEASUREMENTS (PHASE I)                     *                         *
*                                                    *    TH./05076/86/FR/..   *
*                                                    *                         *
*******************************************************************************
* CONTRACTOR :                                       * PROGRAM :               *
*   GERTH                                            *   HYDROCARBONS          *
*   AVENUE DE BOIS PREAU 4        TEL (1)47.52.61.39 *                         *
*   FR - 92502 RUEIL MALMAISON TLX (1)47.52.69.27    *                         *
*                                                    * SECTOR :                *
*                                                    *   SECONDARY AND         *
* PERSON TO CONTACT FOR FURTHER INFORMATION :        *   ENHANCED RECOVERY     *
*   MR. D. DESPAX                 TEL (1)42.91.38.16  *                         *
*                                 TLX TCFP615700      *                         *
*                                                    *                         *
*******************************************************************************
                                                         VERSION : 01/05/87
```

AIM OF THE PROJECT :
****
This project consists of measuring subsidence movements induced by pressure
charges associated to reservoir depletion in view of evaluating that depletion.
These subsurface movements are then converted into reservoir pressure charges
versus space and time using a F.E. model of all relevant layers i.e. overburden,
 reservoir and underbuden. Finally this pone pressure map is converted into a
permeability map in both reservoir and aquifer.

PROJECT DESCRIPTION :
****
Before the method can be applied on a larger scale it is necessary to prove
that the pressure variations in the reservoir provoke movements that can be
detected from the subsurface.
The present project thus consists of :
- a first step, in measuring the subsidence generated by the production of a
well within the VILLEPERDUE reservoir (operated by TOTAL-CFP, partner TRITON),
- a second step, in deducing therby the possible ditribution of the pressure
variations existing around the well.
PHASE 1 - MEASUREMENT OF SUBSIDENCE MOVEMENTS
The VILLEPERDUE field is formed by a carbonate reservoir located in the upper
layers of the DOGGER. The production layer is located at a depth of about 1 850
m.
- a first step, performed on a "versus structure" (Hautefeuille) will allow to
optimize the display of the measuring equipment
- a second step which will allow the measurement of the surface or subsurface
movements during depletion operations. It will confirm that the pressure
variations (150 to 200 bars) in the production layer generate surface movements
that can be detected in the middle of a spurious noise due to environment.
This phase will include the following tasks:
- layout of inclinometres
- positioning of level beacons
- sonic logging of the borehole throughout its depths
- recording of inclinometre results for 3 to 6 months
- high precision level measurement : about 2 measurements per month.

PHASE 2 - ELABORATION OF A NUMERICAL MODEL
This step will include :
- Evaluation of the deformation generated by placing a well on production
- Use of a three dimensional radial symetry numerical programme capable of
coupling the behaviour of the rocks and the flow of a fluid.
PHASE 3 - INTERPRETATION OF MEASUREMENTS
Tasks related to this phase will only begin when a significant subsidence will
have been outlined. These tasks will allow the plotting of a map of the
pressure variations around the well in relation to the time. They will include :
a) Development of a method of numerical resolution of the reserve problem in
which the subsurface movements are the data enabling the pressure variations
within the reservoir to be defined.
b) Determination of the mechanical properties of the entire deformable massif
(analysis of the logs and drilling parameters - laboratory measurements).
c) Interpretation of the inclinometry measurements and the level curves, first
by means of a radial symetry model and second with the aid of a three
dimensional model.

STATE OF ADVANCEMENT :
****
Steps 1 and 2 are underway.

```
********************************************************************************
* TITLE : SCLEROGLUCANE OF EOR QUALITY           *        PROJECT NO          *
*                                                *                            *
*                                                *     TH./05078/86/FR/..     *
*                                                *                            *
********************************************************************************
* CONTRACTOR :                                   * PROGRAM :                  *
*   GERTH                                         *   HYDROCARBONS             *
*   4, AVENUE DE BOIS PREAU    TEL 1 47.52.61.39  *                            *
*   FR - 92502 RUEIL-MALMAISON TLX 203 050        *                            *
*                                                * SECTOR :                   *
*                                                *   SECONDARY AND            *
* PERSON TO CONTACT FOR FURTHER INFORMATION :    *   ENHANCED RECOVERY        *
*   MR. A. DONCHE               TEL 59.83.40.00   *                            *
*                               TLX 560 804       *                            *
*                                                *                            *
********************************************************************************
```

AIM OF THE PROJECT :
****
Today, the injection of polymers to sweep reservoirs is a confirmed technique
in the domain of oil production. In spite of the lack of suitable products,
this method is nevertheless restricted to the exploitation of reservoirs of
mean salinity and a temperature below 70 deg.C. Significant progress has to be
made to ensure the treatment of reservoirs in more difficult environments.
Therefore, Scleroglucane, a polymer produced by fermentation, offers new
prospects: its resistance to temperature and salinity makes it possible to
envisage the treatment of hot and salty reservoirs, still not accessible
technically. Scleroglucane is presently available at industrial scale. The
purpose of the project is to provide the required quality for the injection of
polymers and to define the conditions of implementation of such a product.

PROJECT DESCRIPTION :
****
The scleroglucane that is available today, already presents part of the
characteristics that are required for the development of a hydrosoluble polymer
under EOR. Its solutions are very viscous and highly resistant to shear. Their
viscosity is hardly sensitive to salts and to the ph and little or hardly
affected by the temperature.
The product only lacks a few adaptations so to meet the EOR quality in view of
qualifying for a pilot test.
Therefore, a few obstacles need to be overcome:
- today, scleroglucane presents a poor filtrability which may provoke a risk of
clogging the formation in which is will have been injected.
- conditioning of hydrosoluble polymers and namely polysaccharides requires a
particularly well adapted technique to avoid degradation of the initial
filtrability and viscosity of the product. Whether in concentrate solution or
in powder form, this conditioning is essential for the development of the
scleroglucane and has to be finalized.
- polymer runs many risks of being degraded during its progress in the
reservoir, bacterial degradation being one of the major risks. In this case, it
is necessary to evaluate the risks and define, when necessary, the means of
protecting the polymer.
The project is divided into four phases:

PHASE 1: improving the filtrability of the scleroglucane
The purpose of this activity is to identify the factors that limit the filtrability of the polymer (aggregation factors) to define in a second step the modifications of the manufacturing process of scleroglucane or of other solutions (additives) to improve its filtrability. A detailed study will be carried out on the characteristics of pure polymer, on its structure and on the possible interactions with impurities to identify the endogenous and exogenous aggregation factors.
PHASE 2: adaptation of manufacture process to EOR requirements
Works carried out in Phase 1 will lead to a number of modifications in the scleroglucane manufacturing process (fermentation, finalization treatment, conditioning). These modifications will be systematically tested at the plant on a manufacturing pilot. Special care will be given to the finalization of conditioning techniques which do not alter the rheologic and filtrability characteristics of the scleroglucane obtained after process. Three types of conditioning will be studied, diluted solution directly injectable, concentrate solution and powder solution.
PHASE 3: evaluation of the scleroglucane for EOR
The purpose of this activity is to define the performances of the scleroglucane in porous media and the conditions of its implementation in EOR. Studies will first cover pure scleroglucane under model conditions to evaluate the limits of its utilisation domain then in natural porous media with improved filtrability scleroglucane to analyse its behaviour and efficiency when displacing oil.
PHASE 4: protection against bacterial degradation.

STATE OF ADVANCEMENT :
****
Ongoing

RESULTS :
****
As soon as these different developments and modifications will have been confirmed, an injectivity test will be run on a well beyond the scope of this project.

```
******************************************************************************
* TITLE : RETENTION OF A MULTIPLE CHEMICAL SYSTEM    *      PROJECT NO        *
*         WHILE INCREASING THE RECOVERY FACTOR BY    *                        *
*         CHEMICAL FLOODING                          *     TH./05082/86/DE/.. *
*                                                    *                        *
******************************************************************************
* CONTRACTOR :                                       * PROGRAM :              *
*   PREUSSAG AG                                      *   HYDROCARBONS         *
*   ARNDTSTRASSE 1               TEL 05176/17 247    *                        *
*   DE - 3000 HANNOVER 1         TLX 92655           *                        *
*                                                    * SECTOR :               *
*                                                    *   SECONDARY AND        *
* PERSON TO CONTACT FOR FURTHER INFORMATION :        *   ENHANCED RECOVERY    *
*   DR. D. MENZ                                      *                        *
*                                                    *                        *
*                                                    *                        *
******************************************************************************
                                                      VERSION : 01/01/87
```

AIM OF THE PROJECT :
****

The goal of this development project is to determine retention on the rock
surface under simulated reservoir conditions. In the proposed investigation,
retention data for both the individual surfactants and their combined mixture
as well as data on the multichemical system containing both surfactants and
polymers should be gathered and evaluated. Based on observations on linear
flood tests it is presumed that, by post-flooding using additional surfactants
and polymers, the retention of a surfactant mixture can be considerably reduced.
 This would result in a further reduction of the required concentration and
slug size of the surfactant combination.

PROJECT DESCRIPTION :
****

1. Development and optimization of the analytical process for specificaton of
both the surfactant combinations and the multiple chemical system.
The surfactants which are planned to be used in the field are technical
products which contain a broad spectrum of compounds of varying molecular
weight and reaction groups. In the past, different methods of analysis
(infrared spectrography, UV-spectrography, two phased Titration, etc were
attempted in the flood tests. The main problem is the chemical separation of
the used components like a mixture of crude oil, brine emulsions.
2. Static absorption tests
The identification of absorption isotherms on sand surfaces should supply
information on the absorption characteristics of both the various components
and the multi-chemical system.
Pressure-free-flooding
Using a model sandstone core with a porosity of 20% and a permeability of 1000
md, pressure-free flooding can be done. The testing of the flood and
displacement behaviour of the individual components, the surfactant combination,
 and the multi-chemical system is done in order to determine the retention
behaviour and to show, if present, a chromatographic separation of the
components.
4. The sandpack flood-tests
The sandpack flood-tests serve mostly in showing possible chromatographic
separation.

5. High pressure flooding

The work on the high pressure flooding apparatus can be done as a supplement to the pressure-free flood tests; here the influence of pressure on the flood-behaviour of the chemical system can be tested under simulated reservoir pressures.

STATE OF ADVANCEMENT :
****
Ongoing with phase 1.
The analytic methods have been optimized.Pretests on static absorption have been carried out.

```
********************************************************************************
* TITLE : DEVELOPMENT OF CELLULOSE DERIVATIVES    *        PROJECT NO      *
*         FOR THE USE AS VISCOSITY BUILDERS IN    *                        *
*         FLOODING MEDIA FOR THE ENHANCED OIL     *     TH./05083/86/DE/.. *
*         RECOVERY                                *                        *
********************************************************************************
* CONTRACTOR :                                    * PROGRAM :              *
*   WOLFF WALSRODE AG                             *   HYDROCARBONS         *
*   DE - 3030 WALSRODE        TEL 05161/442838    *                        *
*                             TLX 924 324         *                        *
*                                                 * SECTOR :               *
*                                                 *   SECONDARY AND        *
* PERSON TO CONTACT FOR FURTHER INFORMATION :     *   ENHANCED RECOVERY    *
*   DR. KLAUS BALSER                              *                        *
*                                                 *                        *
*                                                 *                        *
********************************************************************************
```

VERSION : 01/01/87

AIM OF THE PROJECT :
****
The aim of the project is the development of cellulose derivatives that can be
used as viscosifiers in flooding media in enhanced oil recovery. This novel
cellulose polymers should be used in polymer flooding but also as a mobility
control agent in micellar-polymer-processes.

PROJECT DESCRIPTION :
****
The project is divided into three parts.
PART A
"Modification of carboxylmethyl cellulose (CMC)" contains two phases: the
improvement of the injectibility and specific viscosity of CMC and the
insertion of a second substitute onto CMC to improve the electrolyte stability.
PART B
"Development of a new cellulose derivatives" is the main part of the project,
divided into four phases. Preparation of new cellulose derivatives using up to
now industrially not used substitutes. Optimization of the preparation and the
properties of the new cellulose derivatives. Insertion of a second substitute
onto these new derivatives. The use of bulky substitutes as a third substitute
onto new cellulose derivatives.
PART C
"Screening of the products under practical aspects" is continuously done during
the whole proccessing of the project.

STATE OF ADVANCEMENT :
****
The project started on 01.09.85.

RESULTS :
****
The specific viscosity and the injectivity measured in a laboratory test of CMC
could be improved by process optimisation. The insertion of a second
substituent onto CMC lowered the sensitivity of viscosity against electrolytes.
The development of the new cellulose derivatives led to products that showed a
good resistance against electrolytes and also had a good injectivity.

ENVIRONMENTAL INFLUENCE
ON OFFSHORE

```
*******************************************************************************
* TITLE : A FIELD INVESTIGATION INTO THE          *       PROJECT NO        *
*         PERFORMANCE OF A PILED FOUNDATION        *                         *
*         SYSTEM FOR AN OFFSHORE OIL PRODUCTION    *   TH./06014/81/UK/..    *
*         PLATFORM                                 *                         *
*******************************************************************************
* CONTRACTOR :                                     * PROGRAM :               *
*   BRITISH PETROLEUM                              *   HYDROCARBONS          *
*   BRITANNIC HOUSE              TEL 01 920 8000   *                         *
*   MOOR LANE                    TLX 888811        *                         *
*   UK - LONDON ECZY9BU                            * SECTOR :                *
*                                                  *   ENVIRONMENTAL         *
* PERSON TO CONTACT FOR FURTHER INFORMATION :      *   INFLUENCE ON          *
*   MR. D.E. SHARP                                 *   OFFSHORE              *
*                                                  *                         *
*                                                  *                         *
*******************************************************************************
                                                        VERSION : 08/04/87
```

AIM OF THE PROJECT :
****
The aim of the project is to obtain full scale data on the behaviour of the
foundation of an offshore oil production platform.

PROJECT DESCRIPTION :
****
British Petroleum Development Ltd have installed a production platform in 186 m
of water at Magnus field in block 211-12 of the UK sector. North Sea.
The project consists in instrumenting the lower section of one of the four legs
of the Magnus platform and piles supporting that leg. The behaviour of the
instrumented members is recorded and a comparison made between measured pile
loads and those calculated by conventional methods.

STATE OF ADVANCEMENT :
****
Completed

RESULTS :
****
The instrumentation system performed well. All accelerometers, structural
strain gauges and mudmat pressure sensors functioned satisfactorily. The novel
pile instrumentation also performed well. The project has determined the actual
loads imposed on the piling and mudmats by the structural and environmental
forces. In particular:
1. The environmental load acting on the pile group has been measured in
environmental conditions up to a maximum wave height of 21 m.
2. The proportion of the environmental load carried by the mudmat and piles has
been measured.
3. The distribution of load within the pile group and along the length of a
pile has been determined.
4. The actual pile behaviour has been compared with design predictions.
5. The effective stiffness of the pile group has been measured.

```
********************************************************************************
* TITLE : FIELD INVESTIGATION INTO THE SOIL-      *       PROJECT NO          *
*         STRUCT.INTERACTION OF A FOUNDATION      *                           *
*         SYSTEM DURING THE EARLY LIFE OF AN      *    TH./15035/82/UK/..      *
*         OFFSHORE OIL RIG.                       *                           *
********************************************************************************
* CONTRACTOR :                                    * PROGRAM :                 *
*   BRITISH PETROLEUM                             *   HYDROCARBONS            *
*   BRITANNIC HOUSE          TEL 01 920 8000      *                           *
*   MOOR LANE                TLX 888811           *                           *
*   UK - LONDON EC24 9BU                          * SECTOR :                  *
*                                                 *   MISCELLANEOUS           *
* PERSON TO CONTACT FOR FURTHER INFORMATION :     *                           *
*   MR. D.E. SHARP                                *                           *
*                                                 *                           *
*                                                 *                           *
********************************************************************************
                                                   VERSION : 08/04/87
```

AIM OF THE PROJECT :
****
The aim of the project is to obtain early data on the behaviour of the
foundation of an offshore oil production platform.

PROJECT DESCRIPTION :
****
British Petroleum Development Ltd have installed a production platform in 186m
water at Magnus field in block 211/12 of the UK sector, North Sea.
A separate project, 06014/81, provides for the investigation of foundation
behaviour. This project allows the early installation of a temporary data
logger to provide monitoring of foundations at the critical period immediately
following the positioning of the platform tower.

STATE OF ADVANCEMENT :
****
Completed

RESULTS :
****
The majority of results have been reported as part of Project TH./06014/81. In
addition the variation of static load distribution has been measured both as a
function of time and platform structural loads. A continuous record of pile and
mudmat loads has been recorded since platform touchdown or, in the case of the
pile loads, and just prior to an after module installation.

```
********************************************************************************
* TITLE : FIELD INVESTIGATION INTO THE ULTIMATE    *      PROJECT NO        *
*         CAPACITY OF LARGE OFFSHORE PIPE PILES.    *                        *
*                                                   *   TH./15065/84/UK/..    *
*                                                   *                        *
********************************************************************************
* CONTRACTOR :                                      * PROGRAM :              *
*   BRITISH PETROLEUM INTERNATIONAL LIMITED         *   HYDROCARBONS         *
*   BRITANNIC HOUSE            TEL 01 - 920 3845     *                        *
*   MOOR LANE                  TLX 9888811           *                        *
*   UK - LONDON EC2Y 9BU                             * SECTOR :               *
*                                                   *   MISCELLANEOUS        *
* PERSON TO CONTACT FOR FURTHER INFORMATION :       *                        *
*   MR. C.R. MULLIS                                 *                        *
*                                                   *                        *
*                                                   *                        *
********************************************************************************
```
VERSION : 30/01/87

AIM OF THE PROJECT :
****
The aim of the project is to recommend a more fundamental and rigorous approach
to the computation of the ultimate axial load capacity of piled foundations in
clay. The proposed load tests are intended to provide information on the
validity of existing design criteria for predicting skin friction in clays
similar to those found in the North Sea. Associated soil investigations are to
provide data on effective stress changes in the soil around a pile during
installation, clay reconsolidation and loading of the pile to failure.

PROJECT DESCRIPTION :
****
The project consists of two large-scale pile load tests to be performed at
onshore locations in clays similar to those encountered by Hydrocarbon
installation in the North Sea. One test will be carried out in very strong,
heavily over-consolidated clay and the other in stiff, essentially normally-
consolidated clay. The piles will be driven, then loaded to failure via
hydraulic jacks reacting against an anchored frame. The piles and surrounding
soils will be instrumented to monitor their response to pile installation and
subsequent static load testing.

STATE OF ADVANCEMENT :
****
Test sites have been acquired and prepared. Site investigations and laboratory
testing continues. Development of instruments at Queen Mary College, nearing
completion. A driving trial has been successfully completed at the over-
consolidated clay site using a B.S.P. HA40 hammer. It is now proposed to use
the HA40 to drive the instrumented pile. Fabrication of the instrumented piles
continues although delayed due to welding and rolling problems with ROT 701
steel.

RESULTS :
****
No results will be available before the tests completion in Cambridgeshire
(overconsolidated clay) and in Shropshire (consolidated clay).

```
*********************************************************************************
* TITLE : DEVELOPMENT OF TECHNOLOGIES RELATIVE TO   *       PROJECT NO        *
*         IMPROVEMENT OF PILE CAPACITIES BY         *                         *
*         CONTROLLED PRESSURE GROUTING              *    TH./06018/85/BE/..    *
*                                                   *                         *
*********************************************************************************
* CONTRACTOR :                                      * PROGRAM :               *
*   BELGIAN OFFSHORE SERVICES                       *   HYDROCARBONS          *
*   SCHERMERSTRAAT 46          TEL 03 2318770       *                         *
*   BE - 2000 ANTWERPEN        TLX 34129            *                         *
*                                                   * SECTOR :                *
*                                                   *   ENVIRONMENTAL         *
* PERSON TO CONTACT FOR FURTHER INFORMATION :       *   INFLUENCE ON          *
*   MR. S. DECKERS                                  *   OFFSHORE              *
*                                                   *                         *
*                                                   *                         *
*********************************************************************************
```

VERSION : 09/02/87

AIM OF THE PROJECT :
****
The aim of the project is to extend in a reliable manner the technique of
controlled pressure grouting to the improvement of pile capacities of offshore
structures and to develop new pressure grouting technologies in order to
decrease the installation cost of offshore and possibly subsea structures.
So far grouting techniques used in the offshore industry are very conservative
and have not evolved in decades. New innovative technologies will lead to lower
installation costs and improved performances of the pile foundation.

PROJECT DESCRIPTION :
****
The analysis of the state of the art of pressure grouting and of potential
problems existing in the offshore piling industry led to the selection of
pressure grouting techniques which are developed under 3 aspects :
- theoretical analysis of pressure grouting
- detailed study of grouts for offshore applications
- development of new technologies
The theoretical research is related to the phenomenon of soil "cracking" and is
divided in three stages :
- the first stage is dedicated to a naturalistic description of the involved
mechanisms.
- the second stage consists of using an elasto-plastic consolidation finite
element computer program to study the "cracking" phenomenon.
- in the third stage "cracking" laboratory tests are to be performed on small
scale models and interpreted with the help of numerical computations in order
to produce a mathematical model of the "cracking" phenomenon.
The detailed study of grouts for offshore applications consist of :
- checking the suitability of grout used onshore for offshore application
- developing new grouts for specific offshore applications
This part of the project is mostly experimental and it involves intensive
laboratory treating to determine all the grout properties which are essential
for offshore use.

The development of new technologies involves :
- reviewing all the components of existing pressure grouting systems (i.e. pressure grouting units, non return valves, packers, grouting lines...) and adapting them to offshore use. This phase consits in designing, building and testing components.
- defining and testing inventive and low cost pressure grouting techniques and technologies to solve specific offshore piling problems.

STATE OF ADVANCEMENT :
****
Ongoing. The theoretical approach is in progress in the three stages. The laboratory testing equipment is in the building phase. The detailed study of grouts is in the initial experimental stage and the development of new technologies is in advanced realization stage.

RESULTS :
****
THEORETICAL ANALYSIS OF PRESSURE GROUTING
Since the experimental evidences are deduced from laboratory tests, the soil "cracking" study will essentially point out qualitative tendencies. However the support of the numerical computations will enable to make some extrapolations which will be more representative of the in-situ conditions.
Concretely the "cracking" pressure will be studied as a function of : the rate at which the bottomhole pressure rises, the in-situ stress and pore pressure state, the grout viscosity, the soil skeleton deformability (reversible and permanent), the simple tenaille strength of the soil material.
DETAILED STUDY OF GROUTS FOR OFFSHORE APPLICATIONS
New types of grouts which are used onshore for civil engineering purposes have been successfully evaluated for offshore applications :
- MICROSOL grout for fine alluvial soils with a permeability as low as 1 Darcy
- SILACSOL grout which has permeation properties similar to silicate gels and an improved time stability. It is a non polluting chemical grout.
- ACTISOL grout which has higher strength and permeability than conventional cement grouts. It is highly sulfate resistant.
New compositions of cement grouts have been tested for special applications :
- grouts for compaction grouting of calcarenite
- light cement grouts for applications at low temperatures
DEVELOPMENT OF NEW TECHNOLOGIES
A new technique, the grouted driven piles, has been studied, elaborated and tested.
The technology necessary to apply this technique is being developped :
- components for a pressure grouting unit tested
- non return valves which can be used several times with grout were built and successfully tested
- mechanical tests were performed on a 500 mm OD pile equiped with 3 different designs of grouting lines. The equipment austained over 4 300 blows inducing stresses up to 150 MPa without suffering any deterioration.
- a complete onshore test was performed on a 750 OD pile in carbonated sands. The 25 m long pile was driven and grouted.
Static pull out tests were performed before and after 3 grouting steps. The friction of the pile was increased from a few kPa an average of 120 kPa with local values of 240 kPa.

REFERENCES :
****
PAPERS ON GROUTED DRIVEN PILES AND/OR MICROSOL HAVE BEEN OR ARE TO BE PRESENTED AT SEVERAL CONFERENCES AUCH AS NANTES (86), OTC 87, OMAE 87, OFFSHORE BRAZIL (87) AND PERTH 1988.

```
********************************************************************************
* TITLE : THE HYDRA-LOK PILE STRUCTURE CONNECTING   *      PROJECT NO        *
*         SYSTEM FOR LARGE DEEP WATER OFFSHORE      *                        *
*         PRODUCTION FACILITIES                     *    TH./06019/85/UK/..  *
*                                                   *                        *
********************************************************************************
* CONTRACTOR :                                      * PROGRAM :              *
*   BUE HYDRA-LOK LTD                               *   HYDROCARBONS         *
*   WALNEY ROAD                                     *                        *
*   BARROW-IN-FURNESS          TLX 65147            *                        *
*   UK - CUMBRIA LA14 5UX                           * SECTOR :               *
*                                                   *   ENVIRONMENTAL        *
* PERSON TO CONTACT FOR FURTHER INFORMATION :       *   INFLUENCE ON         *
*   MR. J. MARSHALL LOWES                           *   OFFSHORE             *
*                                                   *                        *
*                                                   *                        *
********************************************************************************
                                                    VERSION : 09/03/87
```

AIM OF THE PROJECT :
****
It is the aim of the current project to develop the existing Hydra-Lok
technology into a piece of offshore operational equipment suitable for pile
connection on jackets. This will be achieved by way of a review of the design
of the prototype Hydra-Lok tool and its subsequent use in a field demonstration.
 The field demonstration will involve the installation of a Southern North Sea
jacket from a Derrick Barge. This represents a totally new approach to jacket
installation which has been previously dominated by grouting technology. the
new technology offers a more economical and technically superior connection
system.

PROJECT DESCRIPTION :
****
The project is divided into a number of discrete areas of investigation. These
cover redesign of the existing prototype tools, followed by the building of an
offshore operational tool package, some further small scale testing to verify
certain design parameters and a trial of the system on an offshore installation.
The connection system is a swage pile connection system which serves as a
replacement for more conventional grouting or welding off operations. The
system offers substantial technical and economic advantages since it is
performed by equipment which is reusable. Alternative systems tend to rely on
extensive 'installed equipment' and are therefore very capital cost intensive.
Hydra-Lok avoids these problems with negligible capital equipment being
purchased by the Client.
Initial work centres around the redesign of the tool and the completion of the
small scale testing to verify certain design parameters. This is to be followed
by the building of an offshore usable tool and development of suitable
operating procedures. Subject to the successful completion of these phases the
equipment is to be used in the installation of an offshore jacket with 72"
piles in the Southern North Sea.

STATE OF ADVANCEMENT :
****
Ongoing project which is currently in the early design stages. Project is
scheduled to be completed mid 1988.

RESULTS :
****
To date there are no significant results to report although the project is
proceeding satisfactorily.

```
*********************************************************************************
* TITLE : CLASSIFICATION OF SEA FLOOR ENGINEERING      *        PROJECT NO       *
*         PROPERTIES FOR MARINE FOUNDATIONS AND        *                         *
*         STRUCTURES                                   *     TH./06029/86/IR/..  *
*                                                      *                         *
*********************************************************************************
* CONTRACTOR :                                         * PROGRAM :               *
*   IRISH HYDRODATA LTD                                *   HYDROCARBONS          *
*   RATHMACULLIG WEST           TEL 021 962600         *                         *
*   BALLYGARVAN                 TLX 75850              *                         *
*   IR - CO.CORK                                       * SECTOR :                *
*                                                      *   ENVIRONMENTAL         *
* PERSON TO CONTACT FOR FURTHER INFORMATION :          *   INFLUENCE ON          *
*   MR. T.N. EMERSON                                   *   OFFSHORE              *
*                                                      *                         *
*                                                      *                         *
*********************************************************************************
```

VERSION : 09/03/87

AIM OF THE PROJECT :

****

The main objective of this project is to investigate the interrelationships
between acoustic reflections from the seabed with measured in-situ values and
laboratory determined physical properties. A successful conclusion to the
project will enable compilation of a classification table of results determined
from analysis of acoustic reflections. This will enable engineering properties
and stability of the sea floor to be directly inferred, such that more rapid
and cheaper surveys of areas can be achieved using acoustic methods.

PROJECT DESCRIPTION :

****

The project will compare results obtained from the development of the survey
USP, a seafloor probe for measuring P and S wave velocities in-situ, and
conventional geophysical survey tools which will include coring and grab
sampling.

The project is designed to follow a six phase operation :

Phase 1 - Development of Sea Floor Probe.
          Development of survey version of USP.
  Conventional surveys of test bed areas off Cork/North Wales.

Phase 2
- Joint initial trials of probe, USP and conventional instrumentation.

Phase 3 - Redevelopment of work on probe.
           Redevelopment of work USP.
  Further conventional surveys of test bed areas off Cork/North Sea Wales.

Phase 4
- Joint trials, marine ground truthing and surveys, using USP probe
  and conventional instrumentation including coring/grab sampling.

Phase 5 - Data analysis of USP results.
  Laboratory testing and comparison of probe results.
  Analysis of conventional survey results for characterisation and control
  of sediment distribution in test-bed areas.

Phase 1-5
- Mathematical analysis and numerical modelling of acoustic wave theory.

Phase 6 - Reporting.

STATE OF ADVANCEMENT :

****

Ongoing

AUXILIARY SHIPS AND SUBMERSIBLES

```
********************************************************************************
* TITLE : DEVELOPMENT OF A REMOTELY OPERATED      *       PROJECT NO           *
*         SUBMERSIBLE.                            *                            *
*                                                 *                            *
*                                                 *   TH./07045/82/DE/..       *
*                                                 *   TH./07055/85/DE/..       *
********************************************************************************
* CONTRACTOR :                                    * PROGRAM :                  *
*   ZF-HERION-SYSTEMTECHNIK GMBH                  *   HYDROCARBONS             *
*   POSTFACH 2168              TEL 0711 5209 351  *                            *
*   HOECHENSHASSE 21           TLX 7 254 733      *                            *
*   DE - 7012 FELLBACH                            * SECTOR :                   *
*                                                 *   AUXILIARY SHIPS AND      *
* PERSON TO CONTACT FOR FURTHER INFORMATION :     *   SUBMERSIBLES             *
*   MR. K. WIEMER                                 *                            *
*                                                 *                            *
*                                                 *                            *
********************************************************************************
                                                    VERSION : 11/03/87
```

AIM OF THE PROJECT :
****
The project covers the first phase of a complete program referred to as the
MARS PROJECT which aims to produce a remotely operated submersible vehicle
system to be used for subsea inspection, maintenance and repair work in
situations where to employ a diver would be either too dangerous or
uneconomical.
Phase one of the MARS project was concerned with the collection of data
regarding market requirements, research into critical technical areas which
could later require special attention during further development, preliminary
design and testing of new conceptual ideas and techniques, feasibility
considerations and operational trials in actual offshore conditions.

PROJECT DESCRIPTION :
****
The information and knowledge needed in order to prepare the specification
required for continuation into the second phase of the project was obtained by
the following activities :
MARKET RESEARCH AND DEFINITION OF WORK TASKS :
Data was obtained by establishing and maintaining worldwide contact with oil
companies, diving companies, ROV operating companies, classification
organisations and possible competitors. The contact was made by personal visits
to companies and also attendance at conferences, symposiums and exhibitions.
The information collected was concerned with :
Oil industry policy with regard to inspection, maintenance and repair
activities.
Industrial experience with relevant equipment and technology already in use.
Projected developments for future requirements considering also requirements by
government authorities and classification organisations.
Political factors which could influence the market situation, e.g. oil price
fluctuations.
PRELIMINARY TECHNICAL RESEARCH
Certain research work was necessary in order to ensure that offshore
operational aspects such as launching, recovery, navigation, docking at work
sites, tool operation etc. were fully understood, and also that solutions to
the technical problems involved in achieving the final specification were
reliable in practice.

Functional testing of components using pressure chamber techniques was supplemented by actual offshore trials using existing vehicles as carrier systems for new conceptual ideas and techniques.

PRELIMINARY DESIGN WORK

Considerable and detailed preliminary design work including hardware for testing regarding reliability and feasibility was concern mainly with :
- Vehicle main frame concept
- Truster configuration (layout, drive system)
- Buogancy material arrangement
- Umbilical and termination
- Docking claw arrangement
- Hydraulic system concept
- Manipulator joint development
- Navigational and control system
- Tool modules (water jetting, mud pumping)
- Handling system (crane, depressor)
- Compact umbilical winch.

STATE OF ADVANCEMENT :
****

Completed. From the results of the activities descibed in section 2 above, it has been possible to draw up a complete specification for the MARS System with the knowledge that the solution has been well adapted to future market conditions and is also technically and commercially feasible.The results provide a satisfactory preparation for the following phase of the complete project.

RESULTS :
****

FIRST OFFSHORE TRIALS

Using an existing vehicle, an offshore trial was arranged firstly to establish operational factors involved in working an ROV offshore, secondly to study behaviour and reliability aspects of an ROV and thirdly to define critical areas which would have to be specially considered in the specification and design of the final MARS system.

The first offshore trial took place on the MSV Stadive with the assistance of Shell U.K. Exploration and Production, Aberdeen.

The operational factors were concerned with : influence of deck space requirements, problems of launch and recovery, requirements in terms of knowledge and experience of the operating crew, need for a pre-defined work program and procedures, special requirements for equipment to be suitable for mobilisation offshore.

The system behaviour aspects were concerned with: system performance during subsea operation, general system reliability, difficulties involved in carrying out system service and repair offshore.The critical areas defined for special attention were :

- Overall efficiency with regard to system power input and thruster output.
- Thruster design and layout on vehicle.
- Vehicle frame design with regard to vehicle weight and service.
- Buoyancy material arrangement to allow for trim change according to work module.
- Umbilical specification and Termination.
- Docking device for universal use.
- Hydraulic system with regard to efficiency, reliability and service.
- Handling system suitable for use in severe weather and sea conditions
- Umbilical winch with regard to deck space requirements
- Available tools and tool modules.

PRELIMINARY NEW DESIGN AND FEASIBILITY TESTING

Based on the results of the above trials, new solutions for the points referred to were investigated. Certain components were produced and subjected to initial workshop.

Testing. In order to assess the value of new designs and techniques, the resulting concepts and designs were integrated into an existing vehicle design which was then introduced into the planning for a further offshore evaluation trial with BP Petroleum Development Limited, Aberdeen. The system was planned into the annual inspection programme for the BP-Magnus platform during 1986 for which preparation was carried out.

SPECIFICATION

As a result of the above activity it was possible to prepare the required specification related to the market situation and also to establish operational factors, design concepts and technology for further progress into MARS phase II.

REFERENCES :
****
APART FROM DIRECT CONTACT WITH RELEVANT COMPANIES IN THE OFFSHORE INDUSTRY, THE FOLLOWING EVENTS PROVIDED PARTICULARLY USEFUL INFORMATION :
- OFFSHORE TECHNOLOGY CONFERENCE, HOUSTON USA
- ROV 85 CONFERENCE, SAN DIEGO USA
- OFFSHORE EUROPE 85 CONFERENCE, ABERDEEN UK
- SUBTECH 85 SYMPOSIUM, ABERDEEN UK
- ONS CONFERENCE, STAVANGER, NORWAY
- DOT CONFERENCE, SORRENTO, ITALY.

```
*******************************************************************************
* TITLE : DEVELOPMENT OF A DEEP WATER TETHERED    *       PROJECT NO        *
*         MANNED SUBMERSIBLE                      *                         *
*                                                 *    TH./07046/82/IR/..   *
*                                                 *                         *
*******************************************************************************
* CONTRACTOR :                                    * PROGRAM :               *
*   T.E.ASSOCIATES                                *   HYDROCARBONS          *
*   ELM HOUSE, CLANWILLIAM COU TEL 685222         *                         *
*   LR MOUNT STREET           TLX 30798           *                         *
*   IR - DUBLIN 2                                 * SECTOR :                *
*                                                 *   AUXILIARY SHIPS AND   *
* PERSON TO CONTACT FOR FURTHER INFORMATION :     *   SUBMERSIBLES          *
*   MR. T. EARLS                                  *                         *
*                                                 *                         *
*                                                 *                         *
*******************************************************************************
                                                    VERSION : 12/03/85
```

AIM OF THE PROJECT :
****
Design, manufacture and testing of a tethered one-man submersible capable of
performing sophisticated tasks on hydrocarbon drilling/production installations
in water depths up to 1,500 metres.

PROJECT DESCRIPTION :
****
Three areas of technical innovation related to hull design, underwater
deployment system, and the manipulators. A further area of innovative design
related to the adaption of the vehicle for optional remote control.

STATE OF ADVANCEMENT :
****
Abandoned due to changes in market conditions and development of ROV technology.

RESULTS :
****
Problem areas relating to the hull, tether deployment system and manipulators
were defined and resolved. In the case of the manipulators the second stage of
a planned three-stage development has produced a viable tool which may be
operated by a single joystick control and uses sea water as its hydraulic fluid.
Initial tests of a dual control system have been successfully completed.
This enables the vehicle to be remote-controlled from the surface for
performance of simple tasks or where hazardous conditions might prevail.

```
****************************************************************************
* TITLE : TM 308 - VEHICLE FOR INSPECTION OF    *      PROJECT NO        *
*         OFFSHORE STRUCTURES                    *                        *
*                                                *    TH./07048/82/IT/..  *
*                                                *                        *
****************************************************************************
* CONTRACTOR :                                   * PROGRAM :              *
*   TECNOMARE                                    *   HYDROCARBONS         *
*   S. MARCO 2091            TEL 041 796711      *                        *
*   IT - 30124 VENEZIA       TLX 410484 MAREVI I *                        *
*                                                * SECTOR :               *
*                                                *   AUXILIARY SHIPS AND  *
* PERSON TO CONTACT FOR FURTHER INFORMATION :    *   SUBMERSIBLES         *
*   MR. MAZZON                                   *                        *
*                                                *                        *
*                                                *                        *
****************************************************************************
                                                  VERSION : 01/01/87
```

AIM OF THE PROJECT :
****
Design and development of a tetherless unmanned vehicle for offshore platform
inspection in substitution of divers.

PROJECT DESCRIPTION :
****
The project was divided into 4 stages:
1. Conceptual design
2. Detailed design
3. Construction of prototype
4. Sea tests

STATE OF ADVANCEMENT :
****
Abandoned. The project has been stopped at end of stage 2 (design completed)
having decided to activate an intermediate stage of subsystem development
before initiating vehicle construction.

RESULTS :
****
After detailed analysis of operating requirements and current practices with
divers and evaluation of performances of existing tethered vehicles,
specifications for the tetherless robot have been set up.
The vehicle is a cylindrical-shape vertical body, about 4 m high and 2 m diam.
having the following features:
- ability to navigate within structure based on own position measurement and
route control system.
- acoustic transmission of data (two way) and TV (128 x 128 points, 4 images
per second in navigation).
- ability to dock to inspection mode, by 4 docking arms with limpets.
- 7 degrees of freedom manipulator, which kinematis is designed to maximize the
weld area coverage while maintaining the tools oriented properly.
- ability to execute the three inspection levels required by current rules and
practices (gree, blue, red).
- 10-12 hours autonomy with 80 MP power provided by a closed-circuit diesel
engine diving oil pump and electric generator.
- high pressure water jet, magnetic particle inspection system, TV cameras,
still camera.
The vehicle is controlled by a surface operator by means of a supervision
control architecture and a distributed intelligence multimicro network.
The console is also provided with a 3-D computer graphics system to represent
the operating environment.

REFERENCES :
****
1. TOWARDS TETHERLESS INSPECTION OF PLATFORMS.
M. MAZZON, ROV 1985, SAN DIEGO 1985.
2. TM 308. VEICOLO PER ISPEZIONE E CONTROLLO DI STRUTTURE OFFSHORE IN ACQUE
PROFONDE.
M. MAZZON, ECC SYMPOSIUM ON NEW TECHNOLOGIES FOR HYDROCARBON EXPLOITATION,
LUXEMBOURG, DEC. 1984.
3. CRITERI DI PROGETTO DI SISTEMI SOTTOMARINI CON CAPACITA DI TELEMANIPOLAZIONE.
A. BRIGHENTI, "PROGETTARE", N. 73, JUNE 1986.
4. W. PRENDIN, A. TERRIBILE, SIRI CONFERENCE, MILANO 1986.
5. E. DENTI, S. PIOLA, AIOM CONGRESS, VENEZIA 1986.
6. W. PRENDIN, D. MADDALENA, G. VERONESE, AIOM CONGRESS, VENEZIA 1986.

```
*********************************************************************************
* TITLE : REMOTE OPERATED VEHICLE                *        PROJECT NO          *
*                                                *                            *
*                                                *    TH./07049/83/UK/..      *
*                                                *                            *
*********************************************************************************
* CONTRACTOR :                                   * PROGRAM :                  *
*   BUE SERVICES LTD                             *   HYDROCARBONS             *
*   STONEYWOOD PARK,DYCE        TEL 0224-723415  *                            *
*   ABERDEEN AB2 ODF UK         TLX 73375        *                            *
*                                                * SECTOR :                   *
*                                                *   AUXILIARY SHIPS AND      *
* PERSON TO CONTACT FOR FURTHER INFORMATION :    *   SUBMERSIBLES             *
*   MR CLEGG                    TEL 0224-723415  *                            *
*                               TLX 73375        *                            *
*                                                *                            *
*********************************************************************************
```

VERSION : 09/06/87

AIM OF THE PROJECT :
****
To build an entirely new concept in Remote Operated Vehicles primary designed
to ultimately replace diver intervention in the Platform Inspection and
Cleaning role hence the vehicle name "PIC".

PROJECT DESCRIPTION :
****
Based on a medium sized ROV the vehicle has a payload package carrying cleaning
tools and inspection equipment mounted on a unique attachment device that
enables the vehicle to adhere close to platform nodes. The attachment device
called the Soleplate can rotate in two planes to allow vertical, horizontal and
diagonal members to be addressed.

STATE OF ADVANCEMENT :
****
Completed

RESULTS :
****
The vehicle has been on extensive onshore trials and offshore evaluations with
SHELL EXPRO (UK) and OCCIDENTAL PETROLEUM (CALEDONIA) LTD. These trials proved
the vehicles access capabilities in tight node configurations and its ability
to clean to inspection standards. Further evaluations with MPI systems took
place in October 1985 with BRITISH GAS.

```
*******************************************************************************
* TITLE : ARGYRONETE                            *        PROJECT NO          *
*                                               *                            *
*                                               *   TH./07050/83/FR/..       *
*                                               *                            *
*******************************************************************************
* CONTRACTOR :                                  * PROGRAM :                  *
*   COMEX                                       *   HYDROCARBONS             *
*   36, BOULEVARD DES OCEANS    TEL 091 410170  *                            *
*   B.P. 143                    TLX 410 985     *                            *
*   FR - 13275 MARSEILLE CEDEX                  * SECTOR :                   *
*                                               *   AUXILIARY SHIPS AND      *
* PERSON TO CONTACT FOR FURTHER INFORMATION :   *   SUBMERSIBLES             *
*   MR. Y. DURAND                               *                            *
*                                               *                            *
*                                               *                            *
*******************************************************************************
```

VERSION : 13/03/87

AIM OF THE PROJECT :
****
Development of a long-range autonomous lock-out submersible in order to prove
the validity of new concepts of operations carried out from the seabed and to
demonstrate the capabilities of new technologies essential for the future
development of underwater work.

PROJECT DESCRIPTION :
****
Main characteristics of the submersible :
- Length overall   : 28.06 m
- Submerged displacement   : 545 T
- Maximum depth   : 600 m
- Crew atmospheric compartment : 6 men
- Crew hyperbaric compartment : 6 divers
- Autonomy   : 15 days
Technological innovations of the project :
- Development of composite materials, in particular : high pressure hooped gas
cylinders, buoyant fairing, pressure tanks.
- Cryogenic oxygen storage
- Closed circuit diver breathing system
- Diver heating system

STATE OF ADVANCEMENT :
****
Ongoing. HP gas cylinders (pressure 400 Bar) : models and prototypes tested.
Manufacture of a set of prototypes completed (lengths 1.80, 3.0 and 6.0 m,
diameter 443 mm). Composite pressure tanks under construction. Acceptance tests
scheduled for February 1987. Lightweight fairings manufactured and presently
being assembled onboard. Cryogenic oxygen tanks and associated evaporation
system presently being assembled onboard. Prototype diver breathing system
currently undergoing evaluation tests.

RESULTS :
****
The installation of these innovations onboard the submersible argyronete/saga
will be completed early in 1987. Launching of the submersible is scheduled to
take place mid 1987.
Verification trials and demonstration will take place during the second half of
1987 and during 1988.

```
*********************************************************************************
* TITLE : REMOTE CONTROLLED SYSTEMS FOR          *         PROJECT NO          *
*         INSPECTION WORK-"DAVID V".             *                             *
*                                                *    TH./07051/83/DE/..       *
*                                                *                             *
*********************************************************************************
* CONTRACTOR :                                   * PROGRAM :                   *
*    ZF HERION SYSTEMTECHNIK                      *    HYDROCARBONS             *
*    POSTFACH 2168            TEL 0711-5209-351   *                             *
*    DE - 7012 FELLBACH       TLX 7254 733        *                             *
*                                                * SECTOR :                    *
*                                                *    AUXILIARY SHIPS AND      *
* PERSON TO CONTACT FOR FURTHER INFORMATION :    *    SUBMERSIBLES             *
*    MR. WIEMER                                   *                             *
*                                                *                             *
*                                                *                             *
*********************************************************************************
```

AIM OF THE PROJECT :
****
The project referred to as DAVID V was part of a larger project which was
concerned with the design, manufacture and testing of a diver assistance
submersible system for use in inspection, maintenance and repair tasks
associated with the offshore industry. During the development of the DAVID it
was found that the original specification resulted in certain operational
limitations and it became necessary to make extra provisions to achieve optimum
effectiveness. The required modifications were the subject of the project
DAVID V and the main aim was to make the DAVID suitable for remote operation as
well as for diver assistance.

PROJECT DESCRIPTION :
****
To achieve the extended role of the system it was necessary to modify the
submersible to suit a wider range of tasks and various facilities had to be
arranged in module form so that changes could be made quickly and easily during
operation.
- The buoyancy arrangement was modified so that vehicle trim could be achieved
  irrespective of system configuration.
- The control system was optimised to meet fully the requirements for remote
  control from the surface.
- The vehicle was arranged to accept additional equipment needed for the wider
  range of tasks.
- A lighter, more flexible umbilical cable was designed and fitted.
- A review of the vehicle design was made in order to reduce weight.
- The handling system was reviewed.

RESULTS :
****
The project was completed and as far as possible the results were incorporated
in the existing DAVID prototype.
Initial testing showed that the aims had been achieved in so far as the DAVID
could be operated either as a system remotely controlled from the surface and
also as a diver assistance vehicle. The findings were subsequently verified in
operations offshore.
Certain aspects, for example the requirement to make substantial weight
reduction, were not possible to achieve on the prototype system but sufficient
detailed design work was done to ensure that these changes could be made on
post-prototype systems.

```
********************************************************************************
* TITLE : DEVELOPMENT OF A HEAVY LIFT AND          *        PROJECT NO        *
*         TRANSPORT SYSTEM FOR THE INSTALLATION    *                          *
*         OF OFFSHORE PRODUCTION PLATFORMS         *     TH./07056/85/DE/..    *
*                                                  *                          *
********************************************************************************
* CONTRACTOR :                                     * PROGRAM :                *
*   BLOHM + VOSS AG                                *   HYDROCARBONS           *
*   P.O. BOX 100720            TEL 40 3119-2328    *                          *
*   DE - 2000 HAMBURG 1        TLX 211 047-0       *                          *
*                                                  * SECTOR :                 *
*                                                  *   AUXILIARY SHIPS AND    *
* PERSON TO CONTACT FOR FURTHER INFORMATION :      *   SUBMERSIBLES           *
*   MR. G. O. ANDERSSON                            *                          *
*                                                  *                          *
*                                                  *                          *
********************************************************************************
```

VERSION : 09/03/87

AIM OF THE PROJECT :
****
Installation technique of the topsides of offshore production platforms is by
crane lifting of reasonably sized modules. Considerable cost and time savings
particularly with earlier production would be achieved if the inshore proven
mating technique of platform deck and substructure could be transferred to at
sea application. The present project is a feasibility study of a specialized
transport and lifting barge design for worldwide operation combining
semisubversible motion characteristics and simple dock type lifting capability.
Integrated decks would be picked up by the large at the fabyard, transported to
the site, lifted by deballasting and lowered gently to the preinstalled jacket.
PROJECT DESCRIPTION :
****
In a comparative study the requirements and merits of modularized and
integrated deck installations will be summarized. Various proposals for
offshore mating devices will be analyzed especially with respect to application
in North Sea environment. The chances of a new mating device will be discussed
competing in the market with well established operators of heavy lift crane
vessels. The size of the barge and its form will be determined by an analysis
of form and size of existing jackets and the weight of the topsides they carry.
Operational aspects of load-out, transit amd mating procedure will be looked
into. By weather statistic of a relevant North Sea area the probability of
encountering sufficiently long good weather periods for transport and mating
will be evaluated. Among the studies of the technical feasibility of the barge
will be a discussion of its fabrication on a conventional shipyard. The most
critical load cases will be defined which determine the detail design of form
an structure. Ballast handling during the different loading conditions will be
considered and the compartmentation designed appropriately. Hydrostatic
stability will be checked and damage stability investigated. Motion behaviour
in the decisive loading conditions will be calculated using potential theory.
Model tests will be performed to doublecheck motion characteristics and study
the limiting conditions of the sea state for initiation of the mating procedure.
 Structural design will be controlled by longitudinal strength calculations and
stress analysis using finite element method applied to a beam model of barge
and platform deck structure in the design wave. Internal forces and moments
will be also measured in model tests in regular and irregular waves. Comparison
will be performed with forces obtained from the quasi-steady loaded beam model.

STATE OF ADVANCEMENT :
****
Ongoing. Steps 1 to 4.3 of the work plan are nearly completed; final evaluation
of seekeeping tests and calculation of motion behaviour is ongoing. Step 4.3 -
Structural Analysis was carried out up to the initial design stage.
RESULTS :
****
Review of installation modes (modularized or integrated deck) shows superiority
of integrated type. A vessel for one piece installation of topsides will,
however, envisage strong competition by established heavy lift crane barges.
Therefore, the barge must be also available for platform removal. The vessel
will have U-shape in plan view with similar legs floaters and columns of a
semisubmersible. Open at one side, they are cross-connected by two box shaped
bars at the other side, one below and one above water to achieve good torsional
resistance. With the deck spanned across the legs, conventionally shaped
jackets will be forked in for maching. Environmental forces acting on deck and
U-barge will load the jacket after the first contact is established. Jacket has
to be designed accordingly. Dimensions of U-barge were chosen to serve jackets
of a 60 m width at upper end, carrying a topsides weight of 30,000 t. Length
and beam of the U-barge are 140 m and 120 m, the depth is 28 m. Best method of
fabrication is in a large drydock, but there is a slight though more expensive
chance for conventional shipyards to fabricate the barge in parts and assemble
these in floating condition. Assuming a three day period necessary for transit
from shore to location, positioning of anchors, and mating, it was found from
weather statistics (North Sea, 66 deg. N, 2 deg. E) that such periods with
significant wave heights of up to 1.9 m occur with a frequency of as much as 1.
6 in January/February and 5.9 in May/June. Motion characteristics of the barge -
as found from calculations and model tests - is such that mating is possible
in seaways of up to 2 m significant wave height provided the prevailing wave
periods do not surpass 7 s, maximum 9 s. Model tests also showed that the barge
safely survives the 100 year storm in the North Sea in unloaded condition. The
square arrangement and size of the columns warrant sufficient intact stability
about arbitrary horizontal axes. Damage stability is provided by suitable
selected compartmentation and buoyancy boxes at the four corners of the deck
structure of legs and upper cross bar. For the underwater parts of the barge
strength requirements demand plate thickness of 30 mm to 50 mm, normal strength
steel, above water 20 mm to 25 mm. Only at the transitions between legs and
cross bars high tensile steel and larger plate thiccknesses are required.
Though the tolerable sea state in transit, loaded with a platform deck, must be
limited for safety reasons, the module support frame will be subjected to
additionals loads induced by the motions of the legs against each other. This
will be either lead to an MSF structure properly strengthened or a cross bar
construction designed against large deformations at the leg ends rather than
stresses.
REFERENCES :
****
A PAPER ENTITLED "ALTERNATIVES SCHWERLASTHEBE- UND TRANSPORTSYSTEM FUER DIE
INSTALLATION VON OFFSHORE-PLATTFORMEN" WAS PRESENTED BY G.O. ANDERSON AT THE
81ST ANNUAL MEETING OF THE SCHIFFBAUTECNISCHE GESELLSCHAFT, NOVEMBER 20 AND 21,
1986 IN BERLIN.

```
*********************************************************************************
* TITLE : THE DEVELOPMENT OF INTEGRATED          *        PROJECT NO          *
*          INTERVENTION SYSTEMS FOR SUBSEA        *                            *
*          PRODUCTION EQUIPMENT                   *      TH./07059/86/UK/..     *
*                                                 *                            *
*********************************************************************************
* CONTRACTOR :                                    * PROGRAM :                  *
*   COMEX HOULDER DIVING LTD                      *   HYDROCARBONS             *
*   BUCKSBURN HOUSE, HOWES ROA TEL 0224 714101    *                            *
*   BUCKSBURN                    TLX 73394        *                            *
*   UK - ABERDEEN AB2 9RQ                         * SECTOR :                   *
*                                                 *   AUXILIARY SHIPS AND      *
* PERSON TO CONTACT FOR FURTHER INFORMATION :     *   SUBMERSIBLES             *
*   MR. K. HULLS                                  *                            *
*                                                 *                            *
*                                                 *                            *
*********************************************************************************
                                                      VERSION : 09/03/87
```

AIM OF THE PROJECT :
****
The aim of the project is the development of subsea production equipment
wellheads, trees, manifolds etc.) which are designed on an integrated basis
with remote tooling systems. The tooling systems will be used in subsequent
maintenance operations. The subsea production equipment design will be
concerned with individual components (valves, chokes, control systems etc.) and
the layout of these components on equipment such as trees, manifolds and
templates. The tooling equipment will be designed to attach to the subsea
equipment and be designed to remotely clean, inspect, test, repair, and replace
defective components.

PROJECT DESCRIPTION :
****
The following work will be undertaken.
SUBSEA EQUIPMENT APPRAISAL A detailed study will be performed of the typical
range of satellite wells, template wellheads, template trees, manifolds etc. to
establish the format of equipment used, typical subsea operations and the
physical dimensions of equipment.
On the basis of this study, a subsea system will be specified for intervention
design. Typically this will be a Christmas tree and manifold to include the
range of operations described in Section 4.1.5.
ANALYSIS OF HANDLING SYSTEMS A detailed study will be made of the subsea
handling requirements of the selected system.
ATTACHMENT SYSTEMS Detailed design will be performed of the method of attaching
the parent handling frame to the subsea system.
Detailed design will be performed of the method of attaching tools to the
subsea equipment. The objective is to establish a minimum range of connection
types to perform all intervention tasks.
SUBSEA SYSTEM DESIGN The selected subsea system will be modified and
reconfigured to allow effective intervention.
The attachment system features will be incorporated to allow connection of the
parent handling frame and individual equipment item designs modified for
intervention.
MECHANISMS Detailed design will be undertaken of the individual intervention
tools.

The parent handling frame will be designed with consideration for method of transport, method of deploying tools, power systems and control.
The function and method of operating a range of tools will be studied and detailed design produced for a range of integrated tools.
Typically these tools will consist of :
(a) Gate Valve Override
(b) Hot line stabbing
(c) Testing/Override of Hydraulic Pod Functions
(d) Connector Lock/Release
(e) Dimensional Inspection
(f) Subsea Component Replacement (e.g. hydraulic pod, choke, etc.)
MANUFACTURE Manufacture of a complete subsea intervention system will be undertaken. A fully operational system will include all control functions and surface support modules.
FUNCTIONAL TESTING Full functional testing will be undertaken onshore. This will prove the validity of the handling techniques and the operation of the individual intervention components.
A test bed will be manufactured to provide a simulated Christmas tree installation. This test bed will be in the format of the proposed integrated design and will be capable of mounting the appropriate subsea system components for test purposes.
OFFSHORE TESTING Upon acceptance of the technology, intervention systems may be fully marketed.
Installation of the first system, assumes co-operation of a participant, and will allow testing of the subsea installed equipment and intervention systems offshore.

STATE OF ADVANCEMENT :
****
Ongoing. The project is at the planning stage. It is considered essential for the validity of the project that oil companies are involved and negotiations are taking place to ensure that oil companies participate technically in the programme.

REFERENCES :
****
K. HULLS, M. BOWRING, S. COW - DESIGNING SUBSEA SYSTEMS FOR REMOTE INTERVENTION - PROCEEDINGS OF THE UNDERWATER TECHNOLOGY CONFERENCE 1986: PUBL NUTEC BOX6, BERGEN, NORWAY.
K. HULLS: THE ECONOMICS OF REMOTE INTERVENTION - PROCEEDINGS OF ROV 1987 CONFERENCE.

```
*******************************************************************************
* TITLE : DEVELOPMENT OF KEY SYSTEM FOR AN        *        PROJECT NO        *
*         OFFSHORE SUBMERSIBLE                     *                          *
*                                                  *     TH./07062/86/DE/..   *
*                                                  *                          *
*******************************************************************************
* CONTRACTOR :                                     * PROGRAM :                *
*   THYSSEN NORDSEEWERKE GMBH                       *   HYDROCARBONS           *
*   AM ZUNGENKAI               TEL (4921) 85916     *                          *
*   P.O. BOX 23 51             TLX 27 802           *                          *
*   DE - 2970 EMDEN                                 * SECTOR :                 *
*                                                  *   AUXILIARY SHIPS AND    *
* PERSON TO CONTACT FOR FURTHER INFORMATION :       *   SUBMERSIBLES           *
*   MR. A. FREITAS                                  *                          *
*                                                  *                          *
*                                                  *                          *
*******************************************************************************
                                                        VERSION : 28/04/87
```

AIM OF THE PROJECT :
****
The offshore submersible is planned to be a part of a subsea oil and gas
production system. Its operation profile is very different from that of
conventional naval submarines, being characterized by great operating depth
(over 400 m), long diving periods (up to 21 days), exact 3-dimensional
positioning and low manning levels. As current technology cannot meet these
requirements, new key systems have to be developed. This project aims at the
development and in some cases the prototype testing of 5 key systems of the
submersible and their integration in a basic submersible design.

PROJECT DESCRIPTION :
****
The project covers the development design of the following submersible systems
and their integration in the conceptual design of a prototype vessel :
ENERGY SUPPLY SYSTEM
The energy system is to operate independent of ambient air and is based on a
Cosworth closed cycle diesel engine.
This system uses sea-water for scrubbing the CO_2 from the exhaust gases and a
mixture of CO_2 and argon for the inert part of the suction air.
A special water management system provides fresh sea-water at reduced pressure
and discharges the waste water outboards by using the pressure energy of the
surrounding water.
A test rig for 120 kW prototype system will be built to test the closed cycle
engine under service conditions.
PROPULSION AND DYNAMIC POSITIONING SYSTEM
The propulsion and DP-system must have a highly efficient transmission system
with sensitive power control for economic transit and accurate 3-dimensional
positioning.
At first alternatives for the transmission system are to be investigated and a
system chosen.
Next a conceptual design of the propulsion and DP-system for the prototype
submersible will be prepared to satisfy the requirements of the planned
operation profile.
Finally a prototype thruster will be built and tested under simulated operating
conditions.

LIFE SUPPORT SYSTEM

The extended diving period requires an efficient air purifying system with respect to size and weight. New systems have to be investigated and evaluated. A literature study will be made to identify the possible contaminants in breathing air. Working conditions and safety criteria will be examined. A concept for the life support will be designed and evaluated, and a system specification prepared. Emergency systems will be studied.

MISSION CONTROL SYSTEM

A control and monitoring system for the submersible with a high degree of integration and automation is to be developed in order to increase safety and reduce the manning level.

To begin, system functions, interfaces, kind of signals and type of sensors are to be defined. Concepts for control and monitoring of equipment, navigation and safety will be prepared. Layout for the monitoring and control of the vessel systems will be made. A specification of the required hard and software will be drawn up.

CRANE WORK MODULE

A crane serviced work module with manipulators is to be designed for operation in and from a wet cargo hold. At first the operation procedures and task profiles are to be investigated and defined. On this basis the crane and its components and the multi-function work unit with manipulators will be designed. Special development is required for the machinery operating in a wet environment and for the power supply and control systems.

CONCEPTUAL DESIGN OF THE PROTOTYPE SUBMERSIBLE will be made for a theoretical but realistic operation profile. The work includes the integration of the key systems, the general layout of the submersible and the specification of the main components.

STATE OF ADVANCEMENT :
****
Ongoing, the project is in general, excepting the energy system, in the design phase. The test rig for the closed cycle diesel engine has reached the construction stage.

RESULTS :
****
The results of this project will be a documentation of the development work and of the in-depth investigation on the key systems of an offshore service submersible and a conceptual design of a prototype submersible.

The stages of development of the 5 key systems, which will vary, will be carried out to a level where a descision to build a prototype submersible can be taken :

ENERGY SUPPLY SYSTEM

Full scale testing of a prototype closed cycle diesel engine on a test rig under realistic operating conditions.

PROPULSION AND DYNAMIC POSITIONING SYSTEM

Design of the thrusters and the energy transmission system.

Test of a full scale prototype thruster under operating conditions.

LIFE SUPPORT SYSTEM

Identification of the contaminants that would be present in the submersible atmosphere over long durations and definition of equipment required to purify the air and to supply breathing air under emergency conditions.

MISSION CONTROL SYSTEM

Definition of the operating, navigation and safety concepts with schematic diagramms. Specification of the hard and software for the monitoring and control system.

CRANE WORK MODULE

Working drawings of the crane and work modules and definition of the respective components.

PROTOTYPE SUBMERSIBLE

Main dimensions and general arrangement plan.

Definition of the pressure hull structure.

Preliminary stability and trim data

Preliminary layout of the submersible systems.

Definition of safety and rescue systems.

PIPELINES

```
*******************************************************************************
* TITLE : SELF-DESTROYING INSTRUMENTED VEHICLE      *      PROJECT NO        *
*         FOR THE INSPECTION OF PIPELINES.          *                        *
*                                                   *    TH./10026/81/FR/..  *
*                                                   *                        *
*******************************************************************************
* CONTRACTOR :                                      * PROGRAM :              *
*   SYMINEX                                          *   HYDROCARBONS         *
*   2, BLD DE L'OCEAN           TEL (91) 73 90 03    *                        *
*   FR - 13275 MARSEILLE CEDEX TLX 400563            *                        *
*                                                   * SECTOR :               *
*                                                   *   TRANSPORT            *
* PERSON TO CONTACT FOR FURTHER INFORMATION :        *                        *
*   MESSRS BAUDRY/BISSO                              *                        *
*                                                   *                        *
*                                                   *                        *
*******************************************************************************
```

VERSION : 26/03/87

AIM OF THE PROJECT :
****
The object of the project is to study construction and testing of a low-cost
self-destroying instrumented vehicle destined for the maintenance control of
pipelines (corrosion monitoring).

PROJECT DESCRIPTION :
****
The tool developed by SYMINEX will operate in a 12" diameter pipe in a 10 to 20
km length line. It is made of "syntactic foam" and could be destroyed by a
pyromechanism if it is jammed for a long time in the pipe. In addition, a
magnetic sensor gives information about the corrosion of the pipe. The data
acquisition and storage are managed by an on-board microcomputer.

STATE OF ADVANCEMENT :
****
Completed. The studies to assess the feasibility of the project are finished.
The different tested parts of the vehicle are :
- The automatic fragmentation.
- The electronic for corrosion monitoring.
- A calibration test on a test pipe has been carried out.

RESULTS :
****
Fragmentation tests with a pyrotechnical system have shown that all remaining
parts of the equipment were less than 25% of the pipe diameter.
The electronics for corrosion monitoring are composed of :
* a micro-computer to control the measurement systems and the data processing
* a real time clock for the referencing and which can also be used for
fragmentation programming
* the measurement electronics and analog conditioning
* a high-capacity CMOS RAM module.
Calibration tests were perfomed on a pipe test whose defects were known, a
phase shift of 5 degrees represented an attack of 3% of pipe wall for general
corrosion.

272

```
*******************************************************************************
* TITLE : HYDROELASTIC PHENOMENA ON SUBMARINE    *       PROJECT NO          *
*         PIPELINES.                             *                           *
*                                                *    TH./10038/83/IT/..     *
*                                                *                           *
*******************************************************************************
* CONTRACTOR :                                   * PROGRAM :                 *
*   SNAMPROGETTI                                 *   HYDROCARBONS            *
*   C.P. 12059                  TEL 0721 - 8811  *                           *
*   IT - 20100 MILANO           TLX 560 279      *                           *
*                                                * SECTOR :                  *
*                                                *   TRANSPORT               *
* PERSON TO CONTACT FOR FURTHER INFORMATION :    *                           *
*   DOTT. MATTIELLO                              *                           *
*                                                *                           *
*                                                *                           *
*******************************************************************************
```

VERSION : 01/01/87

AIM OF THE PROJECT :
****
Aim of the project is to improve knowledge of vortex induced oscillations which
occur on submarine pipelines when seabottom unevenness or sediment transport
cause large free spans. Theoretical analysis and experimental work is carried
out in order to achieve new methodologies and experimental data which might be
included in a certification procedure for submarine pipelines.

PROJECT DESCRIPTION :
****
A peculiarity of SVS project is to take into account the overall behaviour
(environment, structural behaviour, hydrodynamics) and to take into account the
scenario of the submarine pipeline resting on the sea bottom.
The project includes four phases :
- theoretical (Phase 1000)
- basin tests (Phase 2000)
- in field tests (Phase 3000)
- processing of experimental data (Phase 4000)
As for the theoretical phase, the physical phenomenon is analyzed and a
critical review of known theoretical and experimental literature is carried out.
Some analytical modelling techniques of phenomenon are analyzed, particularly
concerned of structural behaviour of pipelines freespans relating to specific
equilibrium conditions in presence of unevenness or sedimentological
instability of seabottom. As for the basin test phase, experimental situations
previously not sufficiently investigated are performed, specifically it
concerns : supercritical regime, wave induced vortex shedding, seabottom
proximity free-stream turbulence and flow tridimensionality, cable-type
behaviour of pipes for large freespans and so on; as for the field test phase a
pipeline free-span, installed in vortex shedding induced vibrations favourable
location (as concerned of mean currents velocity and/or wave regimes), is
instrumented, and hydrodynamic response is monitored together with the
meteomarine environment beside meteo-oceanographic situation is carefully
investigated.
As for the processing of experimental data, resulting data from experimental
basin and infield phases are processed and sorted in attempt of inserting them
in a forecast control procedure regarding sealine certification.

STATE OF ADVANCEMENT :
****
Ongoing. Phase 1000 is 75% complete; the state of the art is completed (100%),
the mathematical models are completed for structural parts, in progress for the
environmental and hydrodynamic aspects (68%). Phase 2000 is 59% complete; the
experimental tests are carried out at the Institute and the evaluation activity
of the initial results has started. Phase 3000 is 44% complete. Phase 4000 is
6% complete.

RESULTS :
****
The critical review of available open references has allowed to identify main
parameters which require further investigation.
The simulation of the structural behaviour of an oscillating free span has
allowed to point out the importance of soil-pipe interaction at free span
shoulders.
Reports containing preliminary results are to be put into circulation.

```
*********************************************************************************
* TITLE : DEVELOPMENT OF A FLUORIMETER FOR EARLY    *       PROJECT NO         *
*         DETECTION OF HYDROCARBONS FROM SMALL      *                          *
*         LEAKAGES IN OFF-SHORE PIPELINES           *    TH./09021/85/DK/..    *
*                                                   *                          *
*********************************************************************************
* CONTRACTOR :                                      * PROGRAM :                *
*   WATER QUALITY INSTITUTE                         *   HYDROCARBONS           *
*   SCIENCE PARK AARHUS        TEL 45 6 202000      *                          *
*   10, GUSTAV WIEDSVEJ        TLX 37874            *                          *
*   DK - 8000 AARHUS C                              * SECTOR :                 *
*                                                   *   PIPELINE LAYING        *
* PERSON TO CONTACT FOR FURTHER INFORMATION :       *                          *
*   MR. A. LYNGGAARD-JENSEN                         *                          *
*                                                   *                          *
*                                                   *                          *
*********************************************************************************
```

VERSION : 13/11/86

AIM OF THE PROJECT :
****
The developing of a tool for improvement of the safety of supply of oil/gas
through off-shore pipelines implies the wish for an early detection of small
leakages. An early warning of accidentally occurred cracks in the pipeline will
improve the operation security and thus avoid sudden stop of production.
The traditional methods such as registration of the pressure above the pipeline
will not be sufficient for the detection of small leakages, where as a
determination of the content and type of hydrocarbons in the water along the
pipeline will be a valuable tool.

PROJECT DESCRIPTION :
****
The project is divided into two main phases:
Phase A: Development of a 2nd generation prototype fluorimeter based
          on a laser with firm wavelength and an optical fibre system
          transmitting to one measuring stations.
          The prototype will be tested in the laboratory on different
          reference oil solutions and compared with conventional
          analytical techniques (GC/MS and UVF).
Phase B: Development of a 3rd generation instrument with varying
          wavelength (spectral UVF) including a micro-computer system
          for control of the laser, collection of data and interpreta-
          tion of the fluorescence. The 3rd generation prototype will
          be tested both in the laboratory and in the field.
Each of the two main phases is divided into a number of stages as follows :
A1: Collection of existing knowledge concerning UVF-technology,
    lasers and optical fibres. Specification of a system to be built
    in the laboratory (wavelength, energy-requirement, transmission
    losses in fibres, fibre types etc...
A2: Construction of the first laboratory model of the instrument. The
    model will be constructed in an optical bench at the Institute of
    Physics (University of Aarhus).
    The laboratory model will beconstructed optimized and tested using samples
    of pure hydrocarbons.
A3: Testing of the optimized instrument as a laboratory instrument
    on standard oils, mixtures etc. The instrument will be compared to
    other laboratory instruments.

A4: Construction of a final transportable prototype of the 2nd generation fluorimeter.
B1: Development of a micro-computer system for control of the laser
and collection and storage of the fluorescence.
B2: Development of software for interpretation of the spectral fluorescence
in order to determine concentrations and type of hydrocarbons
in mixture with seawater.
B3: Testing of the spectral UV-fluorimeter in the laboratory(as
stage AA3).
B4: Testing of the 3rd generation instrument in the field. The field
testing will be carried out by towing the sensor head after a
boat and simultaneously taking discrete water samples for laboratory analysis. The tests will be carried out in an area with
varying concentrations of hydrocarbons, for example near an oil
refinery.
STATE OF ADVANCEMENT :
****
The project is in the design and construction phase.
RESULTS :
****
The laboratory now disposes of the two laser systems. One system is a small
battery driven nitrogen laser with a fixed wavelength of 337 nm. The other
system is a large nitrogen laser for pumping a dye laser with a corresponding
frequency doubler. With this system the wavelength may be regulated
continuously from approx. 228nm to 900 nm by selection of suitable dyes and
doubling crystals.
In the test set-up in the laboratory the laser light is coupled directly to a
cuvette with the content to be investigated. For in-situ use on the sea the
idea is to transport light from the dye laser to the transducer via an optical
fibre.
Even small concentrations of oil must be measured, therefore the detector must
be able to see a large part of the interaction area between the laser light and
the oil/water solution. This is carried out by gathering light from an area
with a length of 50 mm. The fluorescence meets an interference filter with
center wavelength lamda-em and a typical FWHM of 5 nm. The transmission of the
center wavelength is normally between 25 and 50% and the blocking outside the
transmission interval will typical be 100.000.
After passage of the filter the light is gathered in a plexiglas rod doped with
a dye, enabling the light to be absorbed and reemitted with a somewhat longer
wavelength. Parts of the light is now caught in the material at a total
internal reflection. Thus the rod acts as a light pipe.
The area from where the fluorescence is caught is larger than if the Photo
Multiplier tube had seen directly towards the interaction area. At the same
time the fluorescence signal is emitted from the plexiglas rods at a higher
wavelength interval at a time.
Tests with known solutions have been made in order to test the measuring
principle. Measurements have been carried out on a dye (bis-MSB), which is
suitable for the wavelengths we had at our disposal at the start.
Experimentally is found a detection limit below 100 ng/l bis-MMSB in cyclohexan.
   This corresponds to 100 ppt. The fluorescence effectivity for a dye is
somewhat higher than for oil, but by optimizing the electronic equipment a
detection limit of 1 ppb for oil should be possible.

```
*****************************************************************************
* TITLE : DEVELOPMENT OF AN INTEGRATED PIPELINE      *      PROJECT NO       *
*         MANAGEMENT SYSTEM (IPMS)                   *                       *
*                                                    *    TH./09024/85/DK/.. *
*                                                    *                       *
*****************************************************************************
* CONTRACTOR :                                       * PROGRAM :             *
*   DHI/R & H PIPEDATA                               *   HYDROCARBONS        *
*   TEKNIKERBYEN 38              TEL 45 2 856500      *                       *
*   DK - 2830 VIRUM             TLX 37108            *                       *
*                                                    * SECTOR :              *
*                                                    *   PIPELINE LAYING     *
* PERSON TO CONTACT FOR FURTHER INFORMATION :        *                       *
*   MR. P.I. HINSTRUP                                *                       *
*                                                    *                       *
*                                                    *                       *
*****************************************************************************
                                                       VERSION : 17/10/86
```

AIM OF THE PROJECT :
****
The objective of the present project is to develop a multiple purpose submarine
pipeline and riser data base management system containing a large variety of
pipeline and riser related data and a multitude of utility programs. The major
purpose of the system is to provide the pipeline operator with an all-inclusive
information base and a range of analysis tools for his current pipeline
integrety evaluations, and inspection/maintenance planning. Project innovative
elements include:
- integration of a large family of databases
- wide range of new analyses methods
- relational database methodology
- high degree of software portability
- posibility for offshore application on survey vessel.

PROJECT DESCRIPTION :
****
The IPMS has been divided into 12 modules to enhance flexibility, viz:
1. Network Modelling
   Description of physical configuration of the pipeline and riser system.
2. Pipeline Design
   Storage of all data generated during pipeline design.
3. Riser Design
   storage of riser design data.
4. Pipeline and Riser Construction
   Data from the construction phase.
5. Inspection Characteristics
   Storage of information related to the execution of the individuel inspection
contracts.
6. Pipeline Inspection
   Storage and analyses of data gathered during pipeline inspection.
7. Position Correction Tables
   Compensation of positioning Inaccuracies
8. Riser Inspection
   Storage and analyses of data gathered during riser inspection
9. Pipeline End Manifold
   Storage and maintenance of data related to a PLEM.

10.Seabed Generel
   Data related to seabed in the vicinity of the pipeline route.
11.Environment
   Storage of relevant hydrographic data.ge of relevant hydrographic data.
 2.External Records
    References to external records.
the project development is divided into seven phases, viz;
a) System Analyses and Definition
b) System design
c) Programming
d) System Documentation
e) Implementation and Testing
f) Installation and Commissioning
g) Training

STATE OF ADVANCEMENT :
****
Two modules are nearing completion. Remaining modules are at different points
in the development process, cf. the above.

RESULTS :
****
The project zil will result in a highly flexible and portable database
management system suitable for installation in a large variety of
hardware/software environments.

REFERENCES :
****
"AN INTEGRATED PIPELINE MANAGEMENT SYSTEM" BY PETER I. HINSTRUP AND T.
SVENSSON
PRESENTED AT THE 1984 OFFSHORE COMPUTERS CONFERENCE IN ABERDEEN

```
*******************************************************************************
* TITLE : NEW TECHNOLOGY FOR INTERNAL IN-SERVICE   *      PROJECT NO         *
*         NDT-INSPECTION OF PIPELINES              *                         *
*                                                  *    TH./09025/85/DK/..   *
*                                                  *                         *
*******************************************************************************
* CONTRACTOR :                                     * PROGRAM :               *
*   A/S NORDISKE KABEL-OG TRAADFABRIKER (NKT)      *   HYDROCARBONS          *
*   BROENDBYVESTERVEJ 95        TEL 452 962688     *                         *
*   DK - 2605 BROENDBY          TLX 33761          *                         *
*                                                  * SECTOR :                *
*                                                  *   PIPELINE LAYING       *
* PERSON TO CONTACT FOR FURTHER INFORMATION :      *                         *
*   MR. J.L. NIELSEN                               *                         *
*                                                  *                         *
*                                                  *                         *
*******************************************************************************
                                                        VERSION : 13/11/86
```

AIM OF THE PROJECT :
****
The aim of the project is to carry out a feasibility study of a proposed new
communication method between an internal instrument carrier, passed through oil
as well as gas pipilineselines and an outside control and recording station.
The system will be based upon the use of an optical fibre connecting the
carrier with a control, recording and evaluation station at the launching end
of the pipeline.
The technological innovation is to use a fibre optical transmission system as
a communication system between the internal instrument carrier and outside
control stations in realistic environments.

PROJECT DESCRIPTION :
****
The project is divided into two major phases :
1. A feasibility study regarding fibre optical communication between the
carrier and the outside station.
2. Construction and test of a demonstration system.
In phase I it is to be investigated whether it is possible at all to transmit
optical signals through the fibre when the pig moves through the pipeline.This
is the ultimate and basic requirement for this project. If the result of phase
I is negative, the project will be terminated.
The work plan for phase I comprises 6 major stages :
1.1 Project planning in detail.
1.2 Calculation of forces acting upon an optical fibre in typical pipelines.
1.3 Fibre-optical transmission system.
1.4 Preparations regarding testing in a pipeline test loop.
1.5 Testing in pipeline test loop.
1.6 Evaluation and reporting.
In phase II the video and scanning equipment is to be developed for the purpose
of demonstration in this technical feasibility study.
The work for phase II comprises 8 major stages :
2.1 Selection and modification of a television camera system.
2.2 Modification of an existing scanning system for corrosion mapping.
2.3 Instrument carrier opto-electronics and interfaces.
2.4 External opto-electronics and interfaces.

2.5 Unit for mounting and testing in pipeline test loop.
2.6 Preparations regarding testing in a pipeline test loop.
2.7 Testing in pipeline test loop.
2.8 Evaluation and reporting.

STATE OF ADVANCEMENT :
****
Ongoing. The project is in the design phase. For the time being calculations of forces acting upon an optical fibre in typical pipelines are carried out.

```
********************************************************************************
* TITLE : TEST OF PROTOTYPE OF A SELF-FRAGMENTING   *      PROJECT NO         *
*         INSTRUMENTED VEHICLE DESTINED FOR THE     *                         *
*         CONTROL OF PIPELINES                      *      TH./09028/85/FR/..  *
*                                                   *                         *
********************************************************************************
* CONTRACTOR :                                      * PROGRAM :               *
*   SYMINEX                                         *   HYDROCARBONS          *
*   2, BOULEVARD DE L'OCEAN      TEL 91 73 90 03    *                         *
*   FR - 13275 MARSEILLE CEDEX TLX 400 563          *                         *
*                                                   * SECTOR :                *
*                                                   *   PIPELINE LAYING       *
* PERSON TO CONTACT FOR FURTHER INFORMATION :       *                         *
*   MR. A. BAUDRY/MR B. BISSO                       *                         *
*                                                   *                         *
*                                                   *                         *
********************************************************************************
                                                       VERSION : 13/11/86
```

AIM OF THE PROJECT :
****
The instrumented vehicle developed by SYMINEX enables the measurements of
corrosion in hydrocarbon pipelines.
The risk of PIG permanently jamming in the pipe is eliminated by a sytem of
automatic fragmentation.
The object of the project was the construction of the vehicle with integration
of the thickness measurement system and the realization of tests on a pipeline
in operation.

PROJECT DESCRIPTION :
****
After studying the various features linked with fragmentation, the use of
syntactic foam for the body of the instrument seemed to be the best solution.
The vehicle length is 0.9 m, its weight with the electronic components is 25 Kg.
It is propelled along the pipe by the fluid flow by means of two urethane cups
at speeds of up to 0.5 m/s.
Fragmentation of the vehicle takes place with the explosion of a pyrotechnical
transmission cord embedded in the body of the vehicle.
Another solution using expansive cement is presently being tested.
Corrosion phenomena are monitored from the measurements of the quantity of
metal in a section of pipe. The operational principle is based on the
deformation undergone by an alternating magnetic field traversing a conducting
surface.
Microprocessor circuits integrated into the equipment enable the measurement of
the phase shift at each period with storage of the results in RAM CMOS.
The sensor/acquisition unit assembly requires little energy and allows complete
autonomy via a distance of 20 Km.

STATE OF ADVANCEMENT :
****
Ongoing.
An industrial prototype was built, with integration of electronics and
transmission and receiving coils.
Tests on site were carried out on a 12" diameter and 136 m length pipeline.

RESULTS :
****
Calibration tests were performed on a pipe test whose defects were know, a phase shift of 5 degrees represented an attack of 3% of pipe wall for general surface pitting.

Tests on site were performed to compare the machine with an industrial environment.

The measurements at different speeds (20 cm/s to 40 cm/s) corresponded well. There was a phase shift in some tubes of 30 degrees corresponding to a 20% reduction in wall thickness.

These measurements were compared to those obtained from tests using an ultrasonic thickness transducer.

Wall thickness was controlled at 500 points per tube, areas of corrosion detected by the PIG corresponded to decrease in wall thickness.

The reliability of the system was established.

```
******************************************************************************
* TITLE : INTERNAL PIPELINE ALIGNMENT CLAMP          *      PROJECT NO        *
*                                                    *                        *
*                                                    *   TH./09030/86/UK/..   *
*                                                    *                        *
******************************************************************************
* CONTRACTOR :                                       * PROGRAM :              *
*   HOULDER OFFSHORE LIMITED                         *   HYDROCARBONS         *
*   53 LEADENHALL STREET        TEL 01 481 2020      *                        *
*   UK - LONDON EC3A 2BR        TLX 884 801          *                        *
*                                                    * SECTOR :               *
*                                                    *   PIPELINE LAYING      *
* PERSON TO CONTACT FOR FURTHER INFORMATION :        *                        *
*   MRS S.M. BEST                                    *                        *
*                                                    *                        *
*                                                    *                        *
******************************************************************************
```

VERSION : 16/03/87

AIM OF THE PROJECT :
****
Pipeline ovality and misalignment are problems which affect the success of any
pipeline connection, whether onshore or subsea in a hyperbaric tie-in. To date,
these problems have been resolved using manually fitted external clamps which
presuppose that the pipe is readily accessible.
We believe that this problem can be overcome by an intelligent pipeline pig
equipped to undertake the task internally. The alignment pig would be inserted
in the pipe prior to the weld such that hydraulically powered radial rams can
be extended to span the weld and correct any misalignment or distorsion in the
pipe, ensuring that welds of consistent high quality are achieved. Its
independent power source allows the pig to operate inside the pipe whilst
controlled by external command signals. In addition, gamma rays of the weld can
be undertaken using an atomic isotope, ensuring improved quality of radiography
as the rays only pass through a single wall thickness of pipe.

PROJECT DESCRIPTION :
****
The proposed design of the pipe alignment pig is as follows :
The unit is made up of a central frame which will house the hydraulics,
electrics and microprocessor in a 60 bar box together with the gamma head for
gamma raying the weld. The unit is equipped with 16 radial rams, these will be
extended to span the weld and further expanded to correct the pipe distorsion
and hold the pipe in position whilst the weld is made.
For transit the pig is fitted with polyurethane wheels to ensure that there is
no metal to metal contact and the unit is designed such that it fails safe into
this position. In addition, driving discs will be fitted to facilitate the
removal of the unit on completion of the operation. The pig will be designed to
negotiate 3D bends whilst withstanding acceleration and deceleration forces of
up to 50 "G".

All internal units will be suitably protected to ensure that no damage is
sustained when recovering the pig into the pig trap.
It is intended to install a gamma ray source of approximately 20 Curies which
will enable a single panoramic scan of the weld to be taken.
This method will allow an improved gamma ray image as the rays only have to
pass through a single wall thickness.
It is intended that the alignment pig should be a totally self-contained unit
requiring no external power source. In addition it is proposed that the
hydraulic requirements will be met using a small positive displacement pump
with a pressure of approximately 700 bar.
The pig will be equipped with an "intelligence" system which will allow all
stages of the operation to be monitored. It is currently proposed that a 32
function control system should be fitted.
The project has been divided into 4 sections to ensure that each stage in the
development progresses with the minimum technical and financial risk.
STAGE 1
Research and development engineering leading to the construction drawings for a
16 inch diameter prototype alignment pig will be undertaken. In conjunction
with this work, a market research programme will be conducted to establish
requirements, trends and possible future developments.
STAGE II
A prototype 16 inch diameter alignment pig will be constructed and tested in a
service test pipeline 200 m in length with a variety of simulated pipeline
bends and conditions. Pressure testing of the unit will aslo be undertaken in
the hyperbaric chamber in Aberdeen. During this stage a marketing video and
brochure will be produced and used for a sales drive to seek customer
commitment for the use of an alignment pig for hyperbaric welds.
STAGE III
Once a client's commitment for the use of an alignment pig has been secured,
the manufacturing drawings for the required diameter of the pig to suit the
clients pipeline will be produced and fully tested as described above. The
first production unit will then be constructed and tested prior to setting to
work in the field with the client.
STAGE IV
Construction drawings for a range of pigs.

STATE OF ADVANCEMENT :
****
Ongoing. The project has reached the completion of Stage I with the design
review taking place at the end of February 1987. It is anticipated that Stage
II will progress through the remainder of 1987.

```
********************************************************************************
* TITLE : TURBO PIG                                 *          PROJECT NO      *
*                                                   *                          *
*                                                   *     TH./10055/86/FR/..   *
*                                                   *                          *
********************************************************************************
* CONTRACTOR :                                      * PROGRAM :                *
*   CHALLENGER S.O.S.                               *   HYDROCARBONS           *
*   49 BIS, AVENUE FRANKLIN RO TEL 1 43.59.12.11    *                          *
*   FR - 75008 PARIS              TLX 642 477       *                          *
*                                                   * SECTOR :                 *
*                                                   *   TRANSPORT              *
* PERSON TO CONTACT FOR FURTHER INFORMATION :       *                          *
*   MR. P. SCEMAMA                                  *                          *
*                                                   *                          *
*                                                   *                          *
********************************************************************************
                                                      VERSION : 06/02/87
```

AIM OF THE PROJECT :
****
Design, build and test a pipe cleaning device (pig) based on new dynamic
concepts; this device will perform specially difficult internal pipe cleaning,
its components being specifically designed according to requirements, and will
incorporate a new and unique by-pass system, avoiding the traditionnal risk of
choking.

PROJECT DESCRIPTION :
****
PHASE 1
This phase is a general design phase, including:
- technical definition
- environmental characterisation
- system general architecture
- specifications
- laboratory simulation and tests
- drawings
PHASE 2
- construction of the prototypes
- field testing

STATE OF ADVANCEMENT :
****
First part of phase 1 (1.1) is complete; the second part (1.2) has begun.

RESULTS :
****
This project is still in its design phase and up to now, no major technical
problem or difficulty has been encountered.

```
********************************************************************************
* TITLE : VOYAGEUR                                  *          PROJECT NO       *
*                                                   *                           *
*                                                   *    TH./10056/86/FR/..     *
*                                                   *                           *
********************************************************************************
* CONTRACTOR :                                      * PROGRAM :                 *
*    CHALLENGER S.O.S.                              *    HYDROCARBONS           *
*    49 BIS, AVENUE FRANKLIN RO TEL 1 43.59.12.11   *                           *
*    FR - 75008 PARIS           TLX 642 477         *                           *
*                                                   * SECTOR :                  *
*                                                   *    TRANSPORT              *
* PERSON TO CONTACT FOR FURTHER INFORMATION :       *                           *
*    MR. P. SCEMAMA                                 *                           *
*                                                   *                           *
*                                                   *                           *
********************************************************************************
```

AIM OF THE PROJECT :
****
Design, build and test a pipeline inspection device (pig) based on new
instrumental concepts; this "pig" will perform traditionnal internal inspection
(gauging/calipering) and also gather all necessary "navigation" data to compute
and draw:
- the pipe path or trajectory
- the critical mechanical constraints (free spans)
providing a full geometrical survey of the pipelines.

PROJECT DESCRIPTION :
****
PHASE 1
This phase is a general design phase, including:
- technical definition
- environmental characterisation
- system general architecture
- specifications
- drawings
PHASE 2
- construction of the prototype
- writing of all necessary computer programs
- laboratory tests
- environmental qualification
PHASE 3
- field testing

STATE OF ADVANCEMENT :
****
Phase 1 has been completed.

RESULTS :
****
First theoretical results are promising:
* expected localization accuracy (computer simulated) is about a few meters,
for a pipe length of ten kilometers; calculations are now in process for pipe
lengths of hundred kilometers and more.
* technological considerations on inertial sensors have led to probably foresee
a special development for this application.

```
*********************************************************************************
* TITLE : PIPELINE INTEGRITY MONITORING EXPERT    *       PROJECT NO        *
*         SYSTEM                                   *                        *
*                                                  *    TH./10057/86/UK/..  *
*                                                  *                        *
*********************************************************************************
* CONTRACTOR :                                     * PROGRAM :              *
*    J.P. KENNY AND PARTNERS LTD                   *    HYDROCARBONS        *
*    BURNE HOUSE              TEL 01 831 6644       *                        *
*    88-89 HIGH HOLBORN       TLX 21823            *                        *
*    UK - LONDON WC1V 6LS                          * SECTOR :               *
*                                                  *    TRANSPORT           *
* PERSON TO CONTACT FOR FURTHER INFORMATION :      *                        *
*    DR. P. TAM                                    *                        *
*                                                  *                        *
*                                                  *                        *
*********************************************************************************
                                                            VERSION : 03/04/87
```

AIM OF THE PROJECT :
****
The main of the project is to provide subsea pipeline operators with a system
carry out accurate assessment of pipeline integrity and the prediction of
remaining useful life. The implementation of such a system will greatly assist
in reducing maintenance costs, improve safety and extend the useful operating
life of submarine pipelines.
The project relies on the application of expert system technology to model
operator expertise in the diagnosis of pipeline condition, based on inspection
records. The use of such methods in a large practical application, interfacing
with operator databases is a highly innovative approach.

PROJECT DESCRIPTION :
****
The project will be performed in three stages. At the end of each stage, an
operational version of the expert system will be delivered, ready for
customisation for pipeline operators. These three versions will be named Marks
I-III.
Mark I will be dedicated to the assessment of the as-built pipeline prior to
commissioning, and will be based on J P KENNY in house data for an actual North
Sea pipeline. This version will perform a fitness for service evaluation,
recommend remedial actions, based on the initial as-built pipeline condition
record (PCR).
Mark I will be a demonstration system, used to evaluate the integrity of a
specific pipeline. Operational aspects will not be considered at this stage and
the system will assume availibility of data in a specific format.
The Mark II system will have additional modules for the monitoring of integrity
using operational data. Firstly, Mark I will be tested using data from
operator(s) on other pipelines, to ensure that the system will operate in a
realistic situation. Then, pipeline operation modules will be added to the
system for the diagnosis and prediction of damage arising from the
transportation of the product.
These modules will take into account established operating practices and data
collection methods.

Mark III will be full system, which will incorporate additional modules for the
planning aspects of pipeline operation. The important functions of assisting
operators to plan and schedule their inspection, maintenance and repair
programmes will be dealt with at this stage. Mark III will include the
following additional features: planning of annual inspection programmes;
scheduling of remedial works; prediction of deterioration related failures;
forecasting of useful life and fitness for new service condition.
Development of the systems will be based on the expert system toolkit "ESSAI",
developed by ESC, Alcatel who will act as sub-contractor to the project by
advising on the use of the toolkit.

STATE OF ADVANCEMENT :
****
The project is currently at the architecture design phase for the Mark I
version of the software.

TRANSPORT

```
*******************************************************************************
* TITLE : LNG-LPG OFFSHORE TRANSFER SYSTEMS           *      PROJECT NO       *
*                                                     *                       *
*                                                     *   TH./10028/81/BE/..  *
*                                                     *                       *
*******************************************************************************
* CONTRACTOR :                                        * PROGRAM :             *
*   F.M.C. EUROPE SA                                  *   HYDROCARBONS        *
*   ROUTE DES CLERIMOIS          TEL 86 65 65 45      *                       *
*   F-89102 SENS CEDEX           TLX 800477           *                       *
*                                                     * SECTOR :              *
*                                                     *   TRANSPORT           *
* PERSON TO CONTACT FOR FURTHER INFORMATION :         *                       *
*   MR.GILLES GRONEAU                                 *                       *
*                                                     *                       *
*                                                     *                       *
*******************************************************************************
```

VERSION : 23/01/87

AIM OF THE PROJECT :
****
The aim of the project is to identify the most critical part of equipment in
LNG-LPG transfer systems, then design, build and test this part in order to
solve specific offshore problems.

PROJECT DESCRIPTION :
****
The most critical part identified is the swivel joint likely to be used in all
kinds of transfer systems (including flexibles). This part is generally used to
cope with combined motion, load and fluid transfer and then is subject to
fatigue, wear, thermal, mechanical problems, the first two parameters being
specific to the offshore environment.
It has been found during the design phase that the major reason for age-related
failures of the conventional swivel joints was that the wear was not equally
spread all over the circumference of the wear surfaces, and the idea developed
was to use, for the first time in this field, an arrangement of heavy-duty ball
bearings and free-wheels which would transform the oscillatory motions of the
piping in internal stepwise-unidirectional motions.
The project also included the manufacturing and testing of a true-scale
prototype. A sixteen inch swivel joint was then manufactured in strict
accordance with the needs of an actual project, fitted with thermal sensors and
installed on a test rig specially for that purpose. This test was designed to
apply oscillatory motions to the swivel joint with an amplitude of 5 degrees
and a period of 1.5 sec (i.e. accelerated four times).
A particular feature of the test is the fact that genuine liquefied natural gas
has been used instead of liquid nitrogen. Life testing has not been restricted
to motions but also included thermal cycles during which the swivel was allowed
to warm-up and then cooled-down again.
Mechanical loads have also been applied to the swivel by means of two cylinders
acting continuously during the motions. These loads were an axial force of
17000 N, a radical force of 70 000 N and a bending moment of 60 000 Nm.

STATE OF ADVANCEMENT :
****
Completed

RESULTS :
****
5 millions of cycles approximately and 50 thermal cycles have been completed.
Neither the ball bearings nor the free-wheels showed visible signs of wear and
the torque recordings have not demonstrated any significant increase of the
friction during the tests, from which it can be concluded that the mechanical
parts could have withstood more cycles than experimented. The seals have been
replaced once during the tests as part of the program but did not show any sign
of wear-out.

REFERENCES :
****
THE DESIGN PART OF THIS PROJECT HAS BEEN PRESENTED AT THE FIFTH INTERSOCIETY
CRYOGENICS SYMPOSIUM ORGANISED BY THE ASME IN DECEMBER 1984 (SESSION CRYOGENIC
PROCESSES AND EQUIPMENT).

```
*****************************************************************************
* TITLE : OFFSHORE LOADING OF LIQUEFIELD GAS        *      PROJECT NO        *
*                                                   *                        *
*                                                   *   TH./10035/82/FR/..   *
*                                                   *                        *
*****************************************************************************
* CONTRACTOR :                                      * PROGRAM :              *
*   GERTH-EMH                                       *   HYDROCARBONS         *
*   AV DE BOIS PREAU 4            TEL (1) 47.52.61.39 *                       *
*   FR - 92502 REIL-MALMAISON   TLX (1) 47.52.69.27 *                        *
*                                                   * SECTOR :               *
*                                                   *   TRANSPORT            *
* PERSON TO CONTACT FOR FURTHER INFORMATION :       *                        *
*   MR. CHAUVIN                   TEL (1) 47.71.91.22 *                       *
*                                 TLX 204586        *                        *
*                                                   *                        *
*****************************************************************************
```

VERSION : 01/01/87

AIM OF THE PROJECT :
****
This project is intented to develop an offshore terminal for refrigerated
liquefied gases and particularly liquefied petroleum gases (LPG). The terminal
envisaged comprises a loading station in the open sea capable either of loading
an LPG tanker (product at - 48 deg.C) from a shore storage facility, or the
reverse.

PROJECT DESCRIPTION :
****
The project studied the following subsystems:
. transfer line between shore storage facility and loading station,
. offshore loading station with single mooring point enabling the vessel to tie
up whilst remaining free to turn completely around the station so as to lie to
the winds and currents,
. transfer line between this single point mooring (SPM) and the ship.

STATE OF ADVANCEMENT :
****
The project was completed by the 31st of July 1985.

RESULTS :
****
Terminal has been designed on the basis of a CALM buoy fully motorized and
remote controlled, with a low temperature bi-fluid joint.
LPG machines : insulation and mechanical protection is obtained through an
outer steel casing with centering anchor points and a polyurethane insulation.
Low temperature tests enabled to verify field joint procedures, thermal and
mechanical behaviour of the concept and tightness to water.
CALM buoy for LPG : Adaptation of an oil loading CALM buoy to LPG transfer has
been carried out (structural steels, pipework configuration, equipment)
Safety requirements have been considered in order to avoid gas accumulation in
closed space (outdoor valves and sealing of the central shaft). Electrical
installations have been designed considering hazard zones defined for the buoy
in agreement with Bureau Veritas.

Bi-fluid swivel joint of the LPG CALM buoy : For transferring LPG at 48 deg.C and boil-off gas at 0 deg.C, a two-stage bi-fluid joint was developed. Design consists of two coaxial single fluid joints. The joint has an internal tube with external insulation protected by a metal casing comprising the bearings. Static and dynamic tests programme and fatigue tests linings have qualified the swivel joint and the sealings linings.

LPG transfer hoses : Rubber hoses composite hoses with concentrical steel turns have been developed by several industrial firms at TOTAL/EMH request and tested following detailed specification of TOTAL/EMH. Qualification trials on 1/2 scale prototypes and static tests on 16" prototypes have been carried out. Long fatigue tests of 16" prototypes at low temperature (refrigerated isopentane at 50 deg.C) have proved feasibility of the selected designs.

REFERENCES :
****
THE CHACAL PROJECT WAS PRESENTED TO GASTECH IN 1985.

```
*******************************************************************************
* TITLE : PROTOTYPE CONSTRUCTION AND PROTOTYPE    *      PROJECT NO         *
*         TESTING OF THE UW-WORK AND PIPELINE     *                         *
*         REPAIR SYSTEM "SUPRA"                   *    TH./10037/82/DE/..    *
*                                                 *                         *
*******************************************************************************
* CONTRACTOR :                                    * PROGRAM :               *
*   ARGE SUPRA                                    *   HYDROCARBONS          *
*   OCEAN CONSULTANT GMBH        TEL (04153)2314  *                         *
*   HALBMOND 30D                                  *                         *
*   D - 2058 LAUENBURG                            * SECTOR :                *
*                                                 *   TRANSPORT             *
* PERSON TO CONTACT FOR FURTHER INFORMATION :     *                         *
*                                                 *                         *
*                                                 *                         *
*                                                 *                         *
*******************************************************************************
```

VERSION : 31/12/86

AIM OF THE PROJECT :
****
Supra is designed to ensure unlimited working autonomy at sea bottom.

PROJECT DESCRIPTION :
****
New and versatile type of underwater work system for 420 m water depth and
integrated into a catamaran-type floatable vessel. It can be operated either
manned and diver-assisted or remotely controlled and unmanned. It is equipped
with underwater television and an obstacle avoidance sonar system. For diver-
assisted UW-works, conventional offshore saturation diving systems are used.

STATE OF ADVANCEMENT :
****
Completed. Supra is expected to be ready for operation in the summer of 1987.

RESULTS :
****
Supra's main applications are:
1. Gripping and aligning of pipelines by the four integrated alignment frames;
2. Supply of hydraulic and electric power for the machining of pipelines or
structures on the sea bottom;
3. Hyperbaric dry welding in the integrated welding habitat;
4. Interconnecting of pipelines by mechanical couplings;
5. Transport of heavy loads to the sea-bottom by a storage platform and
platform and underwater handling thereof by integrated underwater cranes;
6. Inspection and service works at the sea bottom, either in the "dry" habitat
or with the telescope swivel crane.
Supra's versatile applicability together with the vessel's capability to float
and to dive and its independence of large surface barges renders it
particularly economic and attractive to operators and companies which only have
smaller diving and supply vessels.

```
********************************************************************************
* TITLE : SUBSEA OIL LOADING SYSTEM.                    *        PROJECT NO     *
*                                                       *                       *
*                                                       *      TH./15034/82/DE/..*
*                                                       *                       *
********************************************************************************
* CONTRACTOR :                                          * PROGRAM :             *
*   AEG AKTIENGESELLSCHAFT                               *   HYDROCARBONS        *
*   STEINHOEFT 9              TEL (040)36161             *                       *
*   D - 2000 HAMBURG 11       TLX 211868                *                       *
*                                                       * SECTOR :              *
*                                                       *   MISCELLANEOUS       *
* PERSON TO CONTACT FOR FURTHER INFORMATION :           *                       *
*   DR. WILKE                                           *                       *
*                                                       *                       *
*                                                       *                       *
********************************************************************************
                                                         VERSION : 01/01/87
```

AIM OF THE PROJECT :
****
Study of a concept of a subsea oil loading system and in particular for
engineering of the individual components. The aim of the R & D project is
functional and manoeuvring trials of the system.

PROJECT DESCRIPTION :
****
The system can be described as follows: the tanker is manoeuvred into position
and has moonpool close to its manifold. This moonpool enables a re-entry unit
at end of loading hose which is lowered into the water and guided to the
pipeline end manifold and couples with it to make the hose link for oil loading.
 Coupling is effected by remote control. PLEM is connected via a pipeline with
production platform.
The project involves the following steps :
- detailed investigation of re-entry unit, hose, pipeline-end
  manifold and equipment aboard the tanker
- simulation calculations
- manufacture of re-entry unit (full scale model)
- manoeuvring trials/evaluations of results.

STATE OF ADVANCEMENT :
****
Completed and followed by Phase III of the project; manufacture of whinch and
hose and deep water trials (TH 10050/85).

RESULTS :
****
The system is expected to give many advantages: low expense in comparison with
the highly complex offshore loading systems required today; favourable handling
times; high availability even under rough weather conditions; economic
operation at great water depths and in marginal oil fields; maintenance and
service not dependent on weather; elimination of collision loading facility;
rediced storage capability.
Moreover, SOLS can be used for the development of production areas not
accessible to surface loading systems and the addition of the dynamic
positioning system gives the tanker major manoeuvrability benefits. SOLS can
economically pump a tensid fluid into a subsea well and carry back crude oil
from another well some miles away. Application of heavy ballast transportation
appears realistic. The investigations done during this project (phase II) have
confirmed the expected advantages.
The manoeuvring trials and the simulations have shown the good manoeuvrability
of the re-entry unit and the possibility to connect a hose to the pipeline and
manifold in short time.

REFERENCES :
****
- "SUBSEA OIL LOADING SYSTEM FOR TANKER"
  1984, EUROPEAN PETROLEUM CONFERENCE (SPE 12980)
- UNTERWASSER -OELUEBERNAHMESYSTEM FUER TANKER SOLS
  SCHIFF & HAFEN, HEFTE 4/5, 1986
WEST EUROPEAN CONFERENCE ON MARINE TECHNOLOGY (WEMT)
  "ADVANCES IN OFFSHORE TECHNOLOGY" 25.-27. NOV.86
  (HOLLAND OFFSHORE, AMSTERDAM)

```
********************************************************************************
* TITLE : OFFSHORE FLEXIBLE HOSES                    *       PROJECT NO        *
*                                                    *                         *
*                                                    *    TH./10039/83/IT/..   *
*                                                    *                         *
********************************************************************************
* CONTRACTOR :                                       * PROGRAM :               *
*   INDUSTRIE PIRELLI S.P.A.                         *   HYDROCARBONS          *
*   VIALE SARCA 222          TEL 2-64421             *                         *
*   IT - 20126 MILANO        TLX 310135 PIRELLI      *                         *
*                                                    * SECTOR :                *
*                                                    *   TRANSPORT             *
* PERSON TO CONTACT FOR FURTHER INFORMATION :        *                         *
*   MR. MANCOSU                                      *                         *
*                                                    *                         *
*                                                    *                         *
********************************************************************************
```

VERSION : 25/05/87

AIM OF THE PROJECT :
****
Planning of offshore hose through the Finite Element Analysis (FEA), reducing
the experimentation time and optimising the structure for the single
application.

PROJECT DESCRIPTION :
****
1. Design and construction of a test tank for experimental models.
2. Preparation of some structures in 1/3 scale in respect to full size scale.
3. Static and dynamic tets on models with collection of data on behaviour.
4. Realization of a mathematical model for the various structures.
5. Simulation of tests mentioned under point 3.
6. Comparison between experimental data and calculation results.
7. Improvement of the mathematical model through iteration of points 3-4-5-6.
8. Extrapolation of 1/3 model to full size through realisation of a prototpype.
9. Practical test of the prototype.
10. Final adjustment of the mathematical model to be used for new projects.

STATE OF ADVANCEMENT :
****
Steps from 1 to 6 completed. Iterations (experimentation-calculation) are
ongoing on two basic structures.

RESULTS :
****
The threedimensional FEA method for model examination was successfully worked
out. The tank for dynamic tests was used with good results.
Due to delays in designing the tank and in working out the calculation code, it
was not possible to test the new project on an actual prototype, within the
agreed term.

```
*****************************************************************************
* TITLE : DEEP SEA REPAIRS TRIALS.            *      PROJECT NO          *
*                                             *                          *
*                                             *   TH./10042/83/FR/..     *
*                                             *                          *
*****************************************************************************
* CONTRACTOR :                                * PROGRAM :                *
*   GERTH/COMEX/ALSTHOM-ACB                   *   HYDROCARBONS           *
*   4, AVENUE DE BOIS PREAU    TEL (1)47.52.61.39 *                      *
*   FR - 92502 RUEIL-MALMAISON TLX (1)47.52.69.27 *                      *
*                                             * SECTOR :                 *
*                                             *   TRANSPORT              *
* PERSON TO CONTACT FOR FURTHER INFORMATION : *                          *
*   MR. P. FABIANI             TEL (1)42.91.42.55 *                      *
*                              TLX TCFP615700 *                          *
*                                             *                          *
*****************************************************************************
                                                    VERSION : 01/04/87
```

AIM OF THE PROJECT :
****
This project consists of testing in a significant water depth and without the
intervention of divers, a remote controlled modular system for the preparation
of pipeline extremities in view of a subsequent connection.
The object of the project is to validate one of the main phases of a complete
subsea pipeline repair operation, an efficient preparation of pipe extremities
being indispensable whatever the considered connection method.
The connection itself is not included in this project.

PROJECT DESCRIPTION :
****
The following means have been implemented to run this test:
- a pipeline section (layed for this purpose) of about 100 m formed by pipes of
610 mm (24" external diameter) and 25 mm (1") thick in API 5L x 65 steel. This
pipeline was lined with coal pitch. Inside this pipe some deposits simulated
the pipe dirt.
- a dynamic positioning surface support of the "SEACOM" type : approximately
2500 tjb with a lifting capacity of 70 t minimum in single line and 500 m2 of
available deck surface.
- an assistance ship
- a submersible aimed at operating the RMP H-frame, ensuring backup functions
and adequate observations.
- an acoustic signalling system allowing to spot the equipment on the seafloor
- the RMP devices : manufactured and tested in shallow waters within the scope
of contract 09019/79 i.e. :
  - a transfer vehicle to handle and position the equipment (H-frame, table +
girder + module assembly) and transmit the required energy.
  - a H-frame, monitored by the transfer vehicle or the submersible, allowing
to adjust the position of the pipe section and hold it during the operation.
  - a worktable to maintain the pipe extremely in position during the operation
of modules, also serving a support for the girder
  - a module carrying girder which mates onto the worktable equipped with a
module carrying trolley serving as guide to the different modules during their
operation.
  - 3 work modules : external brushing, fine cutting, internal brushing.

STATE OF ADVANCEMENT :
****
A test was carried out offshore the western coast of Scotland at Loch Linnhe at a water depth of 135 m in December 1984.
The deep sea tests in Loch Linnhe proved to be a complete success with a particularly reduced operation time.

RESULTS :
****
The main two objectives of the programme have been completely achieved, these objectives being the proof that the pipe extremily preparation module is efficient, and that the remote controlled and guideline less handling, positioning and recovery of heavy loads in deep waters is possible. The general structure of the system is reliable inspite of the small modifications to be made.
Moreover, the concept of handling without guidelines with the aid of a transfer vehicle is well mastered and opens a field of important tasks in significant depths.
From the commercialization standpoint, it must be noted that the RMP equipment is aimed at repairing deep see pipelines, operations which fortunately occur seldom. Consequently, the overall programme must be considered as a guarantee, like for instance when a fault is detected on a deep sea pipeline, the technology exists and the problem can be solved.
The commercialization of the RMP system may thus be envisaged in the form of a club of operators using deep sea pipelines. A multidiameter RMP system would be kept operational by this club. This alternative presents the advantage of sharing the important costs linked of the manufacture and maintenance of such a system. The risk would thus be shared between these companies by the intermediate of the club which would operates as an insurance company.
Nevertheless, no deep sea development (down till 1000 m) has yet been achieved. There is no deep sea pipeline market; there exists only one pipeline beyond a 300 m depth in the straits of Sicily. On the other hand, developments in intermediate depths between 300 and 500 m are envisaged (North Sea, Gulf of Mexico, California).
The nearest potential market thus exists at this level. It will require the achievement of additional studies so to propose a complete repair, the connection being done either by mechanical connector, or by hyperbaric welding. Indeed, for these depths, one may envisage the intervention of divers, but only for tasks such as welding.

REFERENCES :
****
- EEC SYMPOSIUM DECEMBER 1984
- DOT 85 SORRENTE

```
***********************************************************************
* TITLE : TRIPHASE FLOW AND PIPELINE/RISER      *     PROJECT NO      *
*         COUPLING.                             *                     *
*                                               *   TH./10043/84/FR/..*
*                                               *                     *
***********************************************************************
* CONTRACTOR :                                  * PROGRAM :           *
*   GERTH                                       *   HYDROCARBONS       *
*   AVENUE DE BOIS PREAU 4      TEL 1 47 52 61 39*                     *
*   FR - 92500 RUEIL-MALMAISON TLX 203050       *                     *
*                                               * SECTOR :            *
*                                               *   TRANSPORT          *
* PERSON TO CONTACT FOR FURTHER INFORMATION :   *                     *
*   MR. CORTEVILLE            TEL 1 47.52.69.60  *                     *
*                                               *                     *
*                                               *                     *
***********************************************************************
```

VERSION : 01/01/87

AIM OF THE PROJECT :
****
The project consists in the production of crude oil or gas in wells and in the
transport via flowlines and pipelines under polyphasic flowing conditions of
such hydrocarbons towards processing units.
The general purpose is to finalize the development of a number of computing
methods for these polyphasics flows : these include various flowing patterns
which are still poorly known in spite of the increasing research in this domain
in the past years i.e. transport of multiphase effluents. These models will
allow production cost reductions and development of new production
configurations.
Usually based on correlations restricted to a small validity domain,
traditional computing methods are not accurate enough ; optimization of
production equipment and operating techniques.
PROJECT DESCRIPTION :
****
The project is to be developed in three phases :
Phase 1 : Study of triphasic flows in tubings.
The instrumentation and data processing system on a triphasic flow test loop
have been improved and finalized. The tests were started under diphasic gas-oil
and gas-water conditions in order gradually to introduce the new triphasic flow
parameters and to examine more in detail the part played by each of the liquid
phases. An initial campaign of tests under triphasic conditions was then
undertaken with an oil content of 70 % in the liquid mixture. In addition, a
test campaign under petroleum conditions with a natural gas-water system was
performed on the Boussens diphasic loop in order to study problems of water
carryover in natural gas wells.
Phase 2 : Study of the instabilities causes by pipeline-riser coupling in
diphasic flows.
The coupling between pipeline and riser was covered by a campaign of systematic
tests so as to recognize the flow conditions in two horizontal and vertical
lines, coupled or not coupled. The limit between stable and unstable flow
conditions was specified. A simple model was developed to take into account the
phenomena of instability : the initial results are encouraging.

Phase 3 : Transient diphasic flows in transport pipelines : analysis of the
present data and definition of an experimental and modelling programme
(preliminary study). Transient diphasic flows underwent preliminary study. This
study has enabled a specific research programme on these flow conditions to be
defined.

STATE OF ADVANCEMENT :
****
Completed at the end of 1986.

RESULTS :
****

PHASE 1 : TRIPHASIC FLOWS
A thorough databank including 96 measuring points among which 45 under
triphasic conditions and 51 under diphasic conditions water/air and oil/air has
been developed at the I.M.F.T. and completed by measuring points obtained on
the diphasic test loop of Boussens in diameters of 3" and 6" with the oil/gas
and water/gas system.
Results have shown that the creep relative to these two phases is very low, and
that the triphasic flow gas-oil-water can be assimilated to a diphasic flow gas-
liquid, in which the liquid presents particular rheological properties.
Interpretation of the overall test data has allowed the adaptation of the gas-
oil diphasic computation model developed within the scope of a previous
contract, so to cover precisely the overall gas-liquid polyphasic flows. The
WELLSIM programme, incorporating this model, has been tested on operational
data deriving from oil and water production wells: the average error is more
satisfactory than the results of similar comparisons, with conventional
computing methods.

PHASE 2 : PIPELINE-RISER COUPLING
Mechanisms that govern the formation of instabilities and their frequency have
been recognised during specific experiments run under stable flow (high speeds)
and instable flow (low speeds) conditions. This enabled us to set-up an
equation system which translates the movement mass and quantity balance in an
unsteady pattern. However, difficulties were encountered while carrying out the
numerical resolution of the system comprising a minimum of simplifying
hypotheses. A simplified model called voluminal with numerical resolution of
equations through the method of finite differences has allowed to recover in a
qualitative manner the aspect of pressure or flowrate fluctuations measured at
the riser foot but the accuracy, which depends on the time pitch, proved to be
insufficient. Hence, a second model including a smaller number of simplifying
hypotheses has been set-up using as an equations resolution mode the method of
characteristics. Difficulties of numerical order were then encountered and
additional studies are presently undertaken in view of finalisation so to run
the overall validation tests required. This is being performed beyond the scope
of this project.

PHASE 3 : TRANSIENT DIPHASIC FLOWS
The preliminary studies on transient diphasic flows have been carried out.
Tests and modelling studies have been untertaken in view of setting up a
computation method within the scope of other projects.

REFERENCES :
****
A NUMBER OF PAPERS ARE BEING PREPARED RELATIVE TO THESE SUBJECTS WITH REGARD TO
THEIR PRESENTATION IN DEDICATED PETROLEUM MAGAZINES AND INTERNATIONAL
CONFERENCES.
AN INITIAL ARTICLE "EXPERIMENTAL STUDY OF RISING GAS/LIQUID FLOW WITH TWO AND
THREE PHASES IN A VERTICAL PIPELINE" HAS BEEN PUBLISHED IN THE MAGAZINE OF THE
INSTITUT FRANCAIS DU PETROLE, JANUARY 1986, P. 115-129

```
*******************************************************************************
* TITLE : UNMANNED SUBSEA BOOSTING SYSTEM FOR TWO    *        PROJECT NO        *
*         PHASE FLOW (SBS)                           *                          *
*                                                    *     TH./10045/84/IT/..   *
*                                                    .                          *
*******************************************************************************
* CONTRACTOR :                                       * PROGRAM :                *
*   SNAMPROGETTI SPA                                 *   HYDROCARBONS           *
*   C.P. 12059              TEL 02 5201              *                          *
*   IT - 20100 MILANO       TLX 310246              *                          *
*                                                    * SECTOR :                 *
*                                                    *   TRANSPORT              *
* PERSON TO CONTACT FOR FURTHER INFORMATION :        *                          *
*   ING. S. BURATTI                                  *                          *
*                                                    *                          *
*                                                    *                          *
*******************************************************************************
                                                          VERSION : 01/01/87
```

AIM OF THE PROJECT :
****
Aim of the project is to define and demonstrate the feasibility of a Subsea
System for collection, pumping and dispatching of the production of an oil
field to a gathering station either onshore or on a platform in shallow waters.
The System, which is unmanned, must be suitable for installations up to depth
of 1000 m and at a distance of about 100 Km from the coast.

PROJECT DESCRIPTION :
****
The project includes the following phases:
PHASE 1000 - Definition of operating scenarios and performance of the system.
Activities in this phase include the identification of the reference operating
scenarios, the main performance to be required by the SBS system and a
preliminary economic investigation and analysis.
PHASE 2000 - Design of the system and analysis of components subsystem.
In this phase, the activities aim to define, on the basis of the results of
phase 1000, the architecture of the system, as well as to select the subsystems
and start the reliability analysis.
PHASE 3000 - Two phase pump subsystem
Investigations, experimentation and setting-up of two phase pumps are foreseen
to develop and define a two phase machine which can cover the operating
scenario identified in phase 1000.
PHASE 4000 - Power generation subsystem.
To identify and develop the power generation and transmission subsystem.
PHASE 5000 - Instrumentation and control subsystem
To identify and develop the control subsystem and the data
acquisition/transmission.
PHASE 6000 - Prototype construction and onshore tests
To evaluate the performance of the developed system, the construction of a
complete prototype and it tests in an on-land pilot plant are foreseen in this
phase.
PHASE 7000 - Tests on offshore pilot circuit
An offshore pilot circuit will be built to test the prototype in operating
conditions similar to the actual ones.

STATE OF ADVANCEMENT :
****
The project is in the design phase.
Phase 1000 has been completed, phase 2000 is 80% completed. Phase 3000 is 50% completed.
The progress activities, in the above phases, lower than the one estimated, attests both the actual difficulties encountered when developing technologically innovating system and the fact that it has not yet been possible to choose a reference arrangement for the pumping system.
Phases 4000 and 5000 are 60% completed. A detailed definition of the subsystem has been reached

RESULTS :
****
The main performance of SBS system have been confirmed according to the selected scenario:
- capacity of handling the two phase mixture with gas content variable in time and according to "GOR" value of the field fluid. Therefore two reference operating areas have been found:
  * low GOR: up to 50% of vacuum degree;
  * high GOR: up to 90% of vacuum degree.
- capacity of pumping a flowrate of liquids variable in time relevant to a max production of 30,000 BOPD.
- capacity of giving to the pumped mixture a pressure rise on the order of 90 Kg/cm2 (on the assumption of a depth of 1,000 m, of a distance of 100 Km far from the coast, of a pipeline with diameter 16")."The use of a SBS system in the exploitation of marginal offshore oil fields led to conceive the plant in a modular way in order to guarantee a greater operation flexibility to the system.
The production capacity of the basic unit has been fixed in 10,000 BOPD.
In its turn each basic unit is manufactured with modules structurally independent housing the process equipment and the plant feed and control systems.
The modular design of the basic unit results from the need of making it feasible the installation and recovery operations of the plant by ROV in these water depths.
As far as the development of a two phase pump is concerned, the results of the tests on the centrifugal impeller in air-water circuit make us think that the "centrifugal" alternative cannot cover the operating range identified in phase 1000. Consequently, studies have been started aiming at analyzing the feasibility of pumping alternatives in the area of volumetric pumps.

303

```
****************************************************************************
* TITLE : REPAIR SEA TRIALS ON DEEP WATER       *      PROJECT NO        *
*         PIPELINES (600 M WATER DEPTH)          *                        *
*                                                *      TH./10047/85/IT/.. *
*                                                *                        *
****************************************************************************
* CONTRACTOR :                                   * PROGRAM :              *
*   SNAM SPA                                     *   HYDROCARBONS         *
*   PIAZZA VANONI1              TEL 39 2 5205716  *                        *
*   IT - 20097 SAN DONATO MILA TLX 310 246       *                        *
*                                                * SECTOR :               *
*                                                *   TRANSPORT            *
* PERSON TO CONTACT FOR FURTHER INFORMATION :    *                        *
*   MR. A. LOLLI                                 *                        *
*                                                *                        *
*                                                *                        *
****************************************************************************
                                                    VERSION : 17/10/86
```

AIM OF THE PROJECT :
****
Maintenance and repair of submarine pipelines gets harder and harder when depth
increases.
For this reason SNAM, joint owner of the world's deepest pipeline (600 m), has
developed (with other Companies of ENI Group) new techniques for automatic
maintenance and repair of such pipelines.
The new system is called S.A.S. (Stazione Autonoma Sottomarina) and, at the
present time, its construction is completed.
Although a series of deep water test was already scheduled, it is now
considered that a full demonstration of the capability of the system is
necessary, including the repair of a 20" sealine in deepwater.
The innovations and know-how expected from these trials consist in proving for
the first time the feasibility of such operation at great dephts in complete
automatic mode (all steps are computerized), and safety as the human presence
is not required on the sea bottom.

PROJECT DESCRIPTION :
****
2.A Main project phases
   The project is developed in five main phases:
     1) construction of connection systems
     2) construction of a pipeline section (140 m)
     3) transport of the pipeline offshore and laying at 600 m awater depth
     4) repair of the pipeline on the sea bottom at 600 m
     5) analysis of results and final report.
2.B Project phases
   For each project phase the work schedule can be subdivided into the
following groups of operations:
     B1) construction of connection systems
         . modifications and improvements to the connection system developed by
Nuovo Pignone
         . preliminary tests on reduced and full size scale
         . construction of connection systems
     B2) Preparation of the testing pipeline section
         . setting up of onshore yard for transporting the pipeline offshore
         . purchase of 20" pipe lenghts, welding and installation of floats

B3) Transport of the pipeline offshore and laying 600 m depth
. mobilization and demobilization of two ocean-going tugs
. towing of the pipeline to the site of installation and laying on the
seabed
2.C) Repair of the pipeline
. Mobilization of a dinamically positioned ship and of a support vessel
. Modifications in the dinamically positioned ship for the installation on
board of a part of the repair system
. Leasing of a complete R.O.V. system (SCORPIO class)
. Repair of the pipeline consisting in:
- removing any artificial overlays from the pipeline (to be defined in
details later on)
- raising the pipeline from the seabed
- cutting and cleaning the two ends to be connected
- recovery the damaged pipe section by lifting it to the surface
- transporting the base frame into the seabed and measuring the spool to
be connected
- installation of connectors on the pipe ends, sealing tests
- installation of the spool-piece, locking and sealing test of
components
- recovery of modules from the seabed.

STATE OF ADVANCEMENT :
****
At present time the preparation of the repair intervention plant, selection of
the area, and surface support vessel, preparation of the test pipe is under way.
The deep water repair operation is now scheduled for Summer 87.

REFERENCES :
****
STATUS OF SNAM DEEPWATER AUTOMATICAL PIPELINE REPAIR SYSTEM - D.O.T. 2 VALLETTA
MALTA - OCT. 83
STATUS AND TECHNOLOGICAL FALL-OUT OF THE SAS PROJECT A. CONTER - SNAMPROGETTI -
D.O.T. 3 - SORRENTO ITALY - OCT. 85

```
*******************************************************************************
* TITLE : DEVELOPMENT OF A RANGE TWO-STAGE           *       PROJECT NO       *
*         CENTRIFUGAL PUMPS FOR THE TRANSPORT OF     *                        *
*         CRUDE AND ITS DERIVATES                    *     TH./10049/85/NL/..  *
*                                                    *                        *
*******************************************************************************
* CONTRACTOR :                                       * PROGRAM :              *
*   STORK PUMPEN B.V.                                *   HYDROCARBONS         *
*   LANSINKESWEG 30            TEL 074-454321        *                        *
*   P.O. BOX 55               TLX 44324              *                        *
*   NL - 7550 AB HENGELO (O)                         * SECTOR :               *
*                                                    *   TRANSPORT            *
* PERSON TO CONTACT FOR FURTHER INFORMATION :        *                        *
*   IR. E.J. BUSSEMAKER                              *                        *
*                                                    *                        *
*                                                    *                        *
*******************************************************************************
                                                         VERSION : 13/11/86
```

AIM OF THE PROJECT :
****
To develop a new range of two-stage centrifugal pumps for the transport of
crude oil and derivates thereof. The range will be equipped with a rotor which
is supported by a bearing at each end of the pump. Compared with the
traditional design equipped with an overhung pumprotor, the new pump design
incorporates a number of innovative aspects resulting in the following
characteristics: low rotorvibration, low sealleakage, long life, low power
consumption and low weight.

PROJECT DESCRIPTION :
****
The project is divided into the following six phases:(1) orientation phase, (2)
innovation phase, (3) development phase, (4) testphase, (5) design and
production of initial series casting patterns and (6) production and sales.
.
Description of the phases:
(1) choice of manufacturing techniques, choice of materials,
    evaluation of hydraulic features, planning.
(2) further development of hydraulic principles and manufacturing
    methods, alternative designs.
(3) implementation of results from phases (1) and (2), detailed
    design work, choice and specification of materials.
(4) production and test of prototype pump, analysis of test results and
    and implementation of resulting design changes in the design
    of phase (3).
(5) detailed design drawings for the complete range, production of
initial casting patterns and pumps, initial test of every pump-
type and implementqtion of resulting design changes.
(6) normal production and sales.
.
The main technical aspects of this development project are the following:
optimal hydraulic design using CAE/CAD, introduction of very smooth hydraulic
surfaces, very high efficiency, modular set-up, capability to cope with every
pumped fluid, good mechanical pump behaviour at off-design conditions.

STATE OF ADVANCEMENT :
****
Orientation phase (1) not yet ended, development phase (3) has been started as
far as the bearing design is concerned. Because of limited manhours capacity,
the project is not yet under full steam, because a second development project
has got higher priority. Planning to be adjusted.

RESULTS :
****
Not available due to early stage of progress.

```
*********************************************************************************
* TITLE : SUBSEA OIL LOADING SYSTEM (SOLS) FOR    *        PROJECT NO          *
*         TANKERS                                 *                            *
*                                                 *     TH./10050/85/DE/..     *
*                                                 *                            *
*********************************************************************************
* CONTRACTOR :                                    * PROGRAM :                  *
*   AEG AKTIENGESELLSCHAFT                         *   HYDROCARBONS             *
*   STEINHOEFT 9               TEL 040 3616-1      *                           *
*   DE - 2000 HAMBURG 11       TLX 211 868         *                           *
*                                                 * SECTOR :                   *
*                                                 *   TRANSPORT                *
* PERSON TO CONTACT FOR FURTHER INFORMATION :     *                            *
*   DR.-ING K. WILKE                              *                            *
*                                                 *                            *
*                                                 *                            *
*********************************************************************************
```

                                                          VERSION : 13/11/86

AIM OF THE PROJECT :
****
After a feasibility study has shown the advantages of the SOLS-concept in a
second phase the main component - the re-entry unit with its water jet thruster
control - was fabricated as prototype and successfully tested in shallow water.
During phase III the system is to be completed with loading hose and whinch and
to be tested in deep water to prove operability under realistic conditions.

PROJECT DESCRIPTION :
****
SOLS is to be operated from a dynamically positioned tanker, which carries a
hose storage whinch on deck over a moonpool.
The re-entry unit is lowered through the moonpool hanging on the hose to the
pipeline and manifold (PLEM) on the seabed. The position of the REU can be
controlled by waterjets and a hydroacoustic measuring system. After having
reached the final position inside the cone of the PLEM the valves are to be
switched into loading position so the oil can be pumped from the production
platform through REU and hose into the tanker.
The phases of the project are:
- preparation of re-entry unit and pressure test
- safety analysis
- design and manufacture of PLEM adapter
- design and manufacture of winch
- factory tests and fabrication of hose
- simulations and preparation of tests program
- deep water trials (120 m)
- evaluation of tests and dismounting of equipement
- documentation and project management
After a milestone at this point it is provided to install the system for sea
trials and repeat the test program under different seastate conditions.

STATE OF ADVANCEMENT :
****
Ongoing construction of the components, as winch, hose and completion of the re-
entry unit. Fabrication has started and software is tested under simulations.
The pre-tests will start at the end of the year, deep water trials will start
early next year.

RESULTS :
****
The shallow water trials of phase II have shown the manoeuvrability of the re-
entry unit. Simulations show that the system will be operable under the
seastate conditions designed to.
More results are expected from the deepwater trials.

REFERENCES :
****
- "SUBSEA OIL LOADING SYSTEM FOR TANKER"
1984, EUROPEAN PETROLEUM CONFERENCE (SPE 12980)
- UNTERWASSER-OELUEBERNAHMESYSTEM FUER TANKER SOLS
SCHIFF & HAFEN, HEFTE 4/5, 1986

```
******************************************************************************
* TITLE : QUALIFICATION TRIALS OF A DEEPWATER      *      PROJECT NO        *
*         DEVICE FOR THE CONNECTION OF FLEXIBLE    *                        *
*         LINES (PHASE 1)                          *    TH./10052/86/FR/..  *
*                                                  *                        *
******************************************************************************
* CONTRACTOR :                                     * PROGRAM :              *
*   COFLEXIP/GERTH                                 *   HYDROCARBONS         *
*   4, AVENUE DE BOIS PREAU     TEL 1 47.52.61.39  *                        *
*   FR - 92502 RUEIL-MALMAISON TLX 203 050         *                        *
*                                                  * SECTOR :               *
*                                                  *   TRANSPORT            *
* PERSON TO CONTACT FOR FURTHER INFORMATION :      *                        *
*   MR. H. LECOMTE               TEL 1 47.47.11.42 *                        *
*                                TLX 610 302       *                        *
*                                                  *                        *
******************************************************************************
                                                       VERSION : 10/02/87
```

AIM OF THE PROJECT :
****
The aim of the project is to quantify a specific diverless method for
connecting flexible lines to a subsea structure or another rigid or flexible
line : a heavy duty manipulator arm, fitted on either a crawler vehicle or a
running tool lowered on the subsea structure takes the extremity of the
flexible line on the seabed and brings it to the locking position on the
structure. At its extremity, the arm is fitted with a specific acoustic
detection system which permits to gain the geometric configuration of the two
parts to be connected. Thanks to this system the final connection system is
performed automatically, thus granting the necessary accuracy which could not
be gained manually. The manipulator arm has been manufactured, outside the
scope of this project.
PROJECT DESCRIPTION :
****
The project is to be developed in two main tasks:
Task 1: Engineering and trial preparation
The studies of the acoustic operational detection system are now started,
together with the development of the command of the manipulator arm. This
command has to take into account the various phases of a connection, with their
specific characteristics:
manual approach of the arm toward the end of the flexible line,
gripping the end of the line in robot mode,
installation of gasket in the connector,
preliminary approach towards the structure in manual mode,
final approach and entry of the connector in robot mode,
closure of the connector and hydrostatic test,
unlocking of the arm, contingencies and other phases for disconnection.
The acoustic detection system will be redesigned for operational use and
includes three different acoustic nets, one for the gripping phase, another for
the final approach and the last one for the entry. This system fitted at the
end of the arm includes also video cameras and other ancillaries.
Other works pertaining to this phase include the preparation of all other
equipment and procedures necessary for the performance of the trials, namely
the adaptation of the crawler vehicle, the design and manufacture of the subsea
structure and of auxiliary equipments, power packs, umbilical, instrumentation,
surface equipments... .

Task 2: Extensive trials will be run in a wet dock, the manipulator arm being fitted on an existing crawler. A large number of connections and disconnections will be performed, with various configurations of the flexible line. These trials will first demonstrate the feasibility and interest of this method, as well as confirm the parameters to be taken into account for the preparation and realisation of a connection job.

STATE OF ADVANCEMENT :
****
Ongoing. The specifications for the study and realization of the control system of the arm have been written up and sent to subcontractors. The studies for the modification of the crawler used for the trials have started, as well as the studies of the running toll and the procedures.

RESULTS :
****
- CONTROL-COMMAND OF THE ARM
An analysis of the different tasks of a connection involving the manipulator arm has been performed (taking the end of the flexible, removing the blind flange, installing the gasket, pulling toward the structure, penetration then unlocking of the arm and retrieval). For each of above tasks, the elementary steps have been defined, and the various parameters to be taken into account have been identified. From these, the specifications have been written up, which includes :
- The study of the system safeties and contingencies
- The global architecture of the system
- The realization of the command, hardware and software
- The controls (internal and surface controls)

The command of the arm must include both a feedback on the control of position and one on the control of effort : the efforts on the arm must be accurately known on a real time basis, and the power of the arm minimized to diminish the impact should a collision occur. Trials with the arm only will permit to know if the efforts can be accurately known from the hydraulic pressure in each articulation, or if captors must be installed, leading to more software in the command.

- MODIFICATIONS OF THE CRAWLER VEHICLE
An existing submarine crawler vehicle will be modified to accommodate the manipulator arm and other equipments necessary for the performance of a connection in a wet dock. A rotating (lateral and vertical) boom supports the manipulator arm, a well as a winch, necessary for the first pull of the flexible toward the subsea structure and means for the removal of the blind flange and insertion of the gasket will be installed on this crawler.

- FEASIBILITY STUDIES OF THE RUNNING TOOL
The main characteristics of this running tool ahve been identified, as well as the interface with the structure (guide-posts).

- INSTALLATION PROCEDURES
The end of the flexible line well will be laid about 30 meters from the subsea structure. Due to the characteristics of the sea bed which may lead to a very low trafficability, the following procedure will be choosen, provided that a winch is fitted on the support (vehicle or running tool) :
- The vehicle is lowered on the sea bed and takes place at its final position.
- An ROV takes the cable from the winch and connects it to the end of the flexible. The winch pulls then the flexible.
- The arm grabs the flexible, then the blind flange is removed, the gasket inserted. The connection is then performed.

```
*********************************************************************************
* TITLE : BUNDLE REPAIR HABITAT                          *      PROJECT NO        *
*                                                        *                        *
*                                                        *    TH./10058/86/UK/..  *
*                                                        *                        *
*********************************************************************************
* CONTRACTOR :                                           * PROGRAM :              *
*    APT LTD                                             *    HYDROCARBONS        *
*    3RD FLOOR, TRAFALGAR HOUSE TEL 01  748 4600         *                        *
*    HAMMERMSITH              TLX 262227                 *                        *
*    UK - LONDON W6 8DW                                  * SECTOR :               *
*                                                        *    TRANSPORT           *
* PERSON TO CONTACT FOR FURTHER INFORMATION :            *                        *
*    MR. C. BAXTER                                       *                        *
*                                                        *                        *
*                                                        *                        *
*********************************************************************************
                                                            VERSION : 14/05/87
```

AIM OF THE PROJECT :
****
The aim is to define and design a one atmosphere work habitat to carry out
repairs on pipelines, bundles and umbilicals where hyperbaric repair by diver
is impossible.

PROJECT DESCRIPTION :
****
The purpose of the habitat is to provide an environment in which technicians
can work on the sea bed to carry out repairs to lines containing hydraulic,
electrical or complex flowlines.
The project will define and design the habitat. It will define its operating
technique and working environment. The build and operating costs will be
produced.

STATE OF ADVANCEMENT :
****
Ongoing

NATURAL GAS TECHNOLOGY

```
******************************************************************************
* TITLE : TECHNICO-ECONOMIC OPTIMIZATION OF NAT.    *       PROJECT NO       *
*         GAS LIQUEFACTION PLANTS ON OFFSHORE       *                        *
*         PLATFORMS USING PERMEABLE MEMBRANES TO    *    TH./12006/84/DE/..   *
*         PURIFY GAS                                *                        *
******************************************************************************
* CONTRACTOR :                                      * PROGRAM :              *
*   SALZGITTER AG                                   *   HYDROCARBONS         *
*   POSTFACH 15 06 27          TEL 030 88 42 97 - 15 *                        *
*   DE - 1000 BERLIN 15        TLX 308 611          *                        *
*                                                   * SECTOR :               *
*                                                   *   NATURAL GAS          *
* PERSON TO CONTACT FOR FURTHER INFORMATION :       *   TECHNOLOGY           *
*   DR. PIETSCH                                     *                        *
*                                                   *                        *
*                                                   *                        *
******************************************************************************
```

VERSION : 01/01/87

AIM OF THE PROJECT :
****

This project was understood as a supplement to previous projects, which are dealing with liquefaction of natural gas and associated gas on offshore sites. With said previous projects conventional gas scrubbing with alkanol amine solution followed by adsorptive gas drying were provided for the purification of feed gas. The aim of this project was to ascertain, whether there is a reduction in investment and operating costs, if the conventional gas purification is substituted by one with gas separating membranes. An incentive for this investigation was the fact, that in the case of offshore process plants, the comparatively little space requirement of a membrane battery and the simplicity of its operation are of a greater influence upon the overall economy of LNG production, than in the case of land based process plants.

PROJECT DESCRIPTION :
****

The said previous projects, forming the background of this project, are:
1. a LNG production plant, resting on a tension leg platform
2. a LNG production plant, housed by a spherical concrete casing, which rests on the sea bottom.
The process data which were considered for the 2 cases were respectively:
case 1 - feed rate 71,000 Kg/hr; 93.63% CH_4; 1.79% C_2H_6
case 2 - feed rate 83,000 Kg/hr; 79.60% CH_4; 10.20% C_2H_6
Level of purity which has been set for cryogenic section design: 1 mol-ppm H_2O; 6 mol-ppm H_2S; 150 mol-ppm CO_2.
Since the required purity cannot be achieved by permeation separation alone, a combination of membrane batteries and absorbers was applied, the membranes as a first stage and the adsorbers for the final purification to the above specification. The regenerating gas leaving the adsorbers is fractionated in an extra membrane battery. The residue out of this is recycled to the feed, the permeate is used as fuel gas.
A multitude of possible configurations of membrane batteries, adsorbers and regenerating gas cycles were evaluated for optimization. Optimizable parameters are among others: interface concentration of impurities (between the permeation and the adsorption stage) permeate pressure(s), number and operating sequence of adsorbers, length of time of the operating cycle of the adsorbers, respectively mass of their beds.

A restriction is,that the quantity of
hydrocarbon contained in the produced waste gas must not exceed the fuel gas
demand of the liquefaction process. Otherwise energy would be lost through the
flare.
In the main part of the study only commercially marketed membranes are
considered. Some of such membranes, with improved qualities, which are
described in public or company literature but not yet used in an industrial
scale, are dealt within an extra part of the study.
The optimized permeation/adsorption plants are incorporated into the LNG
production plants and rearranged layouts of the overall plants were prepared,
in order to evaluate the possible reduction of space requirement for each of
the 2 cases. Then the offshore structures – TLP in case 1, subsea concrete
housing in case 2 – were adapted to the new plant layouts, savings in steel
respectively concrete mass evaluated.
STATE OF ADVANCEMENT :
****
The study is completed
RESULTS :
****

Summary: also for offshore process plants, the performance of up to now
commercially available membrane moduls is not sufficient and their price too
high to justify their use for the purification of LNG-plant feed. For the case
1 plant (on TLP) surplus costs of approx. 700,000 DM p.a. were calculated,
compared with the annual costs of the original LNG-plant with conventional gas
purification. Only for the case 2 plant (in subsea housing) which of course
represents a very special application, savings of approx. 250,000 DM p.a. were
calculated.
DETAILS ABOUT CASE 1 (TLP):
The so called superstructure deck (13 m above the main deck) could be abandoned.
 The membrane modules are housed in a room below deck, thus leaving the space
on the main deck, which was previously occupied by the amine contacting plant,
for other equipment. With the original plant a greater part of the subdeck
rooms could not be utilized. The overall dimensions of the platform remained
unchanged, the reduction of the mass of steel is not significant.
DETAILS ABOUT CASE 2 (subsea concrete housing):
The diameter of the spherical housing was reduced by 2 m, resulting in a
reduction of concrete mass from 10,100 m3 to 8,100 m3. Main reason: the
membrane battery, which can be composed of modules of different length, can be
fitted into wedge shaped spaces near the spherical wall of the casing, which
cannot be used for other equipment.
The annual energy costs of the non plant are about 500,000 DM less than those
of the conventional plant, due to the process heat needed for the regeneration
of amine solution. In the case of the TLP-plant this saving does not count,
because process heat is recovered from the gas turbine exhausts, i.e. by waste
heat. The subsea plant has only steam as source of energy, so all extra energy
needs extra fuel.
Evaluation of laboratory state membranes: the permeability data of one such
membrane, as stated by its developer, were used to optimize a combined
permeation/adsorption plant as described above. Accordingly the performance of
this membrane (separation factor CO_2/CH_4 about 7 times, permeation rate for CO_2
nearly 2 times the corresponding values of marketed membranes) would be well
above the breakeven point, provided its price would be of comparable magnitude
as that of current prices of gas separating membranes.

```
************************************************************************
* TITLE : PROCESSING HYDROCARBON GASES WITH      *     PROJECT NO      *
*         MEMBRANES (PHASE I)                     *                    *
*                                                 *   TH./12008/84/FR/..*
*                                                 *                    *
************************************************************************
* CONTRACTOR :                                    * PROGRAM :          *
*   GERTH/C.E.A.                                  *   HYDROCARBONS      *
*   AVENUE DE BOIS PREAU 4      TEL 1 47 52 61 39 *                    *
*   FR - 92500 RUEIL-MALMAISON TLX (1)47.52.69.27 *                    *
*                                                 * SECTOR :           *
*                                                 *   NATURAL GAS       *
* PERSON TO CONTACT FOR FURTHER INFORMATION :     *   TECHNOLOGY        *
*   MR. J.E. VIDAL              TEL (1)42.91.40.00 *                    *
*                              TLX TCFP615700     *                    *
*                                                 *                    *
************************************************************************
                                                   VERSION : 09/06/87
```

AIM OF THE PROJECT :
****
The main object of this project consisted of the choice and the development of
semi-permeable membranes, which, implemented in permeators, would allow the
purification of crude natural gases. The purification processes comprise
respectively a gas dehydration phase and a deacidization phase.

PROJECT DESCRIPTION :
****
The first phase of the project is more specific to offshore applications. It
includes:
- the conversion of the process
  - design of pressure chamber
  - fabrication of prototype mini-permeators
  - study of high and low pressure flows
  - trials on test bench to optimize the flows and the conception of the
chamber
- labboratory tests of the membranes, either in the form of samples, or in the
form of mini-permeators.
   These membranes are evaluated according to criteria of permeability,
selectivity, compatibility and pressure resistance.
The second phase of the project is devoted to the design, the manufacture and
the finalization of a permeator study unit allowing to run membrane ageing
tests in the presence of H2S.
These tests required the construction of a unit complying with the manipulation
of important amounts of H2S. This unit was to be set within the SNEA(P)
premises at Lacq where important flows of sulfurous natural gas can be
processed over periods of several months, while complying with the safety
regulations.

STATE OF ADVANCEMENT :
****

The project was stopped on the 30th June 1986.

RESULTS:
****
Phase I
Laboratory tests to control the performance of existing membranes have been run together with tests under real gas use conditions, first on a mini-pilot at the Nuclear Research Center of Saclay, then on a testing device installed at the TOTAL-CFP laboratory of Beauplan.

Comparisons were made between the characteristics of membranes provided by various suppliers.

Test conclusions were the following:
- the temperature effect on permeabilities was very important. A 30°C temperature increase corresponds to a permeability increase of 200 to 250%;
- good resistance versus steam (50°C) and nitrogen (20°C) cycles;
- the transfer coefficient of permeation fibers has been evaluated at 100%, thus making it eventually interesting to use carrier gas;
- according to the permeametry measured on gas alone, selectivities obtained for carbon dioxide are of 30 to 60 and of 270 for helium in relation to methane but pure carbon dioxide on fibers provokes a permeability decrease versus pure methane which cannot be recovered;
- an analysis of the separation in the case of a water/methane mixture shows that the selectivity is very dependant on pressure and temperature conditions. The permeability of the methane in the mixture is lower than that in pure gas.

Phase II
The first tests proved unsatisfactory as none of the membranes presently existing on the market resist to H2S.

Studies related to the design of mini-permeators can only be used if plane membranes or hollow-fiber membranes exist. Tests proved that none of the existing membranes was fully satisfactory. Hence, it was decided to interrupt the project until studies undertaken outside the scope of this project on new membranes had reached satisfactory results according to criteria of permeability, selectivity and compatibility.

```
**********************************************************************************
* TITLE : NEW PROCESSING TECHNIQUES FOR OFFSHORE   *      PROJECT NO        *
*         GAS LIQUEFACTION                          *                        *
*                                                   *   TH./12009/85/FR/..   *
*                                                   *                        *
**********************************************************************************
* CONTRACTOR :                                      * PROGRAM :              *
*   GERTH                                           *   HYDROCARBONS         *
*   4 AVENUE DE BOIS PREAU       TEL 1 47 52 61 39  *                        *
*   FR - 92502 RUEIL-MALMAISON TLX 203 050          *                        *
*                                                   * SECTOR :               *
*                                                   *   NATURAL GAS          *
* PERSON TO CONTACT FOR FURTHER INFORMATION :       *   TECHNOLOGY           *
*   MR. LARUE                    TEL 1 47.49.02.14  *                        *
*                                                   *                        *
*                                                   *                        *
**********************************************************************************
```

VERSION : 01/01/87

AIM OF THE PROJECT :
****
The object of this project is to study and develop a new integrated gas
processing and liquefaction process adapted to offshore production particularly
on a mobile support. The integration of these two functions in a single process
will allow to achieve simultaneously and in a sole phase, the dehydration,
deacidyzation and extraction of liquids from a natural gas and constitutes the
first step towards the transformation in LNG. Compared to the more conventional
processes of deacidyzation, dehydration and extraction of liquids from a
natural gas, which are generally implemented in a successive and distinct
manner, this new method should provide a gain in investment, in weight and in
surface utilisation. Furthermore the innovations implemented in this new
process make it particularly suitable for offshore production.

PROJECT DESCRIPTION :
****
In theory, the process studied in this project, consists of refrigerating the
gas to be treated with the appropriate solvent, thus making it possible to run
in a single phase, both the treatment (dehydration, deacidyzation) and the
refrigeration (extraction of liquids from natural gas) of the gas.
Technological developments envisaged thus concern the treatment and the cooling
of gas and their integration in a sole process.
The treatment function allows to reduce the water and acid gas contents (CO_2,
H_2S). Therefore these constituents need to be separated owing to problems of
hydrates, crystallization, corrosion and specifications to which they can lead.
Classical dehydration and deacidyzation techniques resort to the use of
solvents in absorption columns (glycol, amines) followed by conditioners, thus
leading to heavy units poorly adapted to offshore production. The proposed
technique resides on a new and optimized implementation of solvent allowing a
greater compactness and a reduction of the number of columns.

The cooling function allows the refrigeration of gas, which leads to the condensation of natural gas liquids. Traditional cooling techniques imply heavy rotary machines (compressors, turbines or motor engines) and are hence poorly adapted to offshore production. Refrigeration techniques studied within the scope of this project and associated with the treatment function are of the static type such as absorption cooling. They allow the production of cold directly from heat which can be supplied by recovering thermal discharges or by combustion. These techniques also increase the operating reliability and flexibility and offer more sophisticated possibilities of integration with the treatment section.

The project is divided into two phases:

PHASE 1: Study of the integrated treatment-liquefaction process. Both, the treatment function (search of appropriate solvents and corresponding data, evaluation of different patterns) and the refrigeration function will be first studied separately. Their integration in a single process will be examined in a second step.

PHASE 2: Experimental tests of the various functions of the process. Experimental units for the treatment and refrigeration functions will be designed and tested over a wide range of operating conditions. A document relative to the process pilot will then be drafted.

STATE OF ADVANCEMENT :
****
The project is presently in its study phase: works performed till now concerned only the study of the gas treatment function (Phase 1)

RESULTS :
****
Results obtained at this level of the project concern the study of the gas treatment function.

A patent has been applied for relative to a new implementation of solvent for the gas treatment. This new technique has been the object of preliminary tests at the laboratory and of simulation calculations, in order to perform a comparative economical and technical evaluation (underway).

```
*********************************************************************************
* TITLE : DEVELOPMENT OF A SCRUBBING PROCESS FOR    *        PROJECT NO        *
*         THE SEPARATION OF (ACID) COMPONENTS       *                          *
*         FROM (NATURAL) GAS                        *    TH./12010/85/NL/..     *
*                                                   *                          *
*********************************************************************************
* CONTRACTOR :                                      * PROGRAM :                *
*   NEDERLANDSE GASUNIE N.V.                         *   HYDROCARBONS           *
*   P.O. BOX 19                TEL 050 219111        *                          *
*   LAAN CORPUS DEN HOORN 102  TLX 53448             *                          *
*   NL - 9700 MA GRONINGEN                           * SECTOR :                 *
*                                                   *   NATURAL GAS            *
* PERSON TO CONTACT FOR FURTHER INFORMATION :        *   TECHNOLOGY            *
*   DR. G.E.H. JOOSTEN                                *                          *
*                                                   *                          *
*                                                   *                          *
*********************************************************************************
                                                         VERSION : 01/01/87
```

AIM OF THE PROJECT :
****
The aim of the project is to develop an efficient and compact scrubing process
for thr removal of specific components from natural gas, like aromatics, CO2
and H2S. The innovative aspect of the project is, that the absorption fluid is
directly injected into the gasstream in cocurrent flow. In contrast to
commercially available countercurrent scrubbers, this allows very high gasflows,
in a compact installation.

PROJECT DESCRIPTION :
****
To meet the objective of the project an absorption solvent is directly injected
into a gasstream. The absorption fluid and the gas flow through one or more
parallel tubes at high velocity in a cocurrent mode, The optimal diameter of
the tubes is a parameter to be determined. The intensity of the gas/liquid
contact and therefore the degree of mass transfer will depend on various
parameters like operating pressure, gas/liquid velocity, gas/liquid ratios, and
spontaneous or pressurised spraying of the liquid phase. After the absorption
stage, the liquid has to be separated from the gasstream. Subsequently, the
absorption liquid can be regenerated and reinjected into the gasstream.
The research in this project will especially focuss on the removal of CO2 from
natural gas. The following phases can be distinguished :
1. Investigation of the hydrodynamic behaviour in the equipment
   - Installation of cocurrent gas/liquid contacting equipment
   - Examination of the dispersion of the liquid in the gasstream under
     various conditions
   - Determination of the efficiency of liquid separation from the gasstream.
2. Investigation of the absorption/desorption behaviour of various absorption
   fluids. In a lab-scale stirred cell reactor the kinetic and equilibrium
   data of various absorption liquids will be studied.
3. Investigation of mass transfer with selected absorption liquids in the
   cocurrently operated contacting equipment.
   - Mass transfer apects in cocurrent operation
   - Regeneration of solvent after the gas/liquid separation

4. Economic evaluation
   A preliminary economic evaluation will be based upon the data obtained
   in phase 3. This evaluation will result in a go/no go decision with
   regard to phase 5.
5. Pilot plant tests
   In a pilot plant the process will be tested under real operating
   conditions. The design data for the commercial application of the process
   will be determined.

STATE OF ADVANCEMENT :
****
The project is still ongoing. A high pressure (40 bar) installation has been
realised within Gasunie's high pressure test facilities. Experiments focussing
on the determination of the hydrodynamic behaviour in the cocurrent gas/liquid
equipment, have been started.
Moreover a labscale absorption/desorption unit has been installed, containing a
300 ml stirred cell reactor. The absorption/desorption measurements for the
collection of kinetic and equilibrium data of absorption fluids are in progress

RESULTS :
****
a) : Hydrodynamic behaviour of the equipment.
Experiments to determine the gas/liquid interfacial area for mass transfer in
the high pressure equipment have not been successful sofar. Two different
techniques have been applied :
- A direct fotographic technique to determine the droplet size and
distribution in the gas/liquid-stream.
- An indirect technique where an average droplet size can be estimated by
measuring at various locations the equipment, the degree of saturation of the
gasstream with water-vapor.
The resolution of the fotographic technique, using a nano-second flash-light,
did not meet the requirement. It was therefore concluded that direct techniques
are not suitable to determine the specific gas/liquid surface area of droplets
moving at high speed in gasstreams.
The moisture-content measurements of the gasstream in the high pressure
equipment showed that the gas was completely saturated under all operating
conditions tested sofar. From these results and additional calculations, this
method was found impraticable due to experimental problems.
An alternative indirect technique will be used to study the mass transfer
process in the high pressure equipment. Water will be used to partially remove
CO_2 from "Groningen gas" (0.89 vol. % CO_2). The changes in the CO_2-content in
the liquid phase will be monitored at various locations in the equipment as a
function of the operating conditions. Equipment will be modified to enable
these measurements : CO_2 regeneration column to be added, liquid injection
system to be changed, CO_2 detection system for liquid phase to be added.
Regarding the gas/liquid separation, the first indicative experiments showed
very good efficiencies of 99% or higher. These experiments were performed at
relative high gasvelocities (25-90 m/sec) and relative low gas/liquid ratios.
The installation has now been slightly modified to enable a more accurate
quantification of the efficiency of the Gasunie scrubber.

b) : Absorption/desorption experiments with various absorption solvents.
The determination of the equilibrium absorption data for various absorption
solvents are in progress. At 20 deg.C and 40 deg.C various mixtures of
alkanolamine solutions in water, and pure water have been used. Sofar only
minor differences in equilibrium data have been observed. In follow-up
experiments a number of commercial absorption solvents for CO_2-removal will be
considered.
The experiments to determine the kinetic data for CO_2, absorption and
desorption are also in progress.
The liquid phase mass transfer coefficient for CO_2-absorption in water was
found to be $(3-4) * 0.00001$ m/s. The enhancement factor for chemical absorption
of CO_2 in the alkanolamine solutions varies between 4 and 15, but only during
the initial stage of the absorption process. At CO_2-loadings above 50% of total
absorption capacity no enhancement occurs. In follow-up experiments commercial
solvents will also tested with respect to their kinetic data.

REFERENCES :
****
PERRY, R.H.; CHILTON, C.H.; CHEMICAL ENGINEERS HANDBOOK, FIFTH ED.
MC GRAW-HILL BOOK COMPANY, NEW YORK
KOHL, A.; RIESENFELD F.; GAS PURIFICATION, SEC. ED. 1974, GPC, HOUSTON, TEXAS,
U.S.A.
MADDOX, R.N. GAS AND LIQUID SWEETENING; CAMPELL PETROLEUM SERIES.
BLAUWHOFF, P.M.M.; DISSERTATION 1982. T.H. TWENTE, THE NETHERLANDS; SELECTIVE
ABSORPTION OF H2S FROM SOUR GASES BY ALKANOLAMINE SOLUTIONS.
WESTERTERP, K.R.; V. SWAAIJ, W.P.M.; BEENACKERS, A.A.C.M.
CHEMICAL REACTOR DESIGN AND OPERATION, 1984.
JOHN WILEY AND SONS, NEW YORK, SEC. ED.

ENERGY SOURCES

```
*******************************************************************************
* TITLE : SURFACE-INDEPENDENT UNDERWATER ENERGY     *      PROJECT NO        *
*         SUPPLY SYSTEM-DIESEL ENGINE WITH CLOSED   *                        *
*         GAS CYCLE                                 *   TH./13006/85/DE/..   *
*                                                   *                        *
*******************************************************************************
* CONTRACTOR :                                      * PROGRAM :              *
*   M.A.N. TECHNOLOGIE GMBH                         *   HYDROCARBONS         *
*   DACHAUER STRASSE 667        TEL 089 1480 3459   *                        *
*   POSTFACH 50 06 20           TLX 523 211         *                        *
*   DE - 8000 MUECHEN 50                            * SECTOR :               *
*                                                   *   ENERGY SOURCES       *
* PERSON TO CONTACT FOR FURTHER INFORMATION :       *                        *
*   DR. H. GEHRINGER                                *                        *
*                                                   *                        *
*                                                   *                        *
*******************************************************************************
```

VERSION : 31/12/86

AIM OF THE PROJECT :
****
The closed-circuit diesel engine using argon as cycle medium was developed
under the completed project TH./15056/84 and successfully tested in a pilot
plant with a 32 kW MAN diesel engine.
The aim of this present development project is the construction of a prototype
surface-independent underwater energy supply system with an argon cycle diesel
engine (100kW), and its testing in a work submergible under practical operating
conditions. In this way the operational suitability of this autonomous energy
supply system as generating plant in submarines, underwater work stations,
diver habitats, etc. in the offshore technology sector is to be demonstrated.
PROJECT DESCRIPTION :

****
The energy supply unit, consisting of the diesel engine, the closed argon gas
circuit and the exhaust gas scrubber, together with the various auxiliarly
systems, is to be designed and manufactured as a complete system for
installation in the test machine room section of a submachine. This test
section is to be developed by Messrs. Bruker and its design data are to be
oriented on the requirements of a small submarine for commercial applications.
The diesel engine will serve as energy source for the craft's hydraulic and
propulsion system and for its generator. Envisaged is a combined operational
mode - a closed argon cycle for submerged operation and an open air cycle for
surface operation.
Design criteria and boundary conditions for the energy supply unit will be
strongly influenced by the submarine's assumed operational profile.
Consequently intensive coordination with Messrs. Bruker is envisaged in the
design phase, in order to achieve optimal adaptation.
Any new findings obtained from tests on the pilot plant will be taken into
account during prototype design and preconstruction stages. Likewise the
requirements for the subsequent underwater test phase will also be taken into
account. Design and pre-construction work will be concluded with the drawing up
of final construction documentation.
Assembly of the energy system with the ordering of the individual components,
manucfacture, assembly and functional testing, will be carried out by MAN.
Then follows installation in the best section at Messrs. Bruker and connection
to the supply tanks.
Testing and trials, both on land and underwater, will be carried out jointly
with Messrs. Bruker.

STATE OF ADVANCEMENT :
****
Ongoing. The construction of the prototype unit for trials at the factory on a
laboratory scale is complete. It has been put into operation and operated in a
closed argon circuit up to a load of 20%. Trials at the factory are being
contimued.

RESULTS :
****
During the construction of the prototype unit various late deliveries by subcontractors led to a delay in completing the construction of the laboratory
scale plant. Various modifications and improvement work had to be carried out
as a result of the functional check of individual components and sub-systems,
all of which resulted in further delay and expense.
The factory trial of the prototype unit is to be continued until the closed
argon operation with the uP regulation has been effected over the full output
range and any problems that thereby occur have been solved. Provisional
estimates indicate that the system can afterwards be installed at Messrs.
Bruker in Karlsruhe in the second quarter of 1987.

```
*******************************************************************************
* TITLE : AN AIR INDEPENDANT POWER SOURCE OF HIGH   *     PROJECT NO        *
*         ENERGY STORAGE DENSITY                    *                       *
*                                                   *   TH./13007/85/DE/..  *
*                                                   *                       *
*******************************************************************************
* CONTRACTOR :                                      * PROGRAM :             *
*   BRUKER MEERESTECHNIK GMBH                       *   HYDROCARBONS        *
*   P.O. BOX 21 02 32            TEL 07 21/59 67-1 80 *                     *
*   DE - 7500 KARLSRUHE 21       TLX 78 25 656      *                       *
*                                                   * SECTOR :              *
*                                                   *   ENERGY SOURCES      *
* PERSON TO CONTACT FOR FURTHER INFORMATION :       *                       *
*   MR. J. HAAS                                     *                       *
*                                                   *                       *
*                                                   *                       *
*******************************************************************************
                                                      VERSION : 09/03/87
```

AIM OF THE PROJECT :
****
Battery powered submarines are suffering from the moderate energy densities of
conventional battery systems and thus limited range and endurance.
The aim of this project is to considerably improve the energy storage capacity
of power sources for autonomous work and research submarines and other subsea
installations requiring electrical, mechanical and/or thermal energy.
A complementary aim of the project is to study smaller, low cost closed cycle
diesel systems to be used, for example, for auxiliary power plants for larger
submarines or as stand alone energy sources for smaller vehicles for shallow
water.

PROJECT DESCRIPTION :
****
The aim shall be achieved by integration of a closed cycle argon diesel engine,
developed by MAN-Technologie GMBH, Munich, into an engine room section of a 50
to Autonomous Inspection Submarine, developed by Bruker Meerestechnik GMBH, and
by carrying out a series of realistic dry and wet tests of the complete engine
plant. In a second stage of the project it is planned to complete the engine
section to an operating inspection submarine to demonstrate the safe and
effective operability of the system.
The power supply unit will consist of a pressure hull section forming the
engine compartment of a minisubmarine, the diesel engine with exhaust gas
recirculation and treatment system, auxiliary systems, storage facilities for
chemical agents, the liquified oxygen storage system with evaporator and
control systems, the fuel storage tanks and the ballast and trim system.

The layout and specification of the submarine, the design, calculation and construction of the pressure hull, the integration of the power plant into the same, the design of the hydraulic and the electric system for the submarine, the liquified oxygen plant, the storage system for chemicals, the trim and ballast system and the fuel system as well is within the responsibility of Bruker Meerestechnik GmbH.

The power package will consist of an MAN Diesel engine type D 2566 of 100 KW rating at 1500 rpm rigidly connnected to a combined DC generator/motor and a hydraulic pressure controlled, variable flow axial piston pump.

The hydraulic pump feeds the hydraulic system with main propeller, thrusters, auxiliary pumps and other auxiliary equipment with hydraulic drives.

For redundance reasons it is foreseen to fit out the submarine with a lead acid battery with reduced capacity for limited subsurface operation when the closed cycle diesel is switched off.

The diesel engine can be operated in both, open and closed circuits. Under closed circuit conditions, no media are leaving or entering the system. Only heat is dissipated. The system is therefore depth independant.

The exhaust gas, when leaving the engine, is cooled down, then chemically scrubbed from CO_2 in two stages and cooled again in a second heat exchanger before evaporated liquified oxygen is fed into the circuit.

Evaporation of the oxygen is carried out in heat exchanger fed out of the engine's cooling system.

The storage system for the chemicals consists of a number of tanks and special valves to carry the fresh and the used chemicals and the condensate as well, drained from the gas coolers.

In the closed circuit operation the air in the system is exchanged against argon to improve the thermodynamical efficiency. During closed cycle operation, no argon is consumed. When shifting to "open", part of the argon is lost and has to be replenished again.

STATE OF ADVANCEMENT :
****
Ongoing. The design of the system and its subsystems has been completed. Components and subsystems have been tested. The main components are available for assembly early in 1987. Assuming the results during assembly of the engine compartment look promising, support for completion to an operational demo-submarine will be applied for.

RESULTS :
****
The 100 KW engine and its auxiliaries and most of the components were tested successfully. The main components are available for immediate assembly and completion of the engine compartment. The system was designed to be used for an autonomous inspection and research submarine of increased range and endurance. According to the preliminary data available, the energy capacity of the inspection submarine with closed cycle diesel engine will increase by a factor of eight compared to a battery powered sub of the same size and displacement.

To study the feasibility of a simplified, low cost closed cycle diesel system, Bruker Meerestechnik has constructed a test-plant with a 20 KW diesel engine driving a hydraulic pump to simulate varying loads. The system works on the CO_2-principle. The exhaust gas is cooled and purified to a certain extend.

Oxygen, stored either in liquefied or compressed form, is fed into the gas circuit before entering the engine again.

The system works on temperature and pressure levels above normal. Except for very shallow dives, excess exhaust gas must be dumped overboard by means of a compressor. The system therefore works depth dependant and will be limited to medium depths.

```
*****************************************************************************
* TITLE : RANKINE CYCLE IMMERSED ENERGY SOURCE       *       PROJECT NO      *
*         WITH HIGH-PRESSURE COMBUSTION CHAMBER      *                       *
*         AND CONSTANT MASS OPERATION                *    TH./15055/84/FR/..  *
*                                                    *                       *
*****************************************************************************
* CONTRACTOR :                                       * PROGRAM :             *
*   BERTIN & CIE                                     *   HYDROCARBONS        *
*   B.P. 3  78373 PLAISIR CEDE TEL 59 64.86.48       *                       *
*   FR - 40220 TARNOS           TLX 570 026          *                       *
*                                                    * SECTOR :              *
*                                                    *   MISCELLANEOUS       *
* PERSON TO CONTACT FOR FURTHER INFORMATION :        *                       *
*   MR. D. GROUSET                                   *                       *
*                                                    *                       *
*                                                    *                       *
*****************************************************************************
```

VERSION : 29/04/87

AIM OF THE PROJECT :
****
The main objective of the current project consists of tuning-up a 60 bar
aeronautical combustion chambre with a thermal power of 400 kW, working with
methanol/oxygen and cooled by injection of combustion products at 25 deg.C.
In a short future, this new combustion chamber is assigned to be the heat
generation system of a constant mass subsea power module incorporating an
organic turbine cycle and a combustion products condensation system. This
latest characteristics involves to work at high pressure in order to use sea-
water for condensation.

PROJECT DESCRIPTION :
****
The project began in July 1985 and has a 3 years duration. It is divided in 4
main phases :
PHASE 1 : Design, sizing, construction drawings, procurement and fabrication of
each component of the test bench.
PHASE 2 : Erection of test facilities : combustion chamber, fluids circuits,
electrical circuits for control, regulation and safety.
Implementation of a data acquisition program for measurement.
PHASE 3 : Tests of the combustion chamber including injection, light-up,
thermal behaviour, flame stability, combustion efficiency and control at
various operating conditions.
PHASE 4 : Synthesis and Integration
partial and final reports give the progress of the work and the synthesis of
the results. The integration of the chamber in the global system will be
approached for various ranges of power according to the industrial needs
evaluated by a prospector research.

The main characteristics of the combustion chamber are :
- nominaly thermal power   =   400 kW
- mass flowrates   =   20 g/s methanol
                                =   30 g/s oxygen
                                =   335 g/s carbon dioxyd
- tube flame   =   0.132 m length
                                =   0.062 m O max
- operating pressure   =   6 MPa
- power density   =   20 MW/m3/atm
- range of working power   =   20 to 120%
- injection by a pneumatic nozzle
- flame stability induced by a swirl
- cooling of the tube flame by external flow and film cooling of CO2
- dilution of the combustion gases by injection of cold CO2
- Light-up by a plasma spark plug and flame control by a photoelectric cell.
- Internal pressure maintained by a sonic valve.

The design of the chamber has been driven by coupled calculations from
independent codes under given the drop size distribution, the pressure drops,
the adiabatique gas temperature, the kinetics of combustion, the wall
temperatures. These first iterative calculations resulted in the sizing of the
chamber. Before construction and tests, a thermomechanical code, FLUENT, was
used to validate the sizing and provide more information about the internal
flows : recirculating zones, film-cooling, dilution gats, droplets
trajectoires, and local wall and gas temperatures.

The first trials are carried out for testing the nozzle and measure the
droplets size distribution. This step is required by the lack of knowledge on
atomization under high pressure.

Lastly, the combustion trials will permit the definition of the thermal
performance and the range of pressure and power values with flame stability and
accepted operating conditions.

STATE OF ADVANCEMENT :
****
Ongoing. At this time, plan 1 is finished, and plan 2 has been initiated
according to schedule.
The test facilities are being set up, the data acquisition programs are being
written. The specific test bench for atomization measurements under high
pressure is designed. The first trials are planned in June or July 87. Finally,
contracts are taken with industry for potential applications.

RESULTS :
****
It is expected that the complete module will produce approximately 80 kW of
usable power and permit operation of unmanned submarine tools for offshore oil
production or deep sea cartographic reconnaissance, with a large degree of
independance.

```
*******************************************************************************
* TITLE : DIESEL ENGINE WITH ARGON CYCLE.           *       PROJECT NO       *
*                                                   *                        *
*                                                   *   TH./15056/84/DE/..   *
*                                                   *                        *
*******************************************************************************
* CONTRACTOR :                                      * PROGRAM :              *
*   M.A.N. TECHNOLOGIE GMBH                          *   HYDROCARBONS         *
*   POSTFACH 500620            TEL 089 14803459      *                        *
*   DE - 8000 MUENCHEN 50      TLX 52321             *                        *
*                                                   * SECTOR :               *
*                                                   *   MISCELLANEOUS        *
* PERSON TO CONTACT FOR FURTHER INFORMATION :        *                        *
*   MR. H. GEHRINGER                                 *                        *
*                                                   *                        *
*                                                   *                        *
*******************************************************************************
                                                        VERSION : 17/11/86
```

AIM OF THE PROJECT :
****
In the field of offshore technology, there is a great need for autonomous
energy systems, particularly for use in deeper waters. Efficient diesel engines
in closed cycle operation have considerable advantages against batteries for
that problem. Aim of the project is therefore the development of a closed
diesel engine cycle with argon as cycle medium and its test on a pilot plant.

PROJECT DESCRIPTION :
****
If argon is used as the cycle medium, the efficiency of the diesel engine can
be increased by approx. 25%. In closed cycle operation the recycle gas argon
has to be scrubbed free of the combustion products H_2O and CO_2 before enriched
with pure oxygen leading back to the diesel engine.
One point main effort therefore is the design, test and optimization of the CO_2
scrubber system under operating conditions. Further points were the design of
the overall system including all of the auxilliary components, and the erection
of the closed argon cycle pilot plant with a 32 kW diesel engine. During the
tests of the pilot plant the function of the overall system was to be
demonstrated.

STATE OF ADVANCEMENT :
****
The project was completed on the 30th of April 1986.

RESULTS :
****

In the preliminary phase of the project parallel tests were carried out with an
M.A.N. diesel engine operating in open argon/oxygen mode, during which 42%
efficiency was achieved and, with a CO_2 scrubber using a caustic potash
solution, a CO_2 reduction from 7.5% to 1% obtained.

The closed argon cycle pilot plant with a 32 kW M.A.N. diesel engine was tested
in numerous trials, the sampled results and experiences led to an improved CO_2
scrubber system by using a self-developed 2-stage rotational scrubber. The
advantages of this 2-stage M.A.N. rotational scrubber are its compact design
and its very high lye absorption rate.

The test carried out with the pilot plant confirmed the function of the closed
argon diesel engine cycle with all of its auxiliary components, and the
increase of the engine efficiency.

The next step of the project is to produce an underwater energy supply section
for submarines with a 100 kW M.A.N. diesel engine by using the sampled results
and experiences and to test it under operation conditions.

STORAGE

```
********************************************************************************
* TITLE : PERMANENT MOORING OF A FLOATING UNIT IN   *      PROJECT NO         *
*         DEEP WATER BY MEANS OF A                   *                         *
*         MULTIARTICULATED COLUMN                    *    TH./14014/82/FR/..   *
*                                                    *                         *
********************************************************************************
* CONTRACTOR :                                       * PROGRAM :              *
*   GERTH-EMH                                        *   HYDROCARBONS         *
*   4, AVENUE DE BOIS PREAU      TEL 1 47.52.61.39   *                         *
*   FR - 92502 RUEIL MALMAISON TLX 203050            *                         *
*                                                    * SECTOR :               *
*                                                    *   STORAGE              *
* PERSON TO CONTACT FOR FURTHER INFORMATION :        *                         *
*   MR. BRANCHEREAU              TEL 1 47.71.91.22   *                         *
*                               TLX 204586           *                         *
*                                                    *                         *
********************************************************************************
```

VERSION : 01/01/87

AIM OF THE PROJECT :
****
Development of the articulated column technique for permanently mooring a
floating unit in depths of water of from 600 to 1,000 metres and environments
of the North Sea or Mediterranean type.

PROJECT DESCRIPTION :
****
General analysis:
- definition of the requirements of the oil companies
- a review of the different mooring systems developed or now undergoing
development for great depths of water
- definition of the representative conditions of environment of the two types
of site (waves, wind, currents, soil)
- definition of typical production schemes for study of fluid transfer systems
- review of the fluid transfer systems available (low and high pressure.
Preliminary projects:
- definition of the general structure of deep water columns
- development of methods for predimensioning and analysing the behaviour of the
structures
Specific studies:
- application of weight-reducing materials for the construction of high
pressure buoys
- analysis of fluid transfer systems along the column.

STATE OF ADVANCEMENT :
****
Ongoing. The project is scheduled to be completed within the first half of 1987.

RESULTS :
****
The general analysis phase has revealed the advantages of permanent mooring
systems on a multi-articulated column for the development of deep water
offshore fields. The reason is that there are many applications, ranging from
simple buffer storage to the accumulation of the support functions of equipment
for controlling and processing production, storage of the oil produced and its
discharge via shuttle tankers moored in pairs or in tandem.
The main alternatives defined both for the architecture of the mooring
structures and the fluid transfer system have demonstrated the existence of a
wide variety of solutions for the technical problems set that will be selected
to suit each specific case studied in the rest of the project.
Detailed studies were performed on :
- dimensioning of baseplates and structures
- development ogf high pressure buoy structures for floats
- study of installation and maintenance procedures for the flow transfer lines
- definition of column structures and performance study with the help of the
MCS computing program in diphasic conditions.
An important series of tests were performed to check the operational behaviour
and the towing and installation operations.
In addition, specific work was devoted to the examination of problems raised by
the eventual immersion of the upper part of the articulated columns in the case
of very violent storms. Arrangements were made to protect the rotary joints in
case of accidental immersion.
The study also concerned the efforts exerted on the mooring arms owing to the
slamming phenomenon.
Additional studies were carried out specific to the definition of the floating
unit, associated to an articulated column.
Feasibility studies have been especially devoted to the sizing of each
structure "600-1000 m-Mediterranean Sea", "600-1000 m-North Sea". Their
operational performances have been analysed in detail with both theoretical
computations and experimental tests.

REFERENCES :
****
D.O.T. 1981 - ARTICULATED IN DEEP WATERS
D.O.T. 1983 TANKER MOORING IN DEEP WATER USING ARTICULATED COLUMNS
COPENHAGEN - JUNE 1984 - SYMPOSIUM ON DESCRIPTION AND MODELLING OF DIRECTIONAL
SEA:
BI ARTICULATED MOORING COLUMN TESTED IN DIRECTIONAL WAVES.
D.O.T. 1985 - MULTI-ARTICULATED COLUMNS FOR PERMANENT MOORING : BOURI FIELD AND
DEEPER WATERS APPLICATION.
JOURNEES FRANCO-MEXICAINES- AVRIL 1986 - CONCEPTS FOR DEEP WATER MOORING AND
PRODUCTION PLATFORM.
OMAE 1986 - THE APPLICATION OF TIME DOMAIN SIMULATION TO DESIGN THE OFFSHORE
LOADING ARTICULATED COLUMNS.
JOURNEES FRANCO-BRESILIENNES RIO JANEIRO NOVEMBER 1986 - OFFSHORE LOADING UNITS
FOR DEEP WATERS.
SNAME (TEXAS AND GULF SECTION)

```
********************************************************************************
* TITLE : NITROGEN-FILLED GEOLOGICAL STRUCTURE AT    *      PROJECT NO        *
*         TONDER FOR STORAGE OF NATURAL-GAS.         *                        *
*                                                    *      TH./14017/83/DK/.. *
*                                                    *                        *
********************************************************************************
* CONTRACTOR :                                       * PROGRAM :              *
*   DANSK OLIE & NATURGAS                            *   HYDROCARBONS         *
*   AGERN ALLE 24-26            TEL 02/571022        *                        *
*   DK 2970 HOERSHOLM           TLX 37322            *                        *
*                                                    * SECTOR :               *
*                                                    *   STORAGE              *
* PERSON TO CONTACT FOR FURTHER INFORMATION :        *                        *
*   MR. HANS OEBRO              TEL 02/571022        *                        *
*                              TLX 37322             *                        *
*                                                    *                        *
********************************************************************************
```

VERSION : 09/01/87

AIM OF THE PROJECT :
****
The objective of the project was to investigate the technical and economical
feasibility of storing natural gas in a geological structure using nitrogen as
part of the cushion gas.

PROJECT DESCRIPTION :
****
The project comprised the following activities :
a) Planning and preparatory work
b) Drilling of two new wells,Toender-4 and -5
c) Geological investigations (logging, coring, mapping)
d) Testing of the new wells,Toender-4 and -5
e) Gas mixing test of the old well, Toender-3.
f) Feasibility study.
Two potential storage reservoirs have been investigated :
- Upper Bunter Sandstone (depth 1650 m) : Nitrogen reservoir
- Lower Bunter sandstone (depth 1800 m): Aquifer containing salt saturated
brine.
Activity d, well testing, comprised longterm pressure testing of both Upper and
Lower Bunter Sandstone. The objective was to investigate the pore volume and
productivity of both reservoirs.
Activity e, gas mixing test, comprised a gas injection test of zone 2 of the
Upper Bunter. The gas was injected in the well Toender-3, using argon as a
tracer gas. The injected gas was then withdrawn and the concentration of the
tracer gas was monitored. The concentration data was used to determine the
degree of gas mixing in the reservoir.
Activity f, feasibility study, comprises reservoir simulations of gas storage
in both Upper and Lower Bunter. The objective is to study the mixing of natural
gas and nitrogen during storage operations. The results were used for a
technical/economical study of the benefits and risks involved in using nitrogen
as part of the cushion gas.

STATE OF ADVANCEMENT :
****
The project was initiated on December 1st, 1982 and completed in August 1985.

RESULTS :
****
The project showed that it is likely that natural gas can be stored in the
Upper Bunter nitrogen reservoir provided that only 60-70% of the natural gas is
cycled. In that case, the concentration of nitrogen in the well stream will be
less than 5-10%.
For the Lower Bunter reservoir up to 10% of the total gas-in-place can be
nitrogen without having serious gas mixing problems.
The cost estimates indicate that use of nitrogen as part of the cushion gas can
reduce the total storage cost by 10-15%.

REFERENCES :
****
"STUDY OF UNDERGROUND GAS STORAGE USING NITROGEN AS CUSHION GAS".
BY HANS OEBRO, SENIOR RESERVOIR ENGINEER, DANSK OLIE & NATURGAS A/S.
PAPER PRESENTED AT THE 1986 INTERNATIONAL GAS RESEARCH CONFERENCE, TORONTO,
SEPTEMBER, 1986.

```
********************************************************************************
* TITLE : STORAGE OF LIQUID AND LIQUEFIED        *       PROJECT NO         *
*         HYDROCARBONS IN LINED HARDROCK CAVERNS  *                          *
*                                                 *   TH./14020/85/DE/..     *
*                                                 *                          *
********************************************************************************
* CONTRACTOR :                                    * PROGRAM :                *
*   SALZGITTER AG                                 *   HYDROCARBONS           *
*   ABTEILUNG FORSCHUNG UND EN TEL 030 88 42 97 21 *                         *
*   POSTFACH 15 06 27          TLX 185 655        *                          *
*   DE - 1000 BERLIN 15                           * SECTOR :                 *
*                                                 *   STORAGE                *
* PERSON TO CONTACT FOR FURTHER INFORMATION :     *                          *
*   DIPL.-ING. W. EBELING                         *                          *
*                                                 *                          *
*                                                 *                          *
********************************************************************************
                                                     VERSION : 13/11/86
```

AIM OF THE PROJECT :
****
Rock caverns have until now been constructed and utilized according to a
storage concept developed in Scandinavia for the storage of crude oil, mineral
oil products such as heating oil, liquid gas (LPG). The application of this
storage concept with a hydrodynamic water seal is severely restricted in the
European inland, as well as in other countries for reasons of environmental and
groundwater protection. In addition contaminations of the storage product,
which fundamentally impair the product stability occur with this storage
principle due to the direct contact between groundwater, rock and storage
product.
This problem could be avoided if the storage were to occur in lined caverns.
The aim of this project is to develop storage concepts, which stand out above
the present-day technologies of their greater geotechnical safety potential,
greater geographical utilization possibilities and more diverse employment
possibilities with respect to different products stored.

PROJECT DESCRIPTION :
****
The aim of this project is the preparation of a largescale workable
construction concept for lined rock caverns for storage of a variety of
hydrocarbon products. An evaluation of the stability and tightness of cavern
storage facilities (with internal lining) is interconnected with questions of
geology, rock mechanics, geohydraulics, geothermics, construction technology,
construction material technology, belowground construction and operation
technology. The stability, tightness and economic viability of this storage
type are predominantly dependent on the geological underground conditions, the
construction process and the product-specific stress characteristics.
In the project's initial phase the bases will be prepared in the individual
subject areas of underground construction. Using rock mechanical and numerical
calculations thr rock caverns' stress/ strain conditions will be examined and
criteria for design and dimensioning prepared. Laboratory tests on material
laws and material characteristics of the construction, sealing and heat
insulation materials will be carried out and incorporated in the theoreticaly
calculated simulation.

The preparation of the construction concepts is carried out giving due consideration to concrete location possiblities but the concepts are to be designed in such a way that an adaptation to various different geological conditions would be possible. A definite pilot project, with respect to location selection is only then possible after finalization of the construction concept and the cost efficiency analysis and after detailed preparation of project proposals with the future operators of the project. Construction concepts developed for lined cavern storage facilities will be the subject of cost efficiency analysis in comparison to existing alternative storage concepts.
 Any concepts which do not appear feasible, wether for economical and/or technical reasons will be dropped.
The last stage of this project is the planning of a pilot plant. After previous preliminary talks with possible operators the following projects have been planned to set up pilot plants.
Storage Type 1: liquid hydrocarbons
Storage Type 2: compressed liquefied hydrocarbons
                location: areas of consumption, large towns.
In the preliminary phase of the planning itemization of the proposed project will be carried out with the operators with respect to storage product, cavity volumes, locations, etc.

STATE OF ADVANCEMENT :
****
Ongoing the project is in the design phase. The work for the subsection "project fundamentals" has to a large extent been completed.

RESULTS :
****
The liquid and liquefied hydrocarbons intented for storage have been described and specificated. The compilation of relevant safety data packages and the determination and description of the process and of significant process elements have been completed. The planning principles for the specified products were also recorded diagrammatically and laid down in basic process flow charts.
A general system of classification for rock is presented and discussed by means of examples. Subsequently, six types of rock were selected as being representative for the geological formations under consideration in Central Europe. The types of rock were described in detail.
The types of rock are very different in their natural state and in their mechanical behaviour. Models in terms of rock mechanics were worked out, which contain the characteristics properties of each rock in a clear form. They are basis for the numerical calculation of the stress and deformation behaviour.
For the economic dimensioning of measures for sealing, drainage, influence of the goundwater on the structure and the environment, the geohydraulic parameters are to be determined: permeability, groundwater relationship, chemistry.
The description of the parameters can be dependent on time and includes examinations during the pre-design stage, construction measures and after completion.
The method of testing in order to determine the individual parameters in the rock therefore plays an important and essential role.
The work on the aspect of geothermics has been completed. General statements are made initially with reference to the thermal characteristics of each type of rock. Values for the six types of rock found in relevant literature were listed in tabular form.

In the section " thermomechanics" a calculation model and calculation method to
determine the stress deformation behaviour as a result of thermal conduction
were described.
As a result of the calculations one receives temperature distribution, thermal
balances and static loads for prestated time periods, such as construction
stages and differing operational conditions. The forces of the individual
calculations can be used as load cases in the static design calculation.
The development of internal sealing systems has been completed. The internal
sealing systems are at present being checked.

REFERENCES :
****
H.J. SCHNEIDER, S. SEMPRICH
REQUIREMENTS MADE OF LINED ROCK CAVERNS FOR STORAGE OF FUELS AND LIQUEFIED
GASES;
INTERNATIONAL SYMPOSIUM "LARGE ROCK CAVERNS", HELSINKI, AUGUST 1986

```
********************************************************************************
* TITLE : INDUSTRIAL PILOT PROJECT FOR          *        PROJECT NO          *
*         UNDERGROUND CRYOGENIC STORAGE OF       *                            *
*         LIQUEFIED GASES                        *     TH./14021/86/BE/..     *
*                                                *                            *
********************************************************************************
* CONTRACTOR :                                   * PROGRAM :                  *
*   DISTRIGAZ SA                                 *   HYDROCARBONS             *
*   AVENUE DES ARTS 31        TEL 230.50.20      *                            *
*   BE - 1040 BRUXELLES       TLX 63 738         *                            *
*                                                * SECTOR :                   *
*                                                *   STORAGE                  *
* PERSON TO CONTACT FOR FURTHER INFORMATION :    *                            *
*   MR. J. DERVILLE                              *                            *
*                                                *                            *
*                                                *                            *
********************************************************************************
                                                  VERSION : 01/01/87
```

AIM OF THE PROJECT :
****
The objective of the proposed project is to build a pilot facility and operate
it under industrial working conditions and over a sufficiently long length of
time in order to prove in practice the reliability of all the innovative
facilities ie:
- Design of the storage cavity and its equipment to resist the constraints due
to cryogenic conditions and depth (specially in the case of clay)
- Operating equipment - and its maintenance - working under such conditions
- Operating process. The pilot should demonstrate the technical feasability and
the advantages of the technology in terms especially of safety and lower
capital andoperating costs.

PROJECT DESCRIPTION :
****
This industrial pilot facility will have a capacity of the order of 3000 m3 and
exhibit all the specific features of a full-scale industrial unit and the
equipment needed for its operation.
The project site will be chosen to minimize costs and will therefore be in
Europe, adjacent to a LNG terminal or peak-shaving plant that can provide the
LNG needed for the initial cooling down phase of the facility and the
subsequent operating phase.
Attention is currently focused on a site at Zeebrugge.
The programme will break down into three phases:
1. Feasibility study:
   Its content will of course depend on information available but mus
   necessarily include:
   1.1. Drilling to achieve a full geological survey of the clay layer and
        take fresh samples
   2.2. Laboratory testing - to determine the clay properties at ambient
        and cryogenic temperatures
   2.3. The feasibility study in itself
   2.4. General design, including for example tunneling methods and the
        main items of operating equipment, as well as capital cost estimates

2. Construction:

Will involve driving a single tunnel of 6 m diameter with a length of clayformation.The access shaft will have a diameter of +- 2m and will be approximately 100 m at a depth of roughly 150 m below ground level in a afterwards converted to an operating shaft.

If required, the tunnel can be supported with concrete segments for mechanical integrity during the excavation work.

This equipment will include:

- LNG pump at ground level
- Spraying line along the tunnel roof
- Submersible pump for lifting LNG to surface
- Boil-off treatment unit
- Operating instrumentation and boil-off return lines
- Monitoring and control equipment and safety systems

3. Cooling-down and operation:

The timing of the cooling down phase will last 5 - 6 months. Once cooled down, the unit will be operated for at least one year to provide an opportunity for simulating various operating phases, observing cavity response, checking the performance of equipment and control systems and optimizing operating performances.

STATE OF ADVANCEMENT :
****

A first drilling has been made down to - 200 meters.

Samples have been taken and are being submitted to a first series of laboratory tests.

REFERENCES :
****

"CAVITE PILOTE DE STOCKAGE CRYOGENIQUE DE SCHELLE" A. BOULANGER, P.V. DE LAGUERIE & W. LUYTEN
(EUROP COMM. SYMPOSIUM, LUXEMBOURG, 5-7/12/84)
"LPG AND LNG TERMINAL ASSOCIATED WITH UNDERGROUND STORAGE" J.P. LAGRON, A. BOULANGER & W. LUYTEN
(GASTECH, NICE, 12-15/11/85)

MISCELLANEOUS

```
*****************************************************************************
* TITLE : DEVELOPMENT OF A RELIABILITY ANALYSIS    *      PROJECT NO       *
*         SYSTEM FOR OFFSHORE STRUCTURES. (RASOS)  *                       *
*                                                  *    TH./15024/81/IR/.. *
*                                                  *                       *
*****************************************************************************
* CONTRACTOR :                                     * PROGRAM :             *
*   I.I.R.S.                                       *   HYDROCARBONS        *
*   BALLYMUN ROAD            TEL 01/370101         *                       *
*   IR - DUBLIN 9            TLX 32501             *                        *
*                                                  * SECTOR :              *
*                                                  *   MISCELLANEOUS       *
* PERSON TO CONTACT FOR FURTHER INFORMATION :      *                       *
*   MR. G. KEANE                                   *                       *
*                                                  *                       *
*                                                  *                       *
*****************************************************************************
```

VERSION : 31/12/86

AIM OF THE PROJECT :
****
To develop a reliability analysis system for offshore structures.

PROJECT DESCRIPTION :
****
The project involved:
-characterisation of stochastic variables for Irish waters.
- development of reliability system (certification procedure and computer
programme).
- development of quality assurance and planned maintenance programmes.
- application of rasos to an offshore installation.

STATE OF ADVANCEMENT :
****
Completed

RESULTS :
****
Comprehensive data base now exists at IIRS, covering environmental parameters,
waves winds, and structural parameter, vibration/acceleration.
Fatigue analysis results are available (confidential) at IIRS.
Theoretical analysis has been carried out on a typical structure, to determine
vibrational modes and displacement levels.
The project was undertaken within the scope of work covering the following:
- information acquisition
- definition of quality assurance criteria and their application
- environmental data analysis to determine external influences operating on an
offshore structure. This involved the analysis of data collected on Kinsale Gas
Platform ALPHA by IIRS since 1980, covering waves, vibrations and derived
displacements.
- fatigue investigation built into the RASOS logic.

```
********************************************************************************
* TITLE : CAST STEEL NODES FOR FIXED OFFSHORE      *        PROJECT NO         *
*         STRUCTURES                               *                           *
*                                                  *    TH./15032/82/UK/..      *
*                                                  *                           *
********************************************************************************
* CONTRACTOR :                                     * PROGRAM :                 *
*   BRITOIL PLC                                    *   HYDROCARBONS            *
*   301 ST VINCENT STREET      TEL (041)2042525    *                           *
*   UK - GLASGOW G2 5DD        TLX 777633          *                           *
*                                                  * SECTOR :                  *
*                                                  *   MISCELLANEOUS           *
* PERSON TO CONTACT FOR FURTHER INFORMATION :      *                           *
*   MR. K.J. WELLS                                 *                           *
*                                                  *                           *
*                                                  *                           *
********************************************************************************
```

VERSION : 01/01/87

AIM OF THE PROJECT :
****
To carry out an independent evaluation of the properties of prototype cast
steel node for offshore structures and to define defect acceptance levels and
mechanical properties for specification purposes.

PROJECT DESCRIPTION :
****
The project encompasses the development work needed to enable cast steel nodes
to be specified with confidence for use in offshore steel structures. Phase 1
consisted of design and production of a full-size prototype launch brace node
as a casting. Phase 2 comprises 100% surface and volumetric non-destructive
testing to determine casting, quality, welding and weld repair demonstration
and evaluation and mechanical testing of a large number of representative
samples taken from different parts of the casting.

STATE OF ADVANCEMENT :
****
Completed
RESULTS :
****
Phase 1 has demonstrated that a large node can be produced as a sound casting
to a high standard and to a realistic production schedule.
Phase 2 has demonstrated that the surface condition of the casting was good and
that ultrasonic testing is in general a satisfactory technique for detecting
and locating buried defects such as shrinkage porosity and hot tears. The stub-
to-tubular weld and weld repair trials have been successfully completed.
Mechanical property values in a large casting have been shown to vary within
wide limits and the data obtained can be utilised for the definition of minimum
property values for specification purposes.

REFERENCES :
****
PARLANE, AJA AND GILLIES, AW - "CAST STEEL LIFTING COMPONENTS TO MEET OFFSHORE
DESIGN AND QUALITY REQUIREMENTS".
THE NINTH ANNUAL ENERGY SOURCES AND TECHNOLOGY CONFERENCE AND EXHIBITION (ASME-
ETCE - OFFSHORE OPERATIONS SYMPOSIUM, 1986).

```
********************************************************************************
* TITLE : THE "MECHANICAL ECHO" DIAGNOSTIC METHOD    *        PROJECT NO       *
*         FOR THE OFFSHORE STRUCTURE "MECCANICO".    *                         *
*                                                    *  TH./15033/82/IT/..     *
*                                                    *                         *
********************************************************************************
* CONTRACTOR :                                       * PROGRAM :               *
*   TECNOMARE                                        *  HYDROCARBONS           *
*   S. MARCO 2091            TEL 041 796711          *                         *
*   IT - 30124 VENEZIA       TLX 410484 MAREVE I     *                         *
*                                                    * SECTOR :                *
*                                                    *  MISCELLANEOUS          *
* PERSON TO CONTACT FOR FURTHER INFORMATION :        *                         *
*   MR. V. BANZOLI                                   *                         *
*                                                    *                         *
*                                                    *                         *
********************************************************************************
                                                    VERSION : 01/01/87
```

AIM OF THE PROJECT :
****
To achieve a better capability to diagnose the actual safety conditions of
steel structures, by means of a new method based on the elastic wave
propagation and reflection.

PROJECT DESCRIPTION :
****
The basic principle of the proposed method is that an elastic wave propagating
along a structure shares its energy between a transmitted wave and a reflecting
one when it reaches a discontinuity in the structure (e.g. a crack). Measuring
the time of return and the energy of the reflected wave, it should be possible
to locate and quantify the discontinuity.
The project is subdivided into the following main subjects:
- acquisition of basic theoretical knowledge
- theoretical and laboratory experimental study of the method
- design and procuring of a subsea portable monitoring system
- field test of the portable system on a jacket model
- sea test of the portable system.

STATE OF ADVANCEMENT :
****
Ongoing. The project is in the testing phase. Shop test of the portable system
has been completed as well as sea test in order to verify, mainly, the
underwater operativity of the system. test on a jacket model, simulating a real
platform, are in course: test are performed both on land and at sea.

RESULTS :
****
Main results so far achieved can be summarised as follows:
- The theoretical models of the waves propagation have furnished useful data
for establishing the best conditions for exciting the structure; however the
developed computer program cannot furnish a forecast of the wave propagation,
as the finite elements calculations are to expensive to be used on a routine
basis.
- A new system for including elastic waves in the structure members has been
studied and developed.
The system is based on the magnetostrictive effects and eliminates all the
problems related to the use of mechanical exciters.
- A software for collecting, analysing and comparing the echoes from the
structures discontinuities (nodes, cracks, etc...) has been developed.
- The theory of the synchronous sampling for the signal analysis has been
further developed.
- A portable instrumentation system, housed in a standard container, has been
designed and built. The system includes, besides the instrumentation systems
(signal anlyzer, computer, amplifiers, etc...), all the equipment necessary for
on field operation (as umbilical cable, cable winch, underwater cage, exciter
and accelerometers, etc...).
- The system has been tested at sea in order to verify its operability by
divers.
- Land and sea test of the system on a jacket model (simulating a real
structure) are in course.
- The results so far achieved demonstrate the capability of the method to
detect damage corresponding to 1/4 cut (or less) of the structural members
sections, when used as comparative local method.

```
******************************************************************************
* TITLE : APPLICATION OF ELECTRON BEAM WELDING    *     PROJECT NO         *
*          FOR LAYING UNDERWATER PIPES .          *                        *
*                                                 *                        *
*                                                 *     TH./15038/82/FR/..  *
*                                                 *                        *
******************************************************************************
* CONTRACTOR :                                    * PROGRAM :              *
*   GERTH                                         *   HYDROCARBONS         *
*   4, AVENUE DE BOIS PREAU     TEL 1 47.52.61.39 *                        *
*   FR - 92502 RUEIL-MALMAISON TLX 203 050        *                        *
*                                                 * SECTOR :               *
*                                                 *   MISCELLANEOUS        *
* PERSON TO CONTACT FOR FURTHER INFORMATION :     *                        *
*   MR. ANDRIER                 TEL 1 47.59.60.00 *                        *
*                                                 *                        *
*                                                 *                        *
******************************************************************************
                                                  VERSION : 01/01/87
```

AIM OF THE PROJECT :
****
The purpose of the project was to adapt the method of welding by electron beams
to conventional laying of subsea pipelines in view to weld pipes with more
elaborate and thicker steel grades.

PROJECT DESCRIPTION :
****
The advantages of the method are its fast welding rate independent of the
thickness and minimum change in the methodological characteristics in the
molten zone, which remains very narrow and which enables high elastic limit
steel to be welded under good conditions.
The difficulties of application of the method lie in keeping the welding zone
in a vacuum and in precisely positioning the beam in the plane of the joint of
the two tubes to be welded. In addition, since the plane of the joint is
vertical, the weld takes place in succession in all positions (flat, up-hand,
under-weld and down-hand), thus requiring the adjustment parameters to be
varied constantly.
A preliminary study showed that the weldable thickness for an under-weld and a
flat weld is limited: the melt falls under the effect of gravity when the
thickness is more than 16 mm. Beyond this thickness, the weld must be made in
two passes with a non-through, one from the inside and one from the outside.
Accordingly, two machines, one external and one internal, have to be designed,
and the welding method developed, allowing for the results of welding tests
made on a simulation bench.
A technical and economic study of the welding method has enabled all the
elements involved in its application to be accounted for: handling, setting up
on barge, projected productivity, future market study and maintenance.

STATE OF ADVANCEMENT :
****
The project began in January 1982 and ended in April 1984.

RESULTS :
****
The technical feasibility of all the components of the method has been
established. However, certain factors must be specified more exactly:
weldability of the steels now used for large subsea pipelines, adaptation of
existing barges.
An other study including welding tests is being performed on eight different
steels which have been actually used for existing subsea pipeline. For steels
that are not directly weldable with this electron beam method various processes
have been developed to make the method applicable. This study is still ongoing
and results are very encouraging.

REFERENCES :
****
PRESENTATION OF THE PROJECT IN THE PROCEEDING OF THE 2ND SYMPOSIUM HELD IN
LUXEMBOURG 5-7 DECEMBER 1984.

```
***********************************************************************
* TITLE : VERTICAL POLYPHASIC FLOW PHASE 1      *      PROJECT NO      *
*                                               *                      *
*                                               *    TH./15039/82/FR/.. *
*                                               *                      *
***********************************************************************
* CONTRACTOR :                                  * PROGRAM :            *
*   GERTH                                        *   HYDROCARBONS       *
*   4, AVENUE DE BOIS PREAU     TEL 1 47.52.61.39 *                    *
*   FR - 92502 RUEIL-MALMAISON TLX 203 050       *                     *
*                                               * SECTOR :             *
*                                               *   MISCELLANEOUS       *
* PERSON TO CONTACT FOR FURTHER INFORMATION :    *                     *
*   MR. CORTEVILLE             TEL 1 47.49.02.14  *                    *
*                             TLX 203 050         *                    *
*                                               *                      *
***********************************************************************
```
VERSION : 15/03/87

AIM OF THE PROJECT :
****
The project was executed from 1982 to 1985. The main object of this project was
to develop a specific calculation method for gas and liquids polyphasic flows
in tubing which equip crude or condensate gas wells as well as in production
risers used offshore.
A new method for predicting pressure losses in these vertical or very inclined
pipes seem indeed necessary, owing to :
- the lack of precision of traditional calculation methods, based mostly on
correlations that cannot be extrapolated to difficult operating conditions
which are more frequent.
- the need for precise evaluations, namely for offshore production, of
quantities of hydrocarbons likely to be extracted from different wells within
one field, whether eruptive or being pumped, so to set-up serious and credible
technical and financial forecasts concerning the volume of the overall
production, then to optimize the pipes and the bottom and surface equipment.
PROJECT DESCRIPTION :
****
This project followed the study on diphasic flows in horizontal or slightly
inclined pipes. (Contract EEC/GERTH nr TH 03080/79), which ended in 1981 in the
finalization of a calculation method called PEPITE. Thanks to this experience,
the project team was able to apply the same experimental methodology and the
same theoretical approach, although vertical or highly inclined diphasic flows
are different and more complex to study.
The project was divided in two main phases :
- STUDY OF OIL-GAS DIPHASIC FLOWS
This most important part resided on experiments carried out under
representative oil industry conditions on the vertical diphasic flow test loop
of Boussens. This loop consisted of two test pipes of a diameter of 3" and 6"
respectively, presenting an effective length of 25 m, and allowing to measure
the main characteristics of the oil/gas diphasic flow up to velocities of about
20 m/s and a pressure of 50 bars. These pipes supported by a metallic crossbeam
mobile around an axle and actuated by jack could be oriented from a horizontal
position to a vertical position so the determine the slope effect. A databank
of pressure gradients, liquid contents, flow patterns, etc... was thus
elaborated in relation to the main parameters : phase velocities, pressure,
gradient, diameter, oil nature, according to the test programme chosen for the
modelization requirements.

The theory of the observed three flow patterns (bubble flow, slug flow and annular flow) and of their transitions has been developed in parallel, and led to computation models which restore with a reasonable accuracy the test data (mean error of about 10%).
Last of all, these models have been tested on a small number of data relative to the exploitation of eruptive wells activated by gas-lift, and proved satisfactoy.ry.
- PRELIMINARY STUDIES ON GAS/OIL WATER TRIPHASIC FLOWS UNDER VERTICAL CONFIGURATION
A triphasic test loop of a diameter of 2" and a length of 12,5 m was built at the Institut de Mecainique des Fluides de Toulouse. This test loop allowed us to visualize the different flow patterns, particularly the slug flows and to prepare the theoretical basic elements and the instrumentation suitable for the development of a future study.

STATE OF ADVANCEMENT :
****
The project is now completed. It is being followed by the Study on triphasic flows and pipeline/riser coupling (contract TH 10043/84) which finalizes the present study.

RESULTS :
****
Results obtained within the scope of this project were restricted for the quantitative elements to the diphasic flows of gas and oil. They could not be applied without reserve in the very frequent case of water producing wells. The preliminary study on triphasic flows that had been carried out did not allow to conclude on the possibility of modelling in the same manner the diphasic and triphasic flows.
The results were nevertheless exploited at the end of the following study which alllowed to develop the modelling of triphasic flows and to finalize a model which represented with similar accuracy both situations. This model was incorporated in a general computing programme for polyphasic flows in eruptive wells or gas-lift wells called WELLSIM (well simulation). This programme which includes the results of the diphasic study has been tested and compared to the traditional methods of Aziz-Govier, Hagedorn-Brown, Orkizewski, Ros, on data provided by 90 eruptive or gas-lift wells, among which more than half presented a water/oil ratio that was nil.
With WELLSIM the mean error is more or less 1% and the standard deviation 12%, whereas the best traditional method provides a mean error of more than 9% with a standard deviation of 19%. The obtained accuracy was thus about two times better than that of the traditional methods.

REFERENCES :
****
A NUMBER OF PAPERS RELATIVE TO THESE SUBJECTS ARE BEING WRITTEN AND WILL BE PUBLISHED IN DEDICATED MAGAZINES OR TO BE DISTRIBUTED AT INTERNATIONAL CONFERENCES.

351

```
********************************************************************************
* TITLE : THERMAL INSULATION OF TUBINGS          *       PROJECT NO         *
*                                                *                          *
*                                                *    TH./15040/82/FR/..    *
*                                                *                          *
********************************************************************************
* CONTRACTOR :                                   * PROGRAM :                *
*   GERTH                                         *   HYDROCARBONS           *
*   4, AVENUE DE BOIS PREAU      TEL 1 47.52.61.39 *                        *
*   FR - 92502 RUEIL-MALMAISON TLX 203 050        *                         *
*                                                * SECTOR :                 *
*                                                *   MISCELLANEOUS          *
* PERSON TO CONTACT FOR FURTHER INFORMATION :    *                          *
*   MR. LESAGE                   TEL 1 47.49.02.14 *                        *
*                                TLX 203 050      *                          *
*                                                *                          *
********************************************************************************
```

VERSION : 20/02/87

AIM OF THE PROJECT :
****
This project aims at improving the enhanced recovery of heavy oils by steam
injection. It has long been recognized that by heating reservoirs containing
heavy oils one can improve ultimate recovery substantially. Although steam
injection is currently the principal thermal recovery method, heat losses
associated with the delivery of the steam from the surface generators to the
oil-bearing formation has limited conventional steam injection to shallow
reservoirs. Extending the injection of high quality steam to greater depths or
increasing the steam quality at bottomhole is of great interest. The objective
of the project is to develop tubing - casing insulation materials enabling the
heat losses to be reduced during the steam injection operations while
minimizing the risk of breakage of the casing under the combined effect of
mechanical and thermal stresses.

PROJECT DESCRIPTION :
****
The project has been divided into three phases :
PHASE 1 - CHOICE OF INSULATING MATERIAL
This phase comprised the choice of the insulating materials and their
application techniques. The materials studied have been :
- Mineral foams
Mineral foams can be formed by using sodium silicates. The solubility of
existing materials is very low even in hot water, and this is a disadvantage
when it is necessary to pull out the tubing coated by foam. The study of
silicates having different compositions of sodium oxide and silica, as well as
the study of the effects of additives may identify some compounds having a good
solubility in water or in appropriate solvents. The rate of dissolution and the
heat transfer coefficient for various foams was measured to characterize the
foam.
- Organic materials

Thixotropic gelatinous oil-base fluids can act as insulating materials. The efficiency of these compounds is related to their low thermal conductivity and to their high viscosity which minimizes heat transfer by convection. The thixotropic fluid consisted of a heavy oil and additives to increase the viscosity and the stability. The effect of the fluid composition on the viscosity, on the maximum temperature to which the gel can be formed and on the thermal stability versus time was examined.

PHASE 2 - TESTS ON LABORATORY CELLS

The materials, which have been found as suitable for thermal insulation during the work carried out in phase 1, were to be tested in laboratory equipments simulating the tubing and the casing in a well. The experiments were performed first at atmospheric pressure in a simplified cell, then up to a pressure of 1 MPa in more sophisticated equipment. The results of the tests were used to compare the efficiency of the different mineral and organic materials, and select the best ones. An evaluation of the cost of the material was planned.

PHASE 3 - TESTS ON WELLS

After having defined the procedure for applying the insulating materials, the products selected in the previous phases were planned to be tested on wells, already subjected to steam injection. Measuring the temperature at the wall of the casing with thermocouples before and after having insulated the tubing were expected to give information on the efficiency of the insulating product tested. A thermal balance and an economical balance on each of the tested process was to be carried out at the end of the project.

STATE OF ADVANCEMENT :
****
Abandoned.
During Phase 1, materials showing good insulating properties (mixture of sodium silicate, thixotropic oils-base fluids) were found. During Phase 2, tests performed with inorganic materials indicated the effectiveness of the silicate foam at moderate pressure, but at high pressure it was difficult to generate the foam. Experiments carried out with thixotropic materials showed insufficient thermal stability of the mixtures. The project was then abandoned before well testing.

RESULTS :
****
PHASE 1 - In the first phase, two types of insulating materials - silicate foam and thixotropic oil-base fluid have been found and patented.
Concerning the mineral foam, study of the various parameters, had led to the formulation of a mixture of two silicates having different ratios between sodium oxide and silica. This mixture gives thermally stable foams that have better insulating properties and are more soluble than conventional polysilicate foams.
Studies of the various formulae for gelled oils have resulted in the development of a thixotropic composition, formed mainly from a oil-base fluid and calcium grease. This product has a good thermal stability up to 300 deg.C, reducing heat losses by convection.
PHASE 2 - In the second phase, the materials previously defined were tested in laboratory cells simulating a small part of a casing and a tubing from a well.

Two cells were designed. The first one works at atmospheric pressure; its
length is 0.5 m and the diameters of the tubing and casing are respectively 7,3
cm and 22,7 cm; the tubing is heated by a thermally stable oil. The second cell
can operate up to a pressure of 10 MPa; its length is 1 m; the casing diameter
is 17,8 cm and it is possible to use tubings having diameters of 7,3 cm, 11,4
cm : the tubing is heated by a thermally stable oil.
Concerning the silica formula, in the tests performed at atmospheric pressure,
the foam is easily formed all along the tubing; its thickness is important and
its insulating effectiveness is good (temperature on the casing is less than
100 deg.C for a tubing temperature of 280 deg.C). Moreover, the foam is
completely soluble in water. Good results are also obtained at a pressure of 5
MPa, but for higher pressures, the foam was not formed in the high pressure
cell. In this case, a gelatinous material, which is an intermediate product
between the liquid silicate and the foam, flows to the bottom of the cell, and
the foam is not formed.
Concerning the gelified oils, several formulae were tested at atmospheric
pressure. All of them showed a good insulating characteristics (temperature on
the casing is less than 100 deg.C for a tubing temperature of 300 deg.C). With
time, there is a separation of the oil from the grease used to make the
gelified oils; in the cell, the fluidity is higher in the top zone than in the
bottom zone. So, due to a very high viscosity of the fluid at the bottom of a
well, it can be difficult to empty the annulus tubing-casing when needed.
Phase 3 - Considering these results, the use of these insulating materials in
wells can be risky. Consequently, this phase was abandoned.

REFERENCES :
****
1 FRENCH PATENT : METHODE D'ISOLATION THERMIQUE D'UN PUITS, NR 2532988
2 FRENCH PATENT : NOUVEAU MATERIAU POUR L'ISOLATION THERMIQUE DES PUITS DE
PRODUCTION D'HUILES LOURDES, NR 2536386
3 J. LESAGE, PROTECTION THERMIQUE DES TUBINGS. SECOND EUROPEAN SYMPOSIUM ON NEW
TECHNOLOGIES FOR EXPLORATION AND EXPLOITATION OF OIL AND GAS RESSOURCES,
LUXEMBOURG 5-7 DECEMBER 1984.

```
******************************************************************************
* TITLE : A HIGH PERFORMANCE ROPE FOR DEEP WATER      *       PROJECT NO     *
*         MOORING.                                    *                      *
*                                                     *    TH./15042/82/FR/..*
*                                                     *                      *
******************************************************************************
* CONTRACTOR :                                        * PROGRAM :            *
*   S.E.P.                                            *   HYDROCARBONS       *
*   LE HAILLAN BP 37            TEL (56)348490        *                      *
*   F - 33165 SAINT MEDARD EN   TLX SEP560678F        *                      *
*                                                     * SECTOR :             *
*                                                     *   MISCELLANEOUS      *
* PERSON TO CONTACT FOR FURTHER INFORMATION :         *                      *
*   YVES APPELL                                       *                      *
*                                                     *                      *
*                                                     *                      *
******************************************************************************
```

VERSION : 12/03/87

AIM OF THE PROJECT :
****
The mooring with synthetic fibre rope system is an attractive solution, due to
the high specific strength, low density and non-corrodable properties of these
materials. The purpose of the project is to investigate the feasibility of
using synthetic fibre ropes for these mooring systems and to identify the
problem areas in order to propose solutions, particularly with regard to the
attachment of these ropes, which is the critical item of the system.

PROJECT DESCRIPTION :
****
The major phases of the programme are:
- documentation research of the existing equipments: materials, rope structures,
  type of terminations, static and fatigue tests
- state of the requirement: service working conditions, specifications
- synthesis of these design phases: choice of a type of material, rope and
termination
- complementary characteristics of the selected rope and termination at a
reduced but representative scale (60 T)
- final design of an optimized link system including synthetic rope,
termination and structure.

STATE OF ADVANCEMENT :
****
Abandoned. Testing phase has been carried out and design for rope termination
up to 600 T has been produced.

RESULTS :
****
For the operation of deep water fields, the mooring of the platform by means of
metal wire ropes is no longer feasible.
It will be necessary to turn to higher performance and lighter in weight
materials : synthetic materials.
The conducted investigation enabled to determine that the material offering the
best characteristics in terms of strength and specific stiffness is an aromatic
polyamide marketed under the name of kevlar.
Among the rope manufacturers using this material we selected a British Supplier
(ICI) of which the range of PARAFIL parallel wire ropes includes a rope that
broke at 1486 tons.
The analysis of the requirement showed that deep water mooring shall require
ropes withstanding maximum loads of at least 150 to 200 tons with a safety
factor ranging from 2.5 to 3. The rope oscillation with respect to the platform
shall amount to a minimum of 5 deg.
Pratically, every tests conducted to date, on PARAFIL or others types of ropes,
were carried out on 6 ton maximum strength ropes and consisted of axial tension
cyclic variation.
We accordingly selected to perform a test representative of the contemplated
application : dynamic fatigue through alternate bending loads under static
tension load. We have chosen for experimental appraisal of the process a rope
with a rupture strength of 60 tons. This choice should not, in our opinion,
lead to a significant scale factor with respect to the ropes intended for the
mooring of platforms.
The target life of such mooring systems being of the order of 20 to 30 years,
we have had to increase the severity of the tests in order to achieve
significant results within acceptable times.
We accordingly decided to work at more than 25% of the rupture strength and to
increase the bending angle of the rope at the outlet of the terminal.
The results show the high sensitivity of the kevlar to such stresses. Although
at significant angles, rapid damaging of the kevlar rope was observed. On the
contrary this same rope exibits excellent behaviour when the oscillations are
limited.
Therefore we propose a mooring approach using laminates, an approach which
allows obtaining a service life of the rope terminal consistent with that of
the rope.
This study ends with three projects of connection subsystems of kevlar ropes
intended for mooring lines of a maximum 600 T load.

```
*********************************************************************************
* TITLE : DATA COLLECTION SYSTEM FOR THE GOBAN    *        PROJECT NO        *
*        SPUR.                                    *                          *
*                                                 *    TH./15043/83/IR/..    *
*                                                 *                          *
*********************************************************************************
* CONTRACTOR :                                    * PROGRAM :                *
*   SEA SURVEYS LTD                               *   HYDROCARBONS           *
*   RATHMACULLIG WEST        TEL 021 962600       *                          *
*   BALLYGARVAN              TLX 75850            *                          *
*   IR - CO. CORK                                 * SECTOR :                 *
*                                                 *   MISCELLANEOUS          *
* PERSON TO CONTACT FOR FURTHER INFORMATION :     *                          *
*   MR. E.D. HANNIGAN                             *                          *
*                                                 *                          *
*                                                 *                          *
*********************************************************************************
```
                                                        VERSION : 20/02/87

AIM OF THE PROJECT :
****
To design, construct and evaluate offshore, a prototype low cost real time
current measuring system to provide data by satellite link. To further develop
a three dimensional software design package for application to riser systems
and capable of incorporating both surface and subsurface environmental data.
Accordingly, important new knowledge will be gained with respect to deep sea
currents and immediately applied to offshore engineering design calculations.
The successful completion of the project will constitute a major advance in an
area of great importance and present uncertainty. The results will also be of
benefit to the Community generally through extrapolation of the current
measurements to other areas and through commercial exploitation of the riser
design package .

PROJECT DESCRIPTION :
****
Current Meter Mooring :
Phase 1 : initial feasibility study and review of equipment and operations.
Phase 2 : design of electronic components, deployment and testing.
Phase 3 : manufacture of final system and mooring design.
Phase 4 : field test of system for 6-12 months.
Phase 5 : post trial evaluation of results.
Riser design :
Phase 6 : extend present two dimensional package to three dimensions.
Phase 7 : include fluid currents and vortex shedding.
Phase 8 : modelling of vessel motions and random directional wave spectrum.
Phase 9 : process British Meteorological data and available observed data.
Phase 10 : execute demonstration riser designs.

STATE OF ADVANCEMENT :
****
Completed. However doubtful that the study will lead to any commercial
proposition.

RESULTS :
****
CURRENT METER MOORING
VHF data transmission link used rather than satellite link to avoid high user
costs. After three month deployment a comparison of internally recorded data
with the data received via the VHF link showed that a reasonable data return
was achieved from the top current meter (80%) but that this reduced
dramatically with increasing distance from the hydrophone. Good quality data
from the two deeper meters some five hundred meters away was found non-existant.
 Acoustic link chosen for the 'in water' data path from current meters to
surface transmitting buoy not powerful enough but system concept proven.
Further investigation proceeded with the current meters and a viable system was
proposed by raising the power output of the acoustic transmissions. By fitting
lithium batteries a one year deployment life could still be achieved. However
towards the end of this year's phase it became clear that American technology
had overtaken us. A prototype acoustic doppler current meter had been developed
and trials seemed to indicate encouraging results.
RISER DESIGN
A software package for three dimensional riser analysis capable of
incorporating environmental data was successfully achieved. The long run times
for the program are a cause for concern and indicate that it may be only suited
for applications on supercomputer. For these computers, the computer code
would need to be vectorised in order to obtain the optimum advantage of
existing parallel processing machines. The general present uncertainty with
respect to the specification of vortex shedding loads in three dimensional
analysis means that riser designers will be less likely to incur the processing
costs required except in exceptionally severe cases.

```
*********************************************************************************
* TITLE : DIMENSIONAL VERIFICATION TO BE CARRIED    *        PROJECT NO         *
*         OUT AT OPEN SEA ON OFFSHORE STRUCTURES    *                           *
*         BY MEANS OF PHOTO GRAMMETRY.              *      TH./15044/83/IT/..    *
*                                                   *                           *
*********************************************************************************
* CONTRACTOR :                                      * PROGRAM :                 *
*   AGIP SPA                                        *   HYDROCARBONS            *
*   C.P. 12069                TEL 62 5201           *                           *
*   IT - 20120 MILANO         TLX 310246 ENI        *                           *
*                                                   * SECTOR :                  *
*                                                   *   MISCELLANEOUS           *
* PERSON TO CONTACT FOR FURTHER INFORMATION :       *                           *
*   MR. BOZZOLATO                                   *                           *
*                                                   *                           *
*                                                   *                           *
*********************************************************************************
                                                      VERSION : 31/12/86
```

AIM OF THE PROJECT :
****
To obtain accurate dimensional parameters of the individual parts and of the
entire structure, necessary for assembling them at open sea and for a general
knowledge concerning construction, useful for technical, legal and insurance
reasons. Such a knowledge is expressed in a series of special measurements, "as
built" data and graphs showing discrepancies (design-actual structure) and
residual errors.

PROJECT DESCRIPTION :
****
Phase 1 : Prefeasibility study and design of equipment, special instruments and
softwares perfecting. Construction of a special metal support truss (nicknamed
"SER") to be installed on a medium-large size helicopter. Manufacture of small
ancillary instruments. Experimental tests of photographing from moving supports
(helicopter, floating craft) and subsequent tests of digital stereoplotting and
mathematical processing.
Phase 2 : In case of successful results in the 1st phase, proceed with the
perfecting of means, equipments, instruments and methods in order to make the
procedure fully operational and economical.

STATE OF ADVANCEMENT :
****
Some experimental tests were carried out from helicopter simulator and from
supply vessel. A contract has been signed by a major helicopters factory, to
built the special support truss (SER). The first prototype of "SER", in its
experimental and still uncomplete configuration was installed on AB412 AGUSTA
helicopter (September 17, 1986). The first official flight test was
successfully made on November 18, 1986. Fabrication is going on, to complete
the configuration with aerodynamical fairings.

RESULTS :
****
More complete results and news about further applications will be available in
1987, after the practical tests of the helicopter (equipped with the special
"SER" truss) on petroleum industrial plants on shore and offshore.

```
*********************************************************************************
* TITLE : MOBILE POLYSACCHARIDES INJECTION UNIT.      *      PROJECT NO        *
*                                                     *                        *
*                                                     *                        *
*                                                     *    TH./15046/83/FR/..  *
*                                                     *                        *
*********************************************************************************
* CONTRACTOR :                                        * PROGRAM :              *
*   GERTH                                             *   HYDROCARBONS         *
*   AVENUE DE BOIS PREAU 4      TEL 1 47 52 61 39     *                        *
*   FR - 92500 RUEIL-MALMAISON TLX 203050             *                        *
*                                                     * SECTOR :               *
*                                                     *   MISCELLANEOUS        *
* PERSON TO CONTACT FOR FURTHER INFORMATION :         *                        *
*   MR. G. BLU/MR. B. MERCIER                         *                        *
*                                                     *                        *
*                                                     *                        *
*********************************************************************************
                                                         VERSION : 12*03*85
```

AIM OF THE PROJECT :
****
The purpose of the project is to inject polysaccharides into reservoirs in
order to enhance the recovery of the oil in place. It consists in finalizing a
mobile unit for dissolving and enzyme clarification of xanthan gums formulated
in the works in powder form or a concentrated brew to prevent any risk of
clogging the reservoir.

PROJECT DESCRIPTION :
****
Laboratory tests on 20 litre and 1 m3 hydrolyzers have enabled the conditions
of application of the enzyme process to be specified, for clarification and
elimination of microgels of raw solutions of these xanthan gums. These tests
have shown that short duration enzyme processing in industrial water is on the
one hand valid for concentrated solutions of polymers in powder and on the
other suitable for concentrated brews. Additional work was also carried out,
namely :
- an evaluation of the efficiency of the process by meand of tests that were
more representative of real-life conditions of use, performed both on filters
and on a variety of porous media;
- a technical and economic study for evaluating the excess cost of the
installation on site of a mobile polyaccharides enzyme processing unit;
- redefinition of the process engineering, allowing for the economic study on
the one hand, and the reductions in the processing durations and variation of
the polymer/enzyme ratio, on the other.
In addition, a feasibility study of processing in the works has enabled an
interesting alternative to processing at the oilfield to be tested, namely the
possibility of producting a brew that has already been concentrated in the
works and which meets the requirementsing a brew that has already been
concentrated in the works and which meets the requirements for injectivity and
circulation inside the reservoir needed for the application envisaged.

STATE OF ADVANCEMENT :
****
The project started on 1st December 1980 and was completed on 30 th September
1985.

RESULTS :
****
Whilst the work has revealed a distinct improvement in the quality of the
products processed, very recent progress at the manufacturing stage in the
works has led to solutions of xanthan gums in the form of brews that can in
practice be filtered without any subsequent treatment. This result reduces the
marketing prospects for the mobile unit, all the more so since the oil
companies are reluctant to resort to methods using powders. Accordingly, it was
decided to abandon the construction of the prototype mobile unit.

```
*********************************************************************************
* TITLE : CAST STEEL NODE-ENGINEERING STUDY.        *      PROJECT NO      *
*                                                    *                      *
*                                                    *                      *
*                                                    *   TH./15047/83/UK/.. *
*                                                    *                      *
*********************************************************************************
* CONTRACTOR :                                       * PROGRAM :            *
*   BRITOIL PLC                                      *   HYDROCARBONS       *
*   ST VINCENT STREET 150        TEL 041 2042566     *                      *
*   UK - GLASGOW G2 5LJ          TLX 776268          *                      *
*                                                    * SECTOR :             *
*                                                    *   MISCELLANEOUS      *
* PERSON TO CONTACT FOR FURTHER INFORMATION :        *                      *
*   MR. J. ANDERSON                                  *                      *
*                                                    *                      *
*                                                    *                      *
*********************************************************************************
                                                       VERSION : 18/03/86
```

AIM OF THE PROJECT :
****
To carry out an independent evaluation of the properties of a prototype cast
steel node for use on offshore structures and to define acceptance levels and
mechanical properties for specification criteria.

PROJECT DESCRIPTION :
****
The project covered the design and manufacture of a 70 ton multi tubular node
typical of those used on tubular jacket structures. The implementation of an
inspection procedure including destructive and non-destructive testing. The
establishment of the mechanical properties throughout the casting. The
reconciliation and verification of the defects shown to be present in the non-
destructive tests by metallurgical examination and the validation of welding
procedures for the repair of casting defects.

STATE OF ADVANCEMENT :
****
Completed.

RESULTS :
****
The project has demonstrated that a large complex node can be produced in a
steel casting to a satisfactory quality and dimensional accuracy and that
buried defects can be accurately located and sized by non destructive testing
techniques. The welding of normal steel fabricated tubulars to the cast members
was accomplished entirely satisfactorily and actual and simulated defects were
repaired by welding without loss of properties. The mechanical properties of
the casting were found to be within the specified target values. The results of
the project give confidence in the ability to produce steel castings to replace
complex fabrications with the resultant improvement in fatigue life.

```
*******************************************************************************
* TITLE : COALESCENCE IN PIPES.                    *        PROJECT NO         *
*                                                  *                           *
*                                                  *    TH./15050/83/FR/..     *
*                                                  *                           *
*******************************************************************************
* CONTRACTOR :                                     * PROGRAM :                 *
*   ALSTHOM NEYRTEC                                *    HYDROCARBONS           *
*   AVENUE DU GENERAL DE GAULL TEL 76 39 55 11     *                           *
*   FR - 38800 LE PONT DE CLAI TLX 320547          *                           *
*                                                  * SECTOR :                  *
*                                                  *    MISCELLANEOUS          *
* PERSON TO CONTACT FOR FURTHER INFORMATION :      *                           *
*   MR. C. BEZARD              TEL 76 39 55 84      *                           *
*                                                  *                           *
*                                                  *                           *
*******************************************************************************
                                                          VERSION : 09/06/87
```

AIM OF THE PROJECT :
****
A mathematical model is to be developed which gives the possibility to predict
liquid/liquid coalescence phenomena in turbulent pipes. The validity of the
model is to be tested with an experimental program.

PROJECT DESCRIPTION :
****
A bibliographic study has been carried out on turbulent dispersions and
collisions probability on one hand and on coalescence and drop breakup
mechanisms on the other hand.
Then a numerical model has been developed to predict the drop size evolution of
an oil in water dispersion along a pipe.
According to the results given by this model, a 100 m long 2" test pipe has
been set on a special rig.
A calibrated oil in water dispersion generation device has been designed and
various types of drop size measurement methods have been tested and evaluated.
Finally a light beam attenuation probe adapted to liquid/liquid interfacial
area measurement has been developed and used during the experiment program.
Tests have been carried out to determine drop size evolution along the pipe
under various flow rate conditions and experimental results compared to
theoretical predictions.

STATE OF ADVANCEMENT :
****
A simplified mathematical model has been developed, which predict evolution of drop population in a pipe when coalescence or breakup phenomena occur.
Experiments carried out on a special oil/water test rig have given results in correlation with numerical predictions.

RESULTS :
****
The drop size evolution along a pipe of an oil in water dispersion has been described by mean of a numerical simulation based on resolution of the drop population balance equation.
Predictions given by the model are that significant coalescence phenomena can take place along industrial pipes which are a few kilometers long.
Tests that have been carried out have confirmed that under standard pipe flow conditions, coalescence phenomena are very slow.
Drop size increase observed during the tests were slightly higher than expected from mathematical model.
This difference empharised the need :
- to settle a new expression for the collision rate that would be better adapted to the description of intermediate drop collision,
- to settle an experimental method to determine the collision efficiency of a given dispersion,
An other result of these tests is that further tests should be carried out with a longer test pipe, 1 km long at least.
A light beam attenuation probe adapted to liquid/liquid interfacial area measurement has been developed that could be used for industrial application.

```
****************************************************************************
* TITLE : FREE SWIMMING RISER PIPE INSPECTION      *      PROJECT NO       *
*         TOOLS.                                   *                       *
*                                                  *   TH./15052/84/NL/..   *
*                                                  *                       *
****************************************************************************
* CONTRACTOR :                                     * PROGRAM :             *
*   R.T.D.  B.V.                                   *   HYDROCARBONS        *
*   DELFTWEG 144 POSTBUS 10065 TEL 010 - 4150200   *                       *
*   NL - 3004 AB - ROTTERDAM    TLX 23366          *                       *
*                                                  * SECTOR :              *
*                                                  *   MISCELLANEOUS       *
* PERSON TO CONTACT FOR FURTHER INFORMATION :      *                       *
*   J.TH. EERING, J.A. DE RAAD                     *                       *
*                                                  *                       *
*                                                  *                       *
****************************************************************************
```

VERSION : 21/01/86

AIM OF THE PROJECT :
****
To develop ultrasonic inspection tools which can be applied under on stream
conditions of oil and gas risers. Major design criteria for oil risers which
will be taken in consideration for the design of the ultrasonic inspection
tools are :
*pipe sizes............: 16" and 20"
*overall length.......: 2,3 m for 16" tool
*pressure.............: 150 bar
*speed................: 4 m/sec
*measuring distance.....: 300 metres
*travelling distance..: 50 Km
*wall thickness range..: up to 40 mm
*accuracy..............: +/- 1 mm
*bi- directional, and capable of passing 3 D - 90 bends, 1 D-T joints and
valves.
For gas risers non contact ultrasonic transducers have to be developed. For the
tool to be developed for gas risers, with similar requirements as mentioned
above, only the sensors will be developed.

PROJECT DESCRIPTION :
****
The intelligent pigs under construction will be optimized and tested at full
speed in a test loop which is representative for all relevant field conditions.
The project includes a dedicated read-out system using a desk top computer for
on site colour enhanced report and analysis of the results.
With the inspection tool for oil risers internal pipe profile and wall
thickness can be measured simultaneously. For the 16" tool 36 probes are in use,
 they measure at intervals of 2.5 mm or a multiple of this interval. For a 20"
tool 48 probes will be used. All values will be stored in a 6 Mbyte solid state
memory, sufficient to store all values over a pipelength of 300 metres.
To pass 3 D bends the tool consists of 3 articulated high pressure resistant
containers which are for that purpose connected with eachother by universal
joints. In these containers all electronics and high energy batteries are
housed.

C-MOS electronics are used to save energy. Electronics are compactly packed
partly as hybrids in order to fit in the scarce space of the containers.
Data reduction facilities enable the measurement of long lengths of pipeline as
well.

STATE OF ADVANCEMENT :
****
The development of ultrasonic sensors suitable for gas pipelines has been
completed by July 1986. By July 1986 development of the tools for oil risers
was completed. Assembling started October 1986. Test loop runs are planned for
April 1987. Application in the field is planned for September 1987.

RESULTS :
****
During the development several unique solutions were engineered with respect to
construction of the actual inspection tool.
In particular a good solution was found for the hinging of the probes which
should be in a concentric and perpendicular position with respect to the
pipewall. Spacecraft technology was introduced to design and construct very
compact and energy saving electronics.
Special ultrasonic probes were designed and built in the frame of the project.
These high pressure resistant ultrasonic sensor modules contain the ultrasonic
probe, its transmitter and preamplifier electronics in one unit. The 6 Mbyte
solid state memory consisting of 32 kbyte RAMs including a powerful set
processors was constructed to store all values at full speed of the inspection
tool.
After the data are retrieved from the memory these are transferred to a
powerful HP 9836 C. desk top computer. With this device supported by
appropriate software several presentation modes of results can be generated.
Colours are used to enhance these results.
Many function tests of parts and assembled part were carried out to prove
proper functioning of the integrated inspection tool.

REFERENCES :
****
DE RAAD, J.A. ETAL
HOLLAND MARITIME, VOL 12, APRIL 1986, P. 14-17
HOLLAND INDUSTRIAL, VOL 10, SEPTEMBER 1986, P 18-21
DE RAAD, J.A.
COMPARISON BETWEEN ULTRASONIC AND MAGNETIC FLUX PIGS
1ST INTERNL SUBSEA PIGGING CONF., SEPTEMBER 23-25, 1986, HAUGESUND, NORWAY.
WILL BE PUBLISHED IN PIPE & PIPELINE INTERNATIONAL, JAN 1987, VOLUME 32, NR
1
VAN DEN BERG, W.H. ETAL
DEVELOPMENT OF AN ELECTROMAGNETIC TRANSDUCER
15TH SYMPOSIUM OF ACOUSTIC IMAGING. HALIFAX, CANADA 14-16 JULY 1986.
DE RAAD, J.A.
DEVELOPMENT OF TOOLS FOR ON-STREAM INSPECTION OF OIL RISERS USING ULTRASONICS.
WEST-EUROPEAN CONF. ON MARINE TECHN, ADVANCES IN OFFSHORE TECHN, AMSTERDAM,
NOVEMBER 25-27, 1986.

```
*******************************************************************************
* TITLE : PARAFIL ROPE DEEPWATER MOORING STUDY      *       PROJECT NO        *
*                                                   *                         *
*                                                   *   TH./15053/84/UK/..     *
*                                                   *                         *
*******************************************************************************
* CONTRACTOR :                                      * PROGRAM :               *
*   BRITISH UNDERWATER ENGINEERING LTD              *   HYDROCARBONS          *
*   3RD FLOOR, TRAFALGAR HOUSE TEL 01 748 46 00     *                         *
*   HAMMERSMITH INTERNATIONAL  TLX 928241           *                         *
*   UK - LONDON W6 8DW                              * SECTOR :                *
*                                                   *   MISCELLANEOUS         *
* PERSON TO CONTACT FOR FURTHER INFORMATION :       *                         *
*   DR. C.F. BAXTER                                 *                         *
*                                                   *                         *
*                                                   *                         *
*******************************************************************************
                                                      VERSION : 31/12/86
```

AIM OF THE PROJECT :
****
The aim of the project is to prove the feasibility of 'Parafil' ropes for
deepwater moorings both for tension leg free floating structures; guyed
structures; and semi-submersible facilities.

PROJECT DESCRIPTION :
****
"Parafil" is a parallel laid rope with high axial stiffness and lightweight.
The project will identify suitable structures and analyse the behaviour of
mooring systems for them made of "Parafil". Methods of installing, maintaining
and monitoring the moorings will be investigated properties of the material,
methods of on-site manufacture and transportation will be defined.

STATE OF ADVANCEMENT :
****
'Parafil' testing has required a longer duration than originally planned as the
material is generally outliving preliminary expectations. An approved extension
for the project period has been granted to accommodate these factors.

RESULTS :
****
Final conclusions have not yet been reached. However there is strong indication
that 'Parafil' will offer an attractive alternative to mooring systems in the
depth range in excess of 1500' (500 metres).
During the course of the ongoing investigations a variety of scenario's have
been examined. These include Tension Leg Platforms (small & large); Tethered
Buoys; Guyed Towers and Semi-Submersible vessles (these utilising Catenary
Moorings). These analyses have considered the mooring system in which 'Parafil'
is a tension carrying component. In each case 'Parafil' can be incorporated
into the mooring system in a manner that produces technological and cost
benefits. The cost benefits are heavily influenced by depth range of the
installation and become more attractive as the water depth increases.
Major programme effort has beenexpended on collecting, collating and compiling
a basis of design which recognizes the mechanical and other properties of
'Parafil' as they apply to Deepwater Moorings. Creep; Bend Resistance
(Handling); Axial Stiffness; Tensional and Fatigue characteristics are under
investigation.
'Parafil' is difficult to inspect by traditional NDT type activities, the
programme has therefore concentrated upon identifying acceptable means of
continuously monitoring the state of health of a rope rather than intermittent
inspection. Future planned investigations will target large scale rope tests
(up to 1500 tonne) and will incorporate testing and improvements to tether
designs.

REFERENCES :
****
A 'PRELIMINARY DESIGN PREMISE'DOCUMENT COVERING DESIGN PARAMETERS INVOLVED IN
THE APPLICATION OF 'PARAFIL'MOORINGS TO THE MARINE STRUCTURES DESCRIBED IN
SECTION 4 ABOVE.
TO DATE THIS DOCUMENT HAS ONLY BEEN PRESENTED TO THE PARTICIPANTS IN THE
PROJECT.

```
**********************************************************************************
* TITLE : DEVELOPMENT OF NEW TECHNIQUES OF        *      PROJECT NO        *
*         OIL/WATER SEPARATION.                   *                        *
*                                                 *     TH./15057/84/NL/..  *
*                                                 *                        *
**********************************************************************************
* CONTRACTOR :                                    * PROGRAM :              *
*   TECHNISCHE HOGESCHOOL DELFT, AFD MIJNBOUWKUNDE *   HYDROCARBONS         *
*   POSTBUS 5028                 TEL 015 781328    *                        *
*   NL - 2600 GA DELFT           TLX 38151         *                        *
*                                                 * SECTOR :               *
*                                                 *   MISCELLANEOUS        *
* PERSON TO CONTACT FOR FURTHER INFORMATION :     *                        *
*   IR. W.M.G.T. VAN DEN BROEK                    *                        *
*                                                 *                        *
*                                                 *                        *
**********************************************************************************
                                                    VERSION : 20/02/87
```

AIM OF THE PROJECT :
****
The aim of the project is the development of a small, light and efficient oil-water separator suitable for a large range of oil-water mixtures.

PROJECT DESCRIPTION :
****
A number of existing oil-water separation techniques (such as plate separation, membrane filtration), as well as some separation techniques not yet used for oil-water separation, will be investigated on a laboratory scale. Then a prototype oil-water separator will be designed and built, based on the most effective separation processes.

STATE OF ADVANCEMENT :
****
Most of the investigated separation techniques are in the laboratory experiment stage. For plate separation a larger scale experimental set-up is available. Furtheron experiments on flow distribution (instream into a separator) are in progress.

RESULTS :
****
An initial investigation on plate separation has been carried out. Critical Reynolds numbers of one up to a few hundred for different corrugated plate forms were measured. Large scale experiments will start very soon. For these experiments a laser-doppler anemometer is available. Calculations of the flow between corrugated plates are in progress.
Primarily experiments on membrane separation were reasonably successful on a small scale. These experiments will be continued. Also primarily experiments on flotation have been completed; however, the results have not yet been evaluated. For experiments on adsorption and magnetic separation no results can be given at this stage. Primarily experiments on flow distribution gave favourable results. These experiments will be continued with a large scale flow distributor.

```
********************************************************************************
* TITLE : DEVELOPMENT OF A SUBSEA CONNECTOR.      *      PROJECT NO          *
*                                                 *                          *
*                                                 *   TH./15058/84/IT/..     *
*                                                 *                          *
********************************************************************************
* CONTRACTOR :                                    * PROGRAM :                *
*   M.I.B. ITALIA SPA                             *   HYDROCARBONS           *
*   C.P. 5                    TEL 049 643099       *                          *
*   IT - 35020 CASALSERUGO (PA TLX 430214         *                          *
*                                                 * SECTOR :                 *
*                                                 *   MISCELLANEOUS          *
* PERSON TO CONTACT FOR FURTHER INFORMATION :     *                          *
*   MR. G. BORMIOLI/MR. R. MAS                    *                          *
*                                                 *                          *
*                                                 *                          *
********************************************************************************
                                                  VERSION : 31/12/86
```

AIM OF THE PROJECT :
****
To develop a subsea emergency disconnector system designed for the special
problems associated with the use of flexible risers and flexible flowlines. The
device to be capable of connection/reconnection at the ocean floor.

PROJECT DESCRIPTION :
****
This development is in two phases : the development of mechanical
disconnection/connector itself and the investigation/development of the
associated handling system if necessary. This involves discussion with oil
companies/end users to establish the operational/design criteria and the
development of a design.
One prototype mechanical disconnection/connector has been manufactured and shop
tested following which, arrangements will be made to undertake a subsea test
together with the development of a handling system. The unit developed is
pressure compensated and will automatically release when subjected to a
predetermined external load. In addition it can ben hydraulically activated so
as to allow hydraulically activated so as to allow connection/reconnection or
disconnection.

STATE OF ADVANCEMENT :
****
Ongoing project; monitoring phase.
Phase executed:
A design has been developed, drawings completed and one 3" prototype unit built.
  Discussions with possible users have allowed to agree on technical solutions,
test procedure and result analysis. Preliminary and detailed shop test have
been carried out, and a report completed.

RESULTS :
****
The test confirmed that all the basic principles were correct, and the unit had
good resistance to applied loads. It was estimated that the "weak bolt"
feature did not offer any particular advantage and could be eliminated.
FOLLOWING TESTS HAVE BEEN SUCCESSFULLY PERFORMED:
- Hydrostatic pressure test and cyclic pressure tests.
- Load test with and without "weak shear bolts."
- Load test with torsion and bending moments.
- Manual disconnection test.
- Cyclic axial load.
- Cyclic bending moment load.
- Disconnection with bending moment.
SUMMARY OF RESULTS
- No unexpected leakage occurred during the tests.
- The pressure compensating principle was correct.
- No damage to the unit occurred during the tests.
- No advantage was gained by the weak bolts.
- High pressure metallic seals showed good performance.
 A test report was issued detailing the test procedure, the results obtained,
the problems encountered and possible improvements or solutions.

```
********************************************************************************
* TITLE : UNDERWATER DRILL FOR LARGE CAPACITY      *        PROJECT NO        *
*         PILES.                                   *                          *
*                                                  *    TH./15059/84/BE/..    *
*                                                  *                          *
********************************************************************************
* CONTRACTOR :                                     * PROGRAM :                *
*   BELGIAN OFFSHORE SERVICES NV                   *   HYDROCARBONS           *
*   SCHERMERSTRAAT 46          TEL 03 231 87 70     *                          *
*   BE - 2000 ANTWERPEN        TLX 34129            *                          *
*                                                  * SECTOR :                 *
*                                                  *   MISCELLANEOUS          *
* PERSON TO CONTACT FOR FURTHER INFORMATION :      *                          *
*   MR. S. DECKERS                                 *                          *
*                                                  *                          *
*                                                  *                          *
********************************************************************************
```

VERSION : 17/03/87

AIM OF THE PROJECT :
****
This project intends to study and build a subsea drilling machine for the
excavation of very large capacity anchoring piles in medium to hard soil. This
drill will be operated from a standard drilling ship in several hundred metres
water depth (3 000 feet maximum). The main advantage of this down hole tool is
that no torque is transmitted from surface, limiting string failures as
encountered in big hole rotary drilling.

PROJECT DESCRIPTION :
****
This equipment is based on the HYDROFRAISE a hydraulically driven drilling tool
developed by SOLMARINE's mother Company SOLETANCHE to excavate trenches and
reinforced piles for Civil Works. The subsea drill consists of :
- a 150 to 200 kN guiding frame held by a string set to the drilling ship.
- 2 parallel cutter wheels, rotating on horizontal axes in opposite directions,
set at bottom of frame.
- a pump set above the wheels to remove cuttings.
- a jack to monitor weight or speed during drilling.
- a casing equipped with reentry funnel is pulled with the drilling tool to
cover overburden layers.
- a 400 HP power pack supplied hydraulic power from the surface through
hydraulic lines.
- deflection is monitored by an inclinometer and maintained below 0.3%.
This machine will be able to drill rectangular shafts (2.40 m x 1.00 m section)
down to 100 metres in soil with a simple compressive strength up to 100 MPa.

STATE OF ADVANCEMENT :
****
Ongoing. Detailed drawings of the rectangular marine Hydrofraise are completed
and the tool can be built at any time. The construction phase would need one
year about prior to trials.
As the rectangular shape is not always adapted to offshore structure design, a
detailed study is in progress for a circular tool. This study is planed to be
completed in early summer 87.

RESULTS :
****
Description of the Rectangular Marine Hydrofraise.
For these studies, we have based the concept on the following fundamentals :
- implementation of simple solutions.
- provisioning of usual components aboard the drill ship.
These fundamentale have led us to an easy and simple conception based on the
use of seawater as power fluid.
This primary energy is converted with a turbine driving an underwater hydraulic
power pack, the whole being mounted inside the A frame.
Description of the Circular Marine Hydrofraise
The Circular Marine Hydrofraise has to answer two operational criteria compared
to rectangular marine Hydrofraise :
- soil stability meanwhile drilling,
- compatibility with the present concept for offshore foundation.
These two constraints have led us to design the Circular Marine Hydrofraise on
the following principles :
- Pile drilling protected with a casing on the whole height of the shaft. This
casing is the definitive casing which is installed into the borehole in one
operation prior to the cementation. This technique is based on the pile driving
principle.
- Undereaming to help casing lowering. The motors and the mills are mounted on
an opening system to allow few centimers over drilling around the casing.

REFERENCES :
****
"UN NOUVEL OUTIL FOND DE TROU NOMME HYDROFRAISE MARINE" PRESENTED IN OCTOBER
1985 TO THE CONFERENCE "ACTUALITE ET AVENIR DE L'HYDRAULIQUE MARITIME"
GRENOBLE, FRANCE.
PAPER WILL BE PUBLISHED IN THE NUMBER 4/5 OF LA HOUILLE BLANCHE".

```
********************************************************************************
* TITLE : DEVELOPMENT OF "LANTERN RING" MOORINGS    *      PROJECT NO        *
*         FOR TANKER CONVERSION IN MARGINAL         *                        *
*         OFFSHORE FIELDS.                          *    TH./15061/84/UK/..  *
*                                                   *                        *
********************************************************************************
* CONTRACTOR :                                      * PROGRAM :              *
*   STOROIL LTD                                     *   HYDROCARBONS         *
*   FETTER LANE 12/15          TEL 01 583 83 44     *                        *
*   UK - EC4A 1EL LONDON       TLX 22143            *                        *
*                                                   * SECTOR :               *
*                                                   *   MISCELLANEOUS        *
* PERSON TO CONTACT FOR FURTHER INFORMATION :       *                        *
*   MR. K. CHARLES                                  *                        *
*                                                   *                        *
*                                                   *                        *
********************************************************************************
                                                          VERSION : 20/01/86
```

AIM OF THE PROJECT :
****
The project aims are to define a floating production vessel based on a turret
mooring using conversion of existing tankers.

PROJECT DESCRIPTION :
****
The project is based on the concept of a single unit to be inserted within a
tanker. This unit is based on the "turret" technology with superstructure
floors to hold the processing equipment. The project will be conducted in
several phases :
- project definition
- model testing, computer design and analysis
- design of systems and equipment
- design interfaces - cost estimates - development schedules
- final report.

STATE OF ADVANCEMENT :
****
Abandoned. Works have been stopped due to lack of interest of oil industries.

RESULTS :
****
The preliminary study has been based on the following parameters : water depth
being 100 to 200 m, recoverable reserves 100 millions of barrels, light crude
36 API, peak production 60,000 barrels per day from 6 wells. A typical tanker
VLCC of 315,00 DWT has been selected. Turret investigation led to a 12 chains
mooring system with a lower turret bearing to support horizontal stresses.

```
*****************************************************************************
* TITLE : AN UNDERWATER MOTION COMPENSATED SPM       *      PROJECT NO       *
*                                                    *                       *
*                                                    *   TH./15062/84/UK/..   *
*                                                    *                       *
*****************************************************************************
* CONTRACTOR :                                       * PROGRAM :             *
*   FLOATECH                                         *   HYDROCARBONS        *
*   GREENFORD HOUSE              TEL 01 5752341      *                       *
*   309 RUISLIP ROAD EAST, GRE TLX 23417            *                       *
*   UK - UB6 9BQ MIDDLESEX                           * SECTOR :              *
*                                                    *   MISCELLANEOUS       *
* PERSON TO CONTACT FOR FURTHER INFORMATION :        *                       *
*   MR. M. CONWAY                                    *                       *
*                                                    *                       *
*                                                    *                       *
*****************************************************************************
```

VERSION : 19/12/86

AIM OF THE PROJECT :
****
The objective and scope of this development programme is to complete the pre-
engineering design of an underwater motion compensated Single Point Mooring
(SPM).

PROJECT DESCRIPTION :
****
Development of an engineering design of an offshore single point mooring for
loading of stabilised crude oil with special application in ice infested
regions.

STATE OF ADVANCEMENT :
****
Ongoing. Computer analytical models have been developed for quasi static and
slow drift wave forces and for the wave frequency analysis. Model test
programme has been defined, with equipment and materials on order. Mechanical
and Structural General Arrangement drawings of the mid-water and seabed mounted
buoy variants and of the component parts are in progress. Discussions have
taken place with potential external consultants and manufacturers.

RESULTS :
****
From the initial work, based upon computer model outputs and development of
general arrangement drawings, it is concluded that to achieve the objectives of
the study and to take advantage of potential sales opportunities two variants
have to be developed.
1. deepwater buoy variant with a mid-water buoy
2. shallow water variant with seabed mounted buoy
The results obtained from the study to date indicate that the Submerged Piston
Mooring is a viable mooring option. The scope of work to completion of the
study will be to confirm this conclusion.

```
********************************************************************************
* TITLE : EXPENDABLE VEHICLE FOR MEASURMENT IN        *        PROJECT NO      *
*         HIGHLY DEVIATED WELLS.                       *                        *
*                                                      *    TH./15063/84/FR/..  *
*                                                      *                        *
********************************************************************************
* CONTRACTOR :                                         * PROGRAM :              *
*   SYMINEX                                            *   HYDROCARBONS         *
*   BOULEVARD DE L'OCEAN 2      TEL 91 73 90 03        *                        *
*   FR - 13275 MARSEILLE CEDEX TLX 400563              *                        *
*                                                      * SECTOR :               *
*                                                      *   MISCELLANEOUS        *
* PERSON TO CONTACT FOR FURTHER INFORMATION :          *                        *
*   MR. A.J. KERMABON                                  *                        *
*                                                      *                        *
*                                                      *                        *
********************************************************************************
```

VERSION : 01/01/87

AIM OF THE PROJECT :
****
The vehicle is a measurement probe with an internal spool of insulated copper
wires. Data acquisition, transmission of temperature, pressure, casing collar
location, etc... is performed by a low power electronic module. The aim of the
project is to assess the limitation of the probe owing to the deviation of the
well and to study a self propelled vehicle able to perform measurements in the
deviated parts of the well.

PROJECT DESCRIPTION :
****
Phase 1 : Feasibility study of a propelling system, laboratory tests.
Phase 2 : Prototype of measuring probe equipped with the propulsion system
developed in phase 1, laboratory and field tests.
Phase 3 : Extension of the principle to more sophisticated measurements,
evaluation of the needs.

STATE OF ADVANCEMENT :
****
Ongoing. A laboratory prototype with two kinds of propellers has been made and
tested in an horizontal well full of water. Now, an operational prototype is
being designed to perform on site tests.

RESULTS :
****
The prototype tests give good results. Speed of the probe is about 1 m/s with
48 VDC/10A. The electronic motor drive is now ready for use in the operational
system. Difficulties are : to find highly deviated well in France.

```
*********************************************************************************
* TITLE : DEVELOPMENT OF A MOBILE LASER SYSTEM       *        PROJECT NO       *
*         FOR THE DETERMINATION AT A DISTANCE OF     *                         *
*         AVERAGE CONCENTRATIONS OF METHANE AND      *    TH./15071/85/IT/..   *
*         ETHANE                                     *                         *
*********************************************************************************
* CONTRACTOR :                                       * PROGRAM :               *
*    AZIENDA ENERGETICA MUNICIPALE                   *    HYDROCARBONS         *
*    CORSO DI PORTA VITTORIA, 4 TEL 02 7720/3459     *                         *
*    IT - 20122 MILANO           TLX 334170          *                         *
*                                                    * SECTOR :                *
*                                                    *    MISCELLANEOUS        *
* PERSON TO CONTACT FOR FURTHER INFORMATION :        *                         *
*    ING. BONFIGLI                                   *                         *
*                                                    *                         *
*                                                    *                         *
*********************************************************************************
```

VERSION : 01/01/87

AIM OF THE PROJECT :
****
The aim of the project is the development of an IR LIDAR system able to detect
small methane and ethane concentrations; sensitivity 1 p.p.m. range 1 km. The
system is based on a correlation technique and utilizes as laser source an
optical parametric oscillator pumped by a ND-YA laser. The system is designed
to be a small, compact and reliable device suitable to perform measurement in
an urban environment from movable platform, such as small lorries or
helicopters.
In case of successful results, it will be sought a cooperation with an
electronic firm, which already operates in the laser field, in order to build a
small series of systems to be put in the market of control and surveillance
equipments.

PROJECT DESCRIPTION :
****
The detection method is based on the measurement of the atmospheric absorption
of a laser beam by the gas to be detected; this technique is called
Differential Absorption LIDAR (DIAL) and is used because it allows to cancel
systematic errors due to target reflectivity and atmospheric scattering.
Conventional DIAL techniques may lead to important errors when used to make
measurements on moving platforms, since two laser shots are required; infact,
time varying parameters, such as topographical target reflectance, may
noticeably change during the time elapsed between the two shots, since the
pulse rate is typically not grater than 10 Hz.
The gas correlation method makes it possible to self-normalize the return
signal obtained from a single laser pulse, thus leaving any time dependent
environmental effect. Main disadvantages are a lower sensitivity and a greater
effect on the measurement of interfering molecules.

STATE OF ADVANCEMENT :
****
Ongoing: the system has been constructed.
Phases 1, 2 and 3 have been completed and phase 4 is now ongoing.

RESULTS :
****
GAS CORRELATION TECHNIQUE
The gas correlation approach to a LIDAR configuration requires a fixed-
wavelength rather than a broadband laser source which wholly overlaps an
absorption line of the target molecule. Instead of shooting the source twice
and using a single detector as in normal DIAL systems, the signal received from
each fixed wavelength pulse is split in half and focused on two photodiodes.
Light passing through the "direct" arm straight overtakes the detector, whereas
in the "correlation" path there is a gas filter correlation cell in front of it.
The cell is filled at atmospheric pressure with a large amount of the target
molecule and some other non interfering gas. In this way the detected energy of
the pulse in the "correlation" arm is nearly unaffected by the presence or
absence of the target gas in the atmosphere and the unbalance between the two
arms is related to its open air concentration. Since a source linewidth larger
than that of the absorption line is used, the transmission deviates from Beer-
Lambert law, thus requiring a preliminary calibration of the ratio
"correlation"/"direct" versus the atmospheric optical depth of the target gas.
Notice that this ratio is independent of topographical target reflectance,
atmospheric turbulence, pulse energy and other time varying phenomena, since
the measurements are simultaneously performed on the same pulse.
LASER SOURCE
To permit tuning on the most suitable absorption line and future use on other
gases, an Optical Parametric Oscillator (OPO) has been chosen as infrared
source. The theoretical tuning range of the OPO, when pumped by Nd:YAG laser at
1.06 micron is from 1.5 to 4 micron. The OPO has a single resonant and single
pass configuration which optimizes the "signal" wave oscillation around 3
micron, where matter has a strong absorption band.
As known, a good pump beam quality is one of the major requirements to
successfully operate OPOs, which is met by using a Nd:YAG laser based on a new
type of resonator cavity, termed SFUR (Self Filtering Unstable Resonator). The
laser emits a near diffraction limited beam which evolves with a near gaussian
behaviour.
The pump beam is about 6 mm in diameter with pulses of about 250 mJ energy, 20
ns duration and 10 pps repetition rate; while OPO pulses are expected to be of
nearly 5 mJ of energy and of similar duration.
RECEIVER
The main components of the receiver are a 30 cm newtonion telescope and a pair
of filters coaled in as photovoltaic detectors, which detects the direct and
correlation signals; other three feltier coaled Pb S detectors are used for
internal reference and chek. The output signals of the detectors are then
amplified, sampled and converted in digital format and finally transmitted to
an Rewlett Packard 9000/310 minicomputer for data averaging, concentration
evaluation and data storage.

REFERENCES :
****
METHANE GAS DETECTION WITH INFRARED CORRELATION LIGHT DETECTION AND RANGING
(LIDAR) SYSTEM. S. DRAGHI, E. GALLETTI, M. GARBI, E. ZANZOTTERA - CISE S.P.A.,
P.O. BOX 12081, 20134 MILANO, ITALY
EUROPEAN CONFERENCE ON OPTICS, FLORENCE 30/9-3/10/86.

```
*********************************************************************************
* TITLE : APPLICATION OF NON-DESTRUCTIVE          *          PROJECT NO          *
*         MONITORING METHODS ON COMPOSITE TUBULAR *                              *
*         EQUIPMENT                               *     TH./15077/86/FR/..       *
*                                                 *                              *
*********************************************************************************
* CONTRACTOR :                                    * PROGRAM :                    *
*   GERTH                                          *   HYDROCARBONS               *
*   4, AVENUE DE BOIS PREAU      TEL 1 47.52.61.39 *                              *
*   FR - 92502 RUEIL-MALMAISON TLX 203 050        *                              *
*                                                 * SECTOR :                     *
*                                                 *   MISCELLANEOUS              *
* PERSON TO CONTACT FOR FURTHER INFORMATION :     *                              *
*   MR. J.J. MASSOT                               *                              *
*                                                 *                              *
*                                                 *                              *
*********************************************************************************
                                                         VERSION : 01/01/87
```

AIM OF THE PROJECT :
****
Development of the utilisation of tubular equipment made of composite materials
is presently faced with the lack of quality guaranteed the equipment that is
implemented. This is of importance in the case of vertical tubings, which are
at the same time very heavy mechanically and for which intervention cost is the
highest.
Therefore, the object of this programme is to create, adapt or finalize non-
destructive control methods based on different physical phenomenal (acoustic,
thermal, electrical, etc...) so to ensure the reproducibility of the produced
equipment.
However, the non-homogeneous nature of these materials, implies in practise,
that the adaptation of operational methods be done in direct with the usage
value of the equipment, in this case the loss of tightness of the pipes and of
the ancillary equipment. Hence, this programme is followed by a study on the
incidence of faults on the utilisation properties of the tubings.

PROJECT DESCRIPTION :
****
Non-destructive control methods for composite materials have been developed
particularly for the aeronautical domain, and concern top quality materials,
most often based on carbon fibers. In the case of more industrial epoxy-glass
fiber tubings, the detection of faults and the estimate of their critical point
must be evaluated simultaneously.
Furthermore, the diversity of existing equipment and methods imply a number of
choices possible for each type of development, as well as the performance
evaluation of reference equipment. The programme has been divided in two main
phases :
a) FEASIBILITY
In this preliminary phase, two different studies will be carried out :
- evaluation of real performances in time of tubings selected as test elements.

This evaluation will be carried out under internal pressure, with and without axial traction, in the presence of water and under temperature. This evaluation will also include an analysis of their creeping properties.
- the inventory and evaluation of the performance of possible methods of non-destructive control, in relation with faults currently observed and with their possibility of being used at the factory or on worksites. This task can be carried out without problem in laboratory test tubes.
This phase will end with the precise definition of applicable methods and their utilisation range. Obviously, only a small number of methods will be chosen for the following phase. Also most of the faults encountered will depend on them. Hence, this phase is an ending point for which a reorientation of the programme could be envisaged.

b) DEVELOPMENT OF SPECIFIC METHODS
This phase will be entirely carried out on tubular elements, as similar as possible with industrial tubings. It will include the following tasks :
- Creation of tubings with calibrated artificial faults similar to those encountered previously
- Evaluation of the limit performances of chosen control methods (size, number, etc..)
- Evaluation of the incidence of these faults on the lifespan of tubings. This incidence will be measured by long duration tests, under static and eventually under fatigue conditions. Considering the special feature of the mechanical operating mode of the tubings, this evaluation will not only be carried out under internal pressure, but also under traction, and anyhow under temperature. Attempts will also be made, when possible, to follow the progress of faults observed initially by the chosen control methods.
Thus, it will be possible, at the end of this phase, to define a coherent control scheme for the tubular elements made of composite materials, which will include the maximum size of the admissible fault and the minimum performances of the control methods to be implemented for an appropriate detection.

STATE OF ADVANCEMENT :
****
The works are beginning in the first days of January 1987.

```
******************************************************************************
* TITLE : THE DEVELOPMENT OF INDUSTRIAL DIVING      *      PROJECT NO        *
*           WITH HYDROGENATED BREATHING MIXTURES    *                        *
*                                                   *      TH./15085/86/FR/.. *
*                                                   *                        *
******************************************************************************
* CONTRACTOR :                                      * PROGRAM :              *
*   COMEX SA                                         *   HYDROCARBONS         *
*   36, BLD DES OCEANS          TEL 91.41.01.70      *                        *
*   FR - 13275 MARSEILLE CEDEX TLX 410985           *                        *
*                                                   * SECTOR :               *
*                                                   *   MISCELLANEOUS        *
* PERSON TO CONTACT FOR FURTHER INFORMATION :       *                        *
*   MR. C. GORTAN                                    *                        *
*                                                   *                        *
*                                                   *                        *
******************************************************************************
```

VERSION : 22/03/87

AIM OF THE PROJECT :
****
The aim of the project is to perfect the techniques necessary for the
development of industrial diving with hydrogenated breathing mixtures :
- Research carried out on laboratory animals for definition of the maximum
acceptable hydrogen concentrations to avoid narcotic effects.
- Design and construction of a gas deshydrogenation system.
- Study of compression and decompression tables for hydrogenated gas mixtures.
- Accomplishment of an experimental dive for selectiion and training of divers
in view of a future actual dive at sea.
- Chemical decompression of the hyperbaric chambers by deshydrogenation of the
diving breathing mixtures.

PROJECT DESCRIPTION :
****
Traditional helium diving is limited by two factors :
- The High Pressure Nervous Syndrome (H.P.N.S.) resulting from hydrostatic
pressure on the central nervous system, which provokes motorial disorders.
- The gas mix density which increases in proportion to depth and makes
breathing increasingly difficult for the diver.
The cumulated effects of these two factors diminish thus considerably the
diver's work capacity.
The narcotic power of hydrogen tends to strongly counteract the development of
H.P.N.S.
As concerns its density, the lowest amongst gases, it allows an important
reduction of the voluminal mass of gas mixtures under pressure which induces
much easier breathing.
The programme will take the folowing four phase :
PHASE 1 : TOXICOLOGICAL RESEARCH
Its purpose is to experimentally determine the maximum acceptable hydrogen
content in respect of narcosis. The research will be conducted on small
laboratory animals under extreme pressure conditions, equivalent to 1200 to
2000 metres. A part of the animals will be sacrified for histological study.

PHASE 2 : DESIGN AND CONSTRUCTION OF GAS DESHYDROGENATION SYSTEM
The system will be designed after a bibliographic study of present state of the art in this subject. The actual operating system will be determined on the basis of laboratory tests results, before a full scale prototype is completed in order to decompress the chambers during the manned simulation dive to 520 metres.
PHASE 3 : STUDY OF COMPRESSION AND DECOMPRESSION TABLES FOR HYDROGENATED GAS MIXTURES
PHASE 4 : SELECTION AND TRAINING OF DIVERS AND REFINING OF DIVE PROCEDURES
Ten divers (two of whom serving as substitutes) will be chosen amongst very experienced professional divers.
Eight of them will perform the experimental dive at 520 metres depth. During this dive, they will be subjected to a complete physiological examination, under normal conditions as well as during their activity in water.

STATE OF ADVANCEMENT :
****
The different phases of the project have been realized in accordance with the described programme.
The recorded results are being analysed.

RESULTS :
****
The analysis of the complete results being in progress, only general information may be provided.
Hydrogen does not seem to have any toxicity of its own. The histological examination does not show specific lesions. Under particularly severe conditions (1800 to 2000 metres), the laboratory animals survival rate is much higher with use of hydrogen than with helium, which emphasizes the hydrogen advantages for deep diving.
The gas deshydrogenation system (for elimination of hydrogen by catalytic oxidization) once developed, has worked perfectly during the decompression of the training dive to 520 metres.
1,250 m3 of hydrogen were eliminated within 10 days at a rate corresponding to the physiological decompression of the divers.
This experimental dive to 520 metres lasted for 28 days. During the dive, divers stay at the bottom and the first part of their decompression, about 20 dives have been performed in the wet chamber, including different activities such as :
- work on a submarine ergometric bicycle (ergocycle) under cardio-resiratory control,
- construction of a complex puzzle made of pipes,
- tests of three diving equipments : two breathing systems and a low flow hot water suit.
The breathing apparatus will need to be specifically developed for being perfectly-suitable with hydrogen use.